Texts in Applied Mathematics 5

Springer Science+Business Media, LLC

Texts in Applied Mathematics

John H. Hubbard   Beverly H. West

Differential Equations: A Dynamical Systems Approach

Ordinary Differential Equations

With 144 Illustrations

 Springer

John H. Hubbard
Beverly H. West
Department of Mathematics
Cornell University
Ithaca, NY 14853
USA

Series Editors

J.E. Marsden
Department of
  Mathematics
University of California
Berkeley, CA 94720
USA

L. Sirovich
Division of Applied
  Mathematics
Brown University
Providence, RI 02912
USA

M. Golubitsky
Department of Mathematics
University of Houston
Houston, TX 77204-3476
USA

W. Jäger
Department of Applied Mathematics
Universität Heidelberg
Im Neuenheimer Feld 294
69120 Heidelberg, Germany

Library of Congress Cataloging-in-Publication Data
Hubbard, John.
  Differential equations: a dynamical systems approach / John
Hubbard, Beverly West.
    p. cm. — (Texts in applied mathematics : 5, 18)
  Contents: pt. 1. Ordinary differential equations—pt. 2. Higher-
dimensional systems.

  1. Differential equations. 2. Differential equations, Partial.
I. West, Beverly Henderson, 1939– . II. Title. III. Series.
QA371.H77 1990
515′.35—dc20   90-9649

Printed on acid-free paper.

9 8 7 6 5 4 3 (Corrected third printing, 1997)

SPIN 10631374

ISBN 978-1-4612-6952-6   ISBN 978-1-4612-0937-9 (eBook)
DOI 10.1007/978-1-4612-0937-9

Series Preface

Mathematics is playing an ever more important role in the physical and biological sciences, provoking a blurring of boundaries between scientific disciplines and a resurgence of interest in the modern as well as the classical techniques of applied mathematics. This renewal of interest, both in research and teaching, has led to the establishment of the series: *Texts in Applied Mathematics (TAM)* .

The development of new courses is a natural consequence of a high level of excitement on the research frontier as newer techniques, such as numerical and symbolic computer systems, dynamical systems, and chaos, mix with and reinforce the traditional methods of applied mathematics. Thus, the purpose of this textbook series is to meet the current and future needs of these advances and encourage the teaching of new courses.

TAM will publish textbooks suitable for use in advanced undergraduate and beginning graduate courses, and will complement the *Applied Mathematical Sciences (AMS)* series, which will focus on advanced textbooks and research level monographs.

Preface

Consider a first order differential equation of form $x' = f(t, x)$. In elementary courses one frequently gets the impression that such equations can usually be "solved," i.e., that explicit formulas for the solutions (in terms of powers, exponentials, trigonometric functions, and the like) can usually be found. Nothing could be further from the truth. In fact, only very exceptional equations can be explicitly integrated—those that appear in the exercise sections of classical textbooks. For instance, none of the following rather innocent differential equations can be solved by the standard methods:

$$x' = x^2 - t,$$
$$x' = \sin(tx),$$
$$x' = e^{tx}.$$

This inability to explicitly solve a differential equation arises even earlier —in ordinary integration. Many functions do not have an antiderivative that can be written in elementary terms, for example:

$$f(t) = e^{-t^2} \text{ (for normal probability distribution)},$$
$$f(t) = (t^3 + 1)^{1/2} \text{ (elliptic function)},$$
$$f(t) = (\sin t)/t \text{ (Fresnel integral)}.$$

Of course, ordinary integration is the special case of the differential equation $x' = f(t)$. The fact that we cannot easily integrate these functions, however, does *not* mean that the functions above do not have any antiderivatives at all, or that these differential equations do not have solutions.

A proper attitude is the following:

> *Differential equations define functions, and the object of the theory is to develop methods for understanding (describing and computing) these functions.*

For instance, long before the exponential function was defined, the differential equation $x' = rx$ was "studied": it is the equation for the interest x' on a sum of money x, continuously compounded at a rate r. We have records of lending at interest going back to the Babylonians, and the formula that they found for dealing with the problem is the numerical Euler approximation (that we shall introduce in Chapter 3) to the solution of the equation (with the step h becoming the compounding period).

Methods for studying a differential equation fall broadly into two classes: qualitative methods and numerical methods. In a typical problem, both would be used. Qualitative methods yield a general idea of how *all* the solutions behave, enabling one to single out interesting solutions for further study.

Before computer graphics became available in the 1980's, we taught qualitative methods by handsketching direction fields from isoclines. The huge advantage now in exploiting the capabilities of personal computers is that we no longer need to consume huge amounts of time making graphs or tables by hand. The students can be exposed to ever so many more examples, easily, and interactive programs such as MacMath provide ample opportunity and inducement for experimentation any time by student (and instructor).

The origin of this book, and of the programs (which preceded it) was the comment made by a student in 1980: "This equation has no solutions." The equation in question did indeed have solutions, as an immediate consequence of the existence and uniqueness theorem, which the class had been studying the previous month. What the student meant was that there were no solutions that could be written in terms of elementary functions, which were the only ones he believed in. We decided at that time that it should be possible to use computers to *show* students the solutions to a differential equation and how they behave, by using computer graphics and numerical methods to produce pictures for qualitative study. This book and the accompanying programs are the result.

Generally speaking, numerical methods approximate as closely as one wishes a single solution for a particular initial condition. These methods include step-by-step methods (Euler and Runge–Kutta, for instance), power series methods, and perturbation methods (where the given equation is thought of as a small perturbation of some other equation that is better understood, and one then tries to understand how the solution of the known equation is affected by the perturbation).

Qualitative methods, on the other hand, involve graphing the field of slopes, which enables one to draw approximate solutions following the slopes, and to study these solutions all at once. These methods may much more quickly give a rough graph of the behavior of solutions, particularly the long term behavior as t approaches infinity (which in real-world mathematical modeling is usually the most important aspect of a solution). In addition, qualitative techniques have a surprising capacity for yielding specific numerical information, such as location of asymptotes and zeroes. Yet traditional texts have devoted little time to teaching and capitalizing on these techniques. We shall begin by showing how rough graphs of fields of can be used to zero right in on solutions.

In order to accomplish this goal, we must introduce some new terminology right at the beginning, in Chapter 1. The descriptive terms "fence," "funnel," and "antifunnel" serve to label simple phenomena that have exceedingly useful properties not exploited in traditional treatments of differential equations. These simple new ideas provide a means of formalizing the observations made by any person *familiar* with differential equations, and they provide enormous payoff throughout this text. They give simple, direct, noniterative proofs of the important theorems: an example is the Sturm comparison and oscillation theorem, for which fences and funnels quickly lead to broad understanding of all of Sturm–Liouville theory. Actually, although the words like fences and funnels are new, the notions have long been found under the umbrella of differential inequalities. However, these notions traditionally appeared without any drawings, and were not mentioned in elementary texts.

Fences and funnels also yield hard quantitative results. For example, with the fences of Chapter 1 we can often prove that certain solutions to a given differential equation have vertical asymptotes, and then calculate, to as many decimal places as desired, the location of the asymptote for the solution with a particular initial condition. Later in Part III, we use fences forming an antifunnel to easily calculate, with considerable accuracy, the roots of Bessel functions. All of these fruits are readily obtained from the introduction of just these three well-chosen words.

We solve traditionally and explicitly few types of first order equations— linear, separable, and exact—in Chapter 2. These are by far the most useful classical methods, and they will provide all the explicit solutions we desire.

Chapter 4 contains another vital aspect to our approach that is not provided in popular differential equations texts: a Fundamental Inequality (expanding on the version given by Jean Dieudonné in *Calcul Infinitésimal;* see the References). This Fundamental Inequality gives, by a constructive proof, existence and uniqueness of solutions *and* provides error estimates. It solidly grounds the numerical methods introduced in Chapter 3, where a fresh and practical approach is given to error estimation.

Part I closes with Chapter 5 on iteration, usually considered as an entirely different discipline from differential equations. However, as another type of dynamical system, the subject of iteration sheds direct light on how stepsize determines intervals of stability for approximate solutions to a differential equation, and to gain understanding (through Poincaré mapping) of solutions to periodic differential equations, especially with respect to bifurcation behavior.

In subsequent volumes, Parts II, III and IV, as we add levels of complexity, we provide simplicity and continuity by cycling the same concepts introduced in Part I. Part II begins with Chapter 6, where we extend $x' = f(t, x)$ to the multivariate vector version $\mathbf{x}' = \mathbf{f}(t, \mathbf{x})$. This is also the form to which a higher order differential equation in a single variable can be reduced. Chapter 7 introduces linear differential equations of the form $\mathbf{x}' = A\mathbf{x}$, where eigenvalues and eigenvectors accomplish transformation of the vector equation into a set of decoupled single variable first order equations. Chapters 8 and 9 deal with nonlinear differential equations and bifurcation behavior.

In Part III, Chapters 10 and 11 discuss applications to electrical circuits and mechanics respectively. Chapter 12 deals with linear differential equations with nonconstant coefficients, $\mathbf{x}' = A(t)\mathbf{x} + \mathbf{q}(t)$, and includes Sturm–Liouville theory and the theory of ordinary singular points. Finally, Part III again fills out the dynamical systems picture, and closes with Chapter 13 on iteration in two dimensions.

In Part IV, partial differential equations and Fourier series are introduced as an infinite-dimensional extension of the same eigenvalue and eigenvector concept that suffuses Part II. The remaining chapters of the text continue to apply the same few concepts to all the famous differential equations and to many applications, yielding over and over again hard quantitative results. For example, such a calculation instantly yields, to one part in a thousand, the location of the seventh zero of the Bessel function J_0; the argument is based simply on the original concepts of fence, funnel, and antifunnel.

Ithaca, New York   John H. Hubbard
February, 1997   Beverly H. West

Acknowledgments

We are deeply indebted to all the instructors, students, and editors who have taught or learned from this text and encouraged our approach.

We especially thank our colleagues who have patiently taught from the earlier text versions and continued to be enthusiastically involved: Bodil Branner, Anne Noonburg, Ben Wittner, Peter Papadopol, Graeme Bailey, Birgit Speh, and Robert Terrell. Additional vital and sutained support has been provided by Adrien Douady, John Martindale, and David Tall. The book *Systèmes Différentiels: Étude Graphique* by Michèle Artigue and Véronique Gautheron has inspired parts of this text.

The students who have been helpful with suggestions for this text are too numerous to mention individually, but the following contributed particularly valuable and sustained efforts beyond a single semester: François Beraud, Daniel Brown, Fred Brown, Frances Lee, Martha MacVeagh, and Thomas Palmer. Teaching assistants Mark Low, Jiaqi Luo, Ralph Oberste-Vorth, and Henry Vichier-Guerre have made even more serious contributions, as has programmer Michael Abato.

The enormous job of providing the final illustrations has been shared by the authors, and Jeesue Kim, Maria Korolov, Scott Mankowitz, Katrina Thomas, and Thomas Yan (whose programming skill made possible the high quality computer output). Homer Smith of ArtMatrix made the intricate pictures of Mandelbrot and Julia sets for Chapter 5.

Anne Noonburg gets credit for the vast bulk of work in providing solutions to selected exercises. However, the authors take complete responsibility for any imperfections that occur there or elsewhere in the text.

Others who have contributed considerably behind the scenes on the more mechanical aspects at key moments include Karen Denker, Mary Duclos, Fumi Hsu, Rosemary MacKay, Jane Staller, and Frederick Yang.

Evolving drafts have been used as class notes for seven years. Uncountable hours of copying and management have been cheerfully accomplished semester after semester by Joy Jones, Cheryl Lippincott, and Jackie White.

Finally, we are grateful to the editors and production staff at Springer-Verlag for their assistance, good ideas, and patience in dealing with a complicated combination of text and programs.

John H. Hubbard   Beverly H. West

Ways to Use This Book

There are many different ways you might use this book. John Hubbard uses much of it (without too much of Chapter 5) in a junior–senior level course in applicable mathematics, followed by *Part II: Higher Dimensional Differential Equations, Part III: Higher Dimensional Differential Equations Continued,* and *Part IV: Partial Differential Equations.* In each term different chapters are emphasized and others become optional.

Most instructors prefer smaller chunks. A good single-semester course could be made from Chapters 1 and 2, then a lighter treatment of Chapter 3; Chapter 4 and most of Chapter 5 could be optional. Chapters 6, 7, and 8 from Part II comprise the core of higher dimensional treatments. Chapter 7 requires background in linear algebra (provided in the Appendix to Part II), but this is not difficult in the two- and three-dimensional cases. Chapter 8 is important for showing that a very great deal can be done today with nonlinear differential equations.

This series of books has been written to take advantage of computer graphics. We've developed a software package for the Macintosh computer called *MacMath* (which includes *Analyzer, DiffEq, Num Meths, Cascade, 1D Periodic Equations*) and refer to it throughout the text.

Although they are not absolutely essential, we urge the use of computers in working through this text. It need not be precisely with the *MacMath* programs. With IBM/DOS computers, readers can, for example, use *Phaser* by Huseyn Koçak or *MultiMath* by Jens Ole Bach. There are many other options. The chapter on numerical methods has been handled very successfully with a spreadsheet program like *Excel*.

Because so much of this material is a new approach for instructors as well as students, we include a set of solutions to selected exercises, as a guide to making the most of the text.

Contents of Part I

Ordinary Differential Equations
The One-Dimensional Theory $x' = f(t, x)$

Contents of Part II

Systems of Ordinary Differential Equations: The Higher-Dimensional Theory $x' = f(t, x)$

Chapter 6. Systems of Differential Equations
Graphical representation; theorems; higher order equations; essential size; conservation laws; pendulum; two-body problem.

Chapter 7. Systems of Linear Equations, with Constant Coefficients $x' = Ax$
Linear differential equations in general; linearity and superposition principles; linear differential equations with constant coefficients; eigenvectors and decoupling, exponentiation of matrices; bifurcation diagram for 2×2 matrices, eigenvalues and global behavior; nonhomogeneous linear equations.

Chapter 8. Nonlinear Autonomous Systems in the Plane
Local and global behavior of a vector field in the plane; saddles, sources, and sinks; limit cycles.

Chapter 8*. Structural Stability
Structural stability of sinks and sources, saddles, and limit cycles; the Poincaré-Bendixson Theorem; structural stability of a planar vector field.

Appendix. Linear Algebra

L1. Theory of Linear Equations: In Practice
Vectors and matrices; row reduction.

L2. Theory of Linear Equations: Vocabulary
Vector space; linear combinations, linear independence and span; linear transformations and matrices, with respect to a basis; kernels and images.

L3. Vector Spaces with Inner Products
Real and complex inner products; basic theorems and definitions; orthogonal sets and bases; Gram–Schmidt algorithm; orthogonal projections and complements.

Contents of Part III

Higher-Dimensional Equations continued, $\dot{\mathbf{x}}' = \mathbf{f}(t, \mathbf{x})$

Contents of Part IV

Partial Differential Equations
As Linear Differential Equations in Infinitely Many
Dimensions: Extension of Eigenvector Treatment
e.g., $x'' = c^2(\partial^2 x/\partial s^2) = \lambda x$

Introduction

Differential equations are the main tool with which scientists make mathematical models of real systems. As such they have a central role in connecting the power of mathematics with a description of the world.

In this introduction we will give examples of such models and some of their consequences, highlighting the unfortunate fact that even if you can reduce the description of a real system to the mathematical study of a differential equation, you may still encounter major roadblocks. Sometimes the mathematical difficulties involved in the study of the differential equation are immense.

Traditionally, the field of Differential Equations has been divided into the linear and the nonlinear theory. This is a bit like classifying people into friends and strangers. The friends (linear equations), although occasionally quirky, are essentially understandable. If you can reduce the description of a real system to a linear equation, you can expect success in analyzing it, even though it may be quite a lot of work.

The strangers (nonlinear equations) are quite a different problem. They are strange and mysterious, and there is no reliable technique for dealing with them. In some sense, each one is a world in itself, and the work of generations of mathematicians may give only very partial insight into its behavior. In a sense which is only beginning to be really understood, it is unreasonable to expect to understand most nonlinear differential equations completely.

One way to see this is to consider a computer; it is nothing but a system of electrical circuits, and the time evolution of an electrical circuit is described by a differential equation. Every time a program is entered, it is like giving the system a set of initial conditions. The time evolution of a computer while it is running a program is a particular solution to that differential equation with those initial conditions. Of course, it would be an enormously complicated differential equation, with perhaps millions of voltages and currents as unknown functions of time. Still, understanding its evolution as a function of the initial conditions is like understanding all possible computer programs; surely an unreasonable task. In fact, it is *known* (a deep theorem in logic) that there is no algorithm for determining whether a given computer program terminates. As such the evolution of this differential equation is "unknowable," and probably most differential equations are essentially just as complicated. Of course, this doesn't mean that there isn't anything to say; after all, there is a discipline called computer science.

Consequently, most texts on differential equations concentrate on linear equations, even though the most interesting ones are nonlinear. There are, after all, far more strangers than friends. But it is mostly one's friends that end up in a photograph album.

There is an old joke, which may not be a great joke but is a deep metaphor for mathematics:

> A man walking at night finds another on his hands and knees, searching for something under a streetlight. "What are you looking for?", the first man asks; "I lost a quarter," the other replies. The first man gets down on his hands and knees to help, and after a long while asks "Are you sure you lost it here?". "No," replies the second man, "I lost it down the street. But this is where the light is."

In keeping with this philosophy, this text, after Chapter 1, will also deal in large part with linear equations. The reader, however, should realize that the really puzzling equations have largely been omitted, because we do not know what to say about them. But, you will see in Chapter 1 that with the computer programs provided it is now possible to see solutions to any differential equation that you can enter in the form $x' = f(t, x)$, and so you can begin to work with them.

For now we proceed to some examples.

1. THE FRIENDLY WORLD

Example 0.1. Our model differential equation is

$$x' = \alpha x. \tag{1}$$

You can also write this

$$dx/dt = \alpha x(t),$$

and you should think of $x(t)$ as a function of time describing some quantity *whose rate of change is proportional to itself.* As such, the solutions of this differential equation (1) describe a large variety of systems, for instance

(a) The value of a bank account earning interest at a rate of α percent;

(b) The size of some unfettered population with birth rate α;

(c) The mass of some decaying radioactive substance with rate of change α per unit mass. (In this case α is negative.)

As you probably recall from elementary calculus, and in any case as you can check, the function

$$x(t) = x_0 e^{\alpha(t-t_0)} \tag{2}$$

satisfies this differential equation, with value x_0 at t_0. Thus we say the equation has been solved, and almost any information you want can be read off from the solution (2). You can predict how much your bank account will be worth in 20 years, if the equation were describing compound interest. Or, if the equation were describing radioactive decay of carbon 14 in some ancient artifact, you may solve for the time when the concentration was equal to the concentration in the astmosphere. ▲

Example 0.1 gives an exaggeratedly simple idea of linear equations. Most differential equations are more complicated. The interest rate of a bank account, or the birth rate of a population, might vary over time, leading to an equation like

$$x' = \alpha(t)x. \tag{3}$$

Other linear differential equations involve higher order derivatives, like

$$ax'' + bx' + cx = 0, \tag{4}$$

which describes the motion of harmonic oscillators (damped, if $b \neq 0$), and also of RLC circuits. It has also been used to model the influence of government spending on the economy, and a host of other things.

The study of equation (4) is quite a bit more elaborate than the study of equation (1), but it still is essentially similar; we will go into it at length in Chapter 7. Again this is a success story; the model describes how to tune radios and how to build bridges, and is of constant use.

These differential equations (1), (3), and (4) are called *linear* differential equations, where each term is at worst the product of a derivative of x and a function of t, and the differential equation is a finite sum of such terms. That is, a linear differential equation is of the form

$$\alpha_n(t)x^{(n)} + \alpha_{n-1}(t)x^{(n-1)} + \cdots + \alpha_2(t)x'' + \alpha_1(t)x' + \alpha_0(t)x = q(t), \tag{5}$$

where $x^{(n)}$ means the nth derivative $d^n x/dt^n$.

The friendly world of linear differential equations is very accessible and has borne much fruit. In view of our next example, it may be interesting to note that the very first mathematical texts we have are Babylonian cuneiform tablets from 3000 B.C. giving the value of deposits lent at compound interest; these values are precisely those that we would compute by Euler's method (to be described in Chapter 3) as *approximations* to solutions of $x' = \alpha x$, the simplest linear equation (1).

2. The Strange World

It is easy to pinpoint the birth of differential equations: they first appeared explicitly (although disguised in geometric language) in 1687 with Sir Isaac Newton's book *Philosophiae Naturalis Principia Mathematica* (Mathematical Principles of Natural Philosophy). Newton is usually credited with inventing calculus, but we think that accomplishment, great as it is, pales by comparison with his discovering that we should focus on the *forces* to which a system responds in order to describe the laws of nature.

In *Principia*, Newton introduced the following two laws:

(a) A body subject to a force has an acceleration in the same direction as the force, which is proportional to the force and inversely proportional to the mass.

$$(\mathbf{F} = m\mathbf{a}).$$

(b) Two bodies attract with a force aligned along the line between them, which is proportional to the product of their masses and inversely proportional to the square of the distance separating them.
(The Universal Law of Gravitation)

and worked out some of their consequences.

These laws combine as follows to form a *differential equation*, an equation for the position of a body *in terms of its derivatives*:

Example 0.2. Newton's Equation. Suppose we have a system of two bodies with masses m_1 and m_2 which are free to move under the influence of the gravitational forces they exert on each other. At time t their positions can be denoted by vectors with respect to the origin, $\mathbf{x}_1(t)$ and $\mathbf{x}_2(t)$ (Figure 0.1).

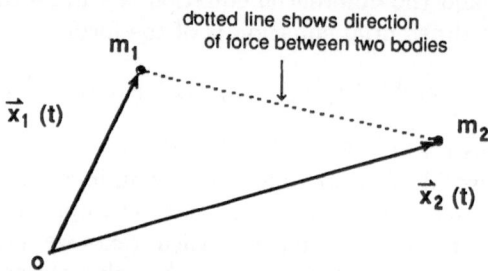

FIGURE 0.1. Representation of two bodies moving in gravitational force field.

The acceleration of the bodies' motions are therefore the second derivatives, $\mathbf{x}_1''(t)$ and $\mathbf{x}_2''(t)$ respectively, of the positions with respect to time.

Combining Newton's Laws (a) and (b) gives the force on the first body as

$$m_1 \mathbf{x}_1''(t) = G \frac{m_1 m_2}{\|\mathbf{x}_2 - \mathbf{x}_1\|^2} \underbrace{\frac{\mathbf{x}_2 - \mathbf{x}_1}{\|\mathbf{x}_2 - \mathbf{x}_1\|}}, \tag{6}$$

this ratio gives the
unit vector from the
first body to the second

and the force on the second body as

$$m_2 \mathbf{x}_2''(t) = G \frac{m_1 m_2}{\|\mathbf{x}_1 - \mathbf{x}_2\|^2} \underbrace{\frac{\mathbf{x}_1 - \mathbf{x}_2}{\|\mathbf{x}_1 - \mathbf{x}_2\|}}. \tag{7}$$

this ratio gives the
unit vector from the
second body to the first

The gravitational constant G of proportionality is *universal*, i.e., independent of the bodies, or their positions.

Equations (6) and (7) form a *system* of differential equations. To be sure, a system is more complicated than a single differential equation, but we shall work up to it gradually in Chapters 1–5, and we can still at this point discuss several aspects of the system.

Most notably, this system of differential equations is *nonlinear;* the equations cannot be written in the form of equation (5), because the denominators are also functions of the variables $\mathbf{x}_1(t)$ and $\mathbf{x}_2(t)$.

As soon as more than two bodies are involved, equations (6) and (7) are simply extended, as decribed by

$$m_i \mathbf{x}_i''(t) = G \sum_{j \neq i} m_i m_j \frac{\mathbf{x}_j - \mathbf{x}_i}{\|\mathbf{x}_j - \mathbf{x}_i\|^3}, \quad \text{for } i = 1, \ldots, n, \tag{8}$$

to a larger nonlinear system which is today in no sense solved, even though it was historically the first differential equation ever considered as such. Regard as evidence the surprise of astronomers at finding the braided rings of Saturn. These braided rings are a solution of Newton's equation which no one imagined until it was observed, and no one knows whether this sort of solution is common or exceptional; in fact no one even knows whether planets usually have rings or not.

Nevertheless, Newton's equation (8) is of great practical importance; for instance, essentially nothing else is involved in the guidance of satellites and other space missions. Newton was able, in the case where there are precisely two bodies, to derive Kepler's laws describing the orbit of each planet around the sun. This derivation can be found in Chapter 6 (Volume II). Its success really launched Newton's approach and completely revolutionized scientists' ways of thinking.

It is instructive to consider what Newton's approach replaced: the Greek and Babylonian theory of *epicycles,* which received its final form in the

Almagest of Ptolemy (2nd century AD). The ancient astronomers had attempted to describe the apparent motion of the planets on the heavenly sphere, and had found that epicycles, a kind of compound circular motion, were a good tool. Imagine a point moving on a circle, itself the center of a circle on which a point is turning, and so on some number of times. Each of these points is moving on a (more and more complex) epicycle, as indicated in Figure 0.2.

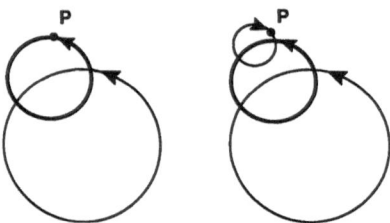

FIGURE 0.2. Epicycles.

It would be quite wrong to ridicule this earlier theory of epicycles; it provides quite good approximations to the observed motions (in fact as good as you wish, if you push the orders of the epicycles far enough). Moreover, if you consider that the sun moves around the earth very nearly on a circle (the right point of view to describe the apparent motions as seen from the earth), and then that the planets turn around the sun also very nearly on circles, epicycles seem a very natural description. Of course, there is the bothersome eccentricity of the ellipses, but in fact an ellipse close to a circle can be quite well approximated by adding one small epicycle, and if you require a better precision, you can always add another. Still, the epicycle theory is involved and complicated; but then again the motions of the planets really are complicated. In fact, *there is no simple description of the motions of the planets.* ▲

The main point of Newton's work was the realization that

the forces are simpler than the motions.

There is no more forceful motivation for the theory of differential equations than this. Historically, Newton's spectacular success in describing mechanics by differential equations was a model for what science should be; in fact Newton's approach became the standard against which all other scientific theories were measured. And the modus operandi laid out by Newton is still largely respected: all basic physical laws are stated as differential equations, whether it be Maxwell's equations for electrodynamics, Schrödinger's equation for quantum mechanics, or Einstein's equation for general relativity.

The philosophy of forces being simpler than motions is very general, and when you wish to make a model of almost any real system, you will want to describe the forces, and derive the motions from them rather than describing the motions directly. We will give an example from ecology, but it could as well come from economics or chemistry.

Example 0.3. Sharks and sardines. Suppose we have two species, one of which preys on the other. We will try to write down a system of equations which reflect the assumptions that

(a) The prey population is controlled only by the predators, and in the absence of predators would increase exponentially.

(b) The predator population is entirely dependent on the prey, and in its absence would decrease exponentially.

Again we shall put this in a more mathematical form. Let $x(t)$ represent the size of the prey population as a function of time, and $y(t)$ the size of the predator population. Then in the absence of interaction between the predators and the prey, the assumptions (a) and (b) could be coded by

$$x'(t) = ax(t)$$
$$y'(t) = -by(t),$$

with a and b positive constants. If the predators and prey interact in the expected way, meetings between them are favorable to the predators and deleterious to the prey. The product $x(t)y(t)$ measures the number of meetings of predator and prey (e.g., if the prey were twice as numerous, you could expect twice as many meetings; likewise if the predators were twice as numerous, you could also expect twice as many meetings). So with c and f also positive constants, the system *with* interaction can be written

$$x'(t) = ax(t) - cx(t)y(t)$$
$$y'(t) = -by(t) + fx(t)y(t). \tag{9}$$

If you have specific values for a, b, c, and f, and if at any time t_0 you can count the populations $x(t_0)$ and $y(t_0)$, then the equations (9) describe exactly how the populations are changing at that instant; you will know whether each population is rising or falling, and how steeply.

This model (much simpler to study than Newton's equation, as you shall see in Chapter 6) was proposed by Vito Volterra in the mid-1920's to explain why the percentage of sharks in the Mediterranean Sea increased when fishing was cut back during the first World War, and is analyzed in some detail in Section 6.3. The modeling process is also discussed there and in Section 2.5. ▲

The equations (9) form another *nonlinear* system, because of the products $x(t)y(t)$. It will be much easier to extract information about the solutions than to find them explicitly.

Basically, what all these examples describe is how a system will be pulled and pushed in terms of where it is, as opposed to stating explicitly the state of the system as a function of time, and that is what every differential equation does. To imagine yourself subject to a differential equation: start somewhere. There you are tugged in some direction, so you move that way. Of course, as you move, the tugging forces change, pulling you in a new direction; for your motion to solve the differential equation you must keep drifting with and responding to the ambient forces.

The paragraph above gives the idea of a *solution* of a differential equation; it is the path of motion under those ambient forces. Finding such solutions is an important service mathematicians can perform for other scientists. But there are major difficulties in the way. Almost exactly a century ago, the French mathematician Poincaré showed that solving differential equations in the elementary sense of finding formulas for integrals, or in the more elaborate sense of finding constants of motion, is sometimes impossible. The King of Sweden had offered a prize for "solving" the 3-body problem, but Poincaré won the prize by showing that it could not be done. More recently, largely as a result of experimentation with computers, mathematicians have grown conscious that even simple forces can create motions that are extremely complex.

As a result, mathematicians studying differential equations have split in two groups. One class, the numerical analysts, tries to find good algorithms to approximate solutions of differential equations, usually using a computer. This is particularly useful in the "short" run, not too far from the starting point. The other class of mathematicians practice the qualitative theory, trying to describe "in qualitative terms" the evolution of solutions in the "long" run, as well as in the short run.

In this book, we have tried constantly to remember that explicit solutions are usually impossible, and that techniques which work without them are essential. We have tried to ally the quantitative and the qualitative theory, mainly by using computer graphics, which allow you to grasp the behavior of many solutions of a differential equation at once. That is, although the computer programs are purely quantitative methods, the graphics make possible qualitative study as well.

We advise the reader, without further waiting, to go to the computer and to run the program *Planets*. Many facts concerning differential equations *and the difficulties in finding solutions* are illustrated by this program, including some phenomena that are almost impossible to explain in text.

The *Planets* program does nothing but solve Newton's equations of motion for whatever initial conditions you provide, for up to ten bodies. Try it first with the predefined initial conditions KEPLER, to see the elliptic orbits which we expect. There are within the program a few other predefined initial conditions which have been carefully chosen to provide systems with some degree of stability.

But entering different initial data for 3 or more bodies is quite a different matter, and you will see that understanding the solutions of the differential equations for Newton's Laws in these cases cannot be easy. You should try experimenting with different masses, positions, and velocities in order to see how unusual it is to get a stable system. In fact, the question of whether our own solar system is stable is still unanswered, despite the efforts of mathematicians like Poincaré.

We shall not further analyze systems of differential equations until Chapter 6; Volume II, where we will use the *Planets* program for computer exploration and discuss specific data. For now we shall move to Chapter 1 and begin with what can be said mathematically about the simplest case, a single differential equation.

1

Qualitative Methods

A *differential equation* is an equation involving derivatives, and the *order* of a differential equation is the highest order of derivative that appears in it. We shall devote this first volume of our study of differential equations to the simplest, the *first order equation*

$$\frac{dx}{dt} = x' = f(t, x),$$

where $f(t, x)$ is a continuous function of t and x. We shall consistently throughout this text use t as the independent variable and x as a dependent variable, a scheme which easily generalizes to higher dimensional systems.

As you shall see in Volume II, *higher* order equations such as $x'' = f(t, x, x')$ can be expressed as a system of *first* order equations, so this beginning indeed underlies the whole subject.

A *solution* of a differential equation is a differentiable function that satisfies the differential equation. That is, for $x' = f(t, x)$,

$$u = u(t) \quad \text{is a } solution \text{ if} \quad u'(t) = f(t, u(t)).$$

We shall behave as if differential equations *have* solutions and discuss them freely; in Chapter 4 we shall set this belief on a firm foundation.

At a point $(t, u(t))$ on a solution curve, the *slope* is $f(t, u(t))$. The core of the qualitative methods for studying differential equations is the "slope field" or "direction field."

1.1   Field of Slopes and Sketching of Solutions

A *direction field* or *slope field* for a differential equation $x' = f(t, x)$ is the direction or slope at every point of the t, x-plane (or a portion of the plane). We shall demonstrate several ways to sketch the slope field by making a selection of points and marking each with a short line segment of slope calculated by $f(t, x)$ for that point.

1. *Grid method.* For the differential equation $x' = f(t, x)$ we first think (roughly) of a rectangular grid of points (t, x) over the entire t, x-plane, and then determine (roughly) the slope of the solutions through each.

Example 1.1.1. Consider $x' = -tx$.

You can sketch the slope field for the solutions to this particular differential equation with a few simple observations, such as

(i) If $t = 0$ *or* if $x = 0$, then $f(t, x) = 0$. So the slope of a solution through any point on either axis is zero, and the direction lines (for solutions of the differential equation) along both axes are all horizontal.

(ii) The direction lines (and therefore the solution curves) are *symmetric* about the origin and about both axes, because the function $f(t, x)$ is antisymmetric about both axes (that is, f changes sign as either variable changes sign). Therefore, considering the first quadrant in detail gives all the information for the others.

(iii) For fixed positive t, the slopes (negative) get steeper as positive x increases.

(iv) For fixed positive x, the slopes (negative) get steeper as positive t increases.

The resulting field of slopes is shown in Figure 1.1.1, with one solution to the differential equation drawn, following the slopes. ▲

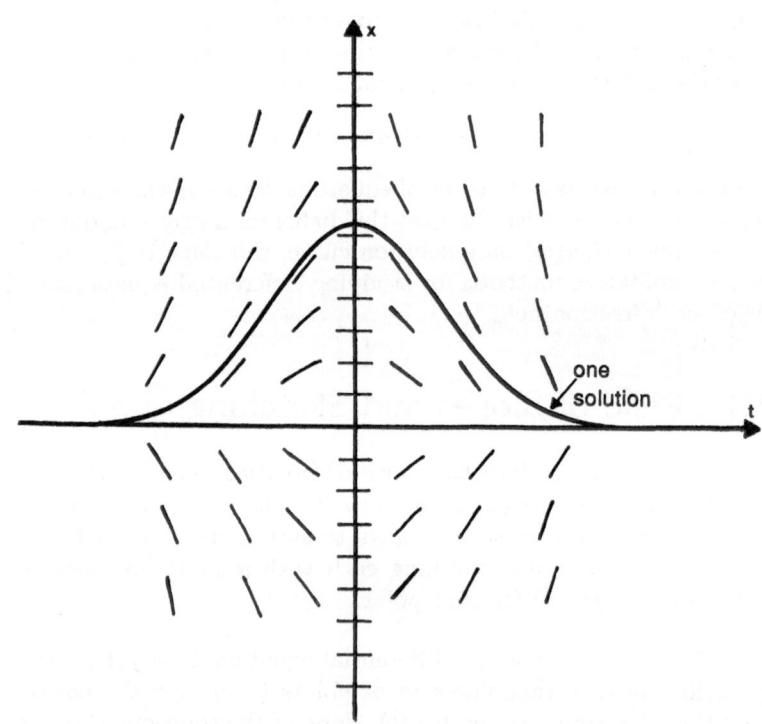

FIGURE 1.1.1. $x' = -tx$. Hand sketch of slope field.

Every solution to the differential equation must follow the direction field, running tangent to every little slope mark that it grazes. (Drawing solutions by hand may take some practice, as you will try in the exercises; a computer does it very well.)

Any point in the plane corresponds to an *initial condition* (t_0, x_0) through which you can draw a solution passing through x_0 at time t_0.

We shall show in Chapter 2 that the particular equation $x' = -tx$ of Example 1.1.1 can be solved analytically by separation of variables, yielding solutions of the form $x = x_0 e^{-t^2/2}$. The solutions that can be drawn in the slope field (Figures 1.1.1, 1.1.3, and 1.1.4) are indeed the graphs of equations of this form, with a different solution for each value of x_0. We call this a *family* of solutions.

In Chapter 4 we shall discuss requirements for existence and uniqueness of solutions. For now we shall simply say that graphically the interpretations of these concepts are as follows:

> *Existence* of solutions means that *you can draw and see them on the direction field.*

> *Uniqueness* means that only one solution can be drawn through any given point or set of initial conditions. Uniqueness holds for the vast majority of points in our examples. An important implication of uniqueness is that *solutions will not meet or cross.*

In Example 1.1.1 the family consists entirely of unique solutions; all except $x = 0$ approach the t-axis asymptotically, never actually meeting or crossing.

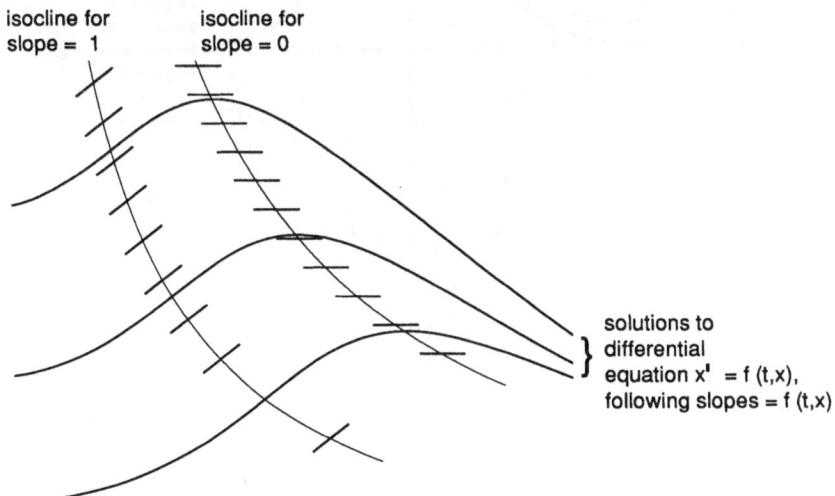

FIGURE 1.1.2. Isoclines.

2. *Isocline method.* Another way (often faster for hand calculation, and more quickly enlightening) to construct a slope field for a differential equation is to find *isoclines,* which are curves on which the solutions to the differential equation have given slope. Set the slope $f(t, x) = c$; usually for each c this equation describes a curve, the isocline, which you might draw in a different color to avoid confusing with solutions. Through any point on the isocline, a solution to the differential equation crosses the isocline with slope c, as shown in Figure 1.1.2. Note that the isocline is simply a locus of "equal inclination"; it is *not* (with rare exceptions) a solution to the differential equation.

Example 1.1.2. Consider again the equation of Example 1.1.1, $x' = -tx$.

The isoclines are found by setting $-tx = c$, so they are in fact the coordinate axes (for $c = 0$) and a family of hyperbolas (for $c \neq 0$). Along each hyperbola we draw direction lines of the appropriate slope, as in Figure 1.1.3.   ▲

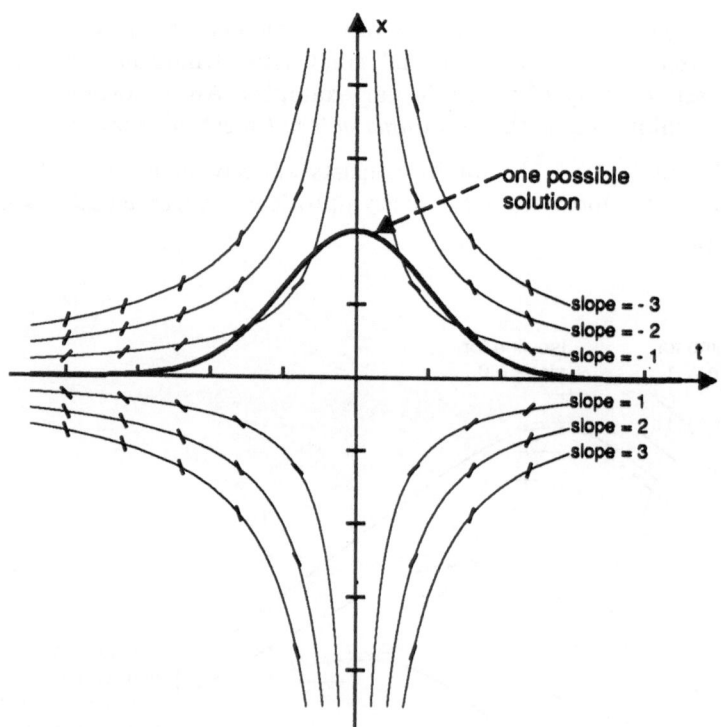

FIGURE 1.1.3. $x' = -tx$. Slope marks on isoclines.

The important thing to realize is that Figure 1.1.3 represents the same direction field as Figure 1.1.1, but we have chosen a different selection of points at which to mark the slopes.

3. *Computer calculation.* A slope field can also be illustrated by computer calculation of $f(t, x)$ for each (t, x) pair in a predetermined grid. Figure 1.1.4 shows the result of such a calculation, for the same equation as Examples 1.1.1 and 1.1.2, with many solutions drawn on the slope field.

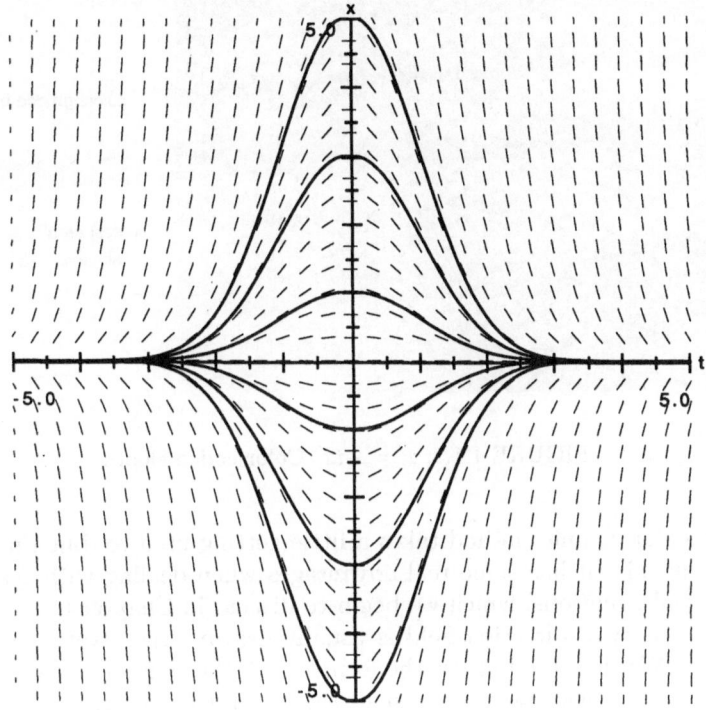

FIGURE 1.1.4. $x' = -tx$. Slope marks on a grid.

The computer program *DiffEq* will draw the slope field for any equation of the form $x' = f(t, x)$ that you enter. Then by positioning the cursor at any point in the window, you are determining an initial condition, from which the computer will draw a solution. (The computer actually draws an *approximate* solution, to be discussed at length in Chapter 3.)

Different programs on different computers may represent the slope field in different ways. The Macintosh version of *DiffEq* makes the familiar little slope marks as shown in Figure 1.1.4. The IBM program for *DiffEq* uses color instead of slope marks to code for direction. Different colors mark regions of different slope ranges; consequently the boundaries between the colors represent isoclines. Figure 1.1.5 gives a sample for the same differential equation as Figure 1.1.4.

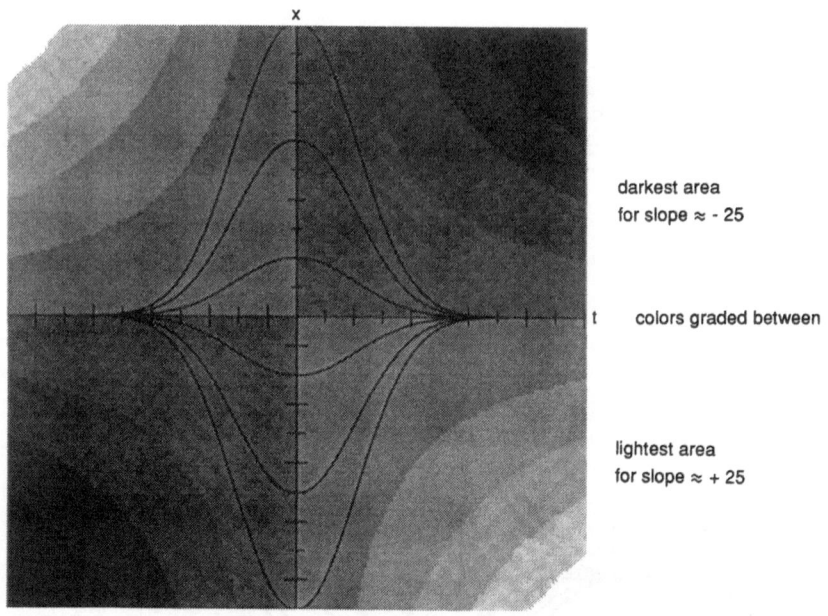

FIGURE 1.1.5. $x' = -tx$. Color-coded slopes.

This color-coding method takes a little getting used to, but that comes rather quickly; it has some real advantages when dealing with systems of differential equations, which we begin to discuss in Chapter 6.

In summary, a direction field is usually drawn either over a grid or by isoclines. We shall tend to use the former with the computer and the latter for hand drawings, but as you shall see in later examples, you will need both. To analyze a direction field you will often need to sketch by hand at least a part of it.

At this point, some different examples are in order.

Example 1.1.3. Consider $x' = 2t - x$.

First we can find the isoclines by setting $2t - x = c$. Hence the isoclines are straight lines of the form $x = 2t - c$, and the slope field looks like the top half of Figure 1.1.6. Solutions can be drawn in this field, as shown in the bottom half of Figure 1.1.6.

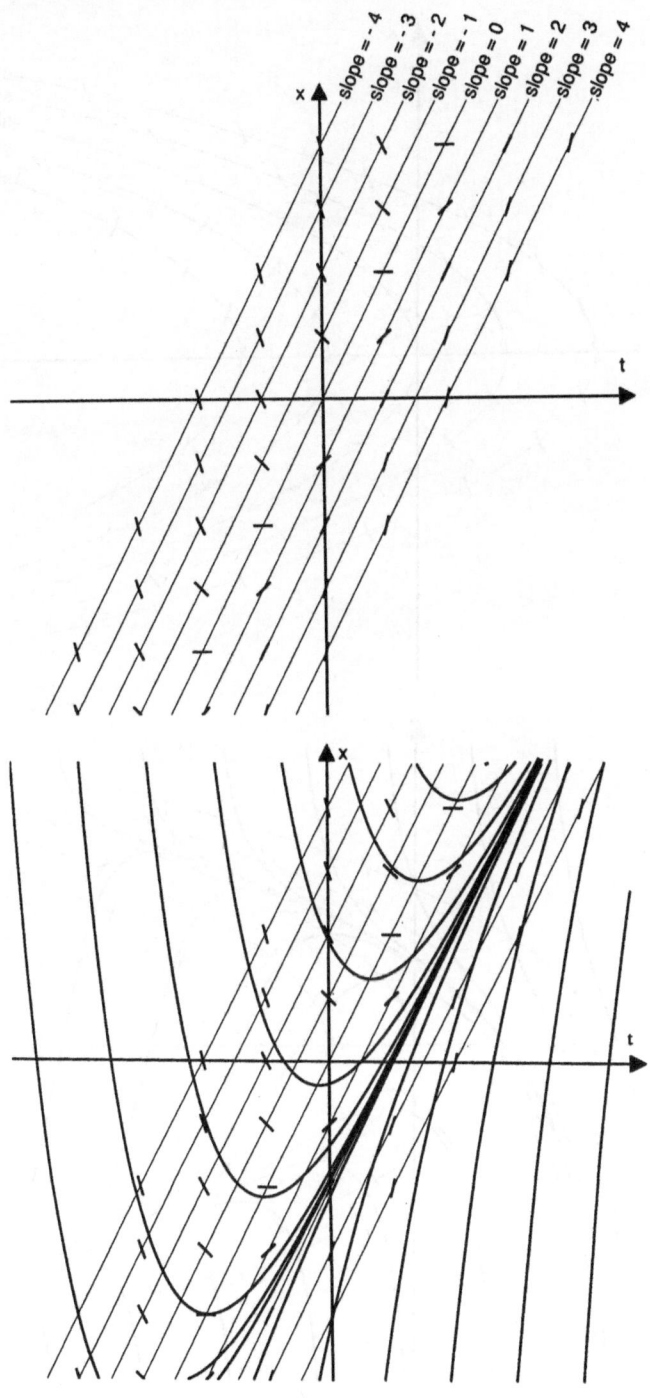

FIGURE 1.1.6. $x' = 2t - x$.

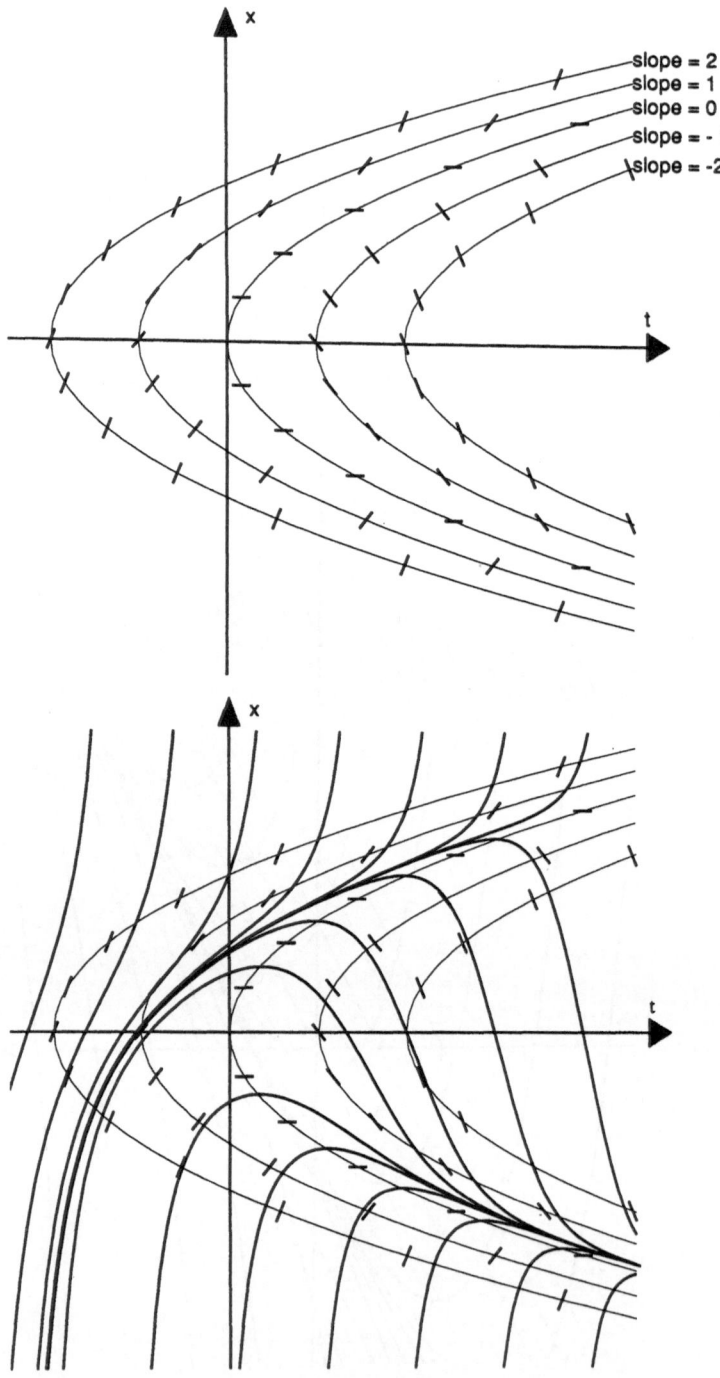

FIGURE 1.1.7. $x' = x^2 - t$.

The solutions are of the algebraic form $x = ke^{-t} + 2t - 2$. You can confirm that these are solutions by substituting this function into the differential equation. You can also see that $x = 2t - 2$ is an asymptote for all the solutions where $k \neq 0$. Furthermore, this line itself is a solution (for $k = 0$), one of those rare cases where an isocline is also a solution. ▲

Example 1.1.4. Consider $x' = x^2 - t$.

The isoclines are of the form $x^2 - t = c$, so they are all parabolas. The direction field is shown at the top of Figure 1.1.7, and solutions are drawn upon it at the bottom. Some solutions fly up, some fall down; exactly one exceptional solution does neither and separates the other two behaviors. Details of these phenomena will be explored later in Section 1.5. ▲

The differential equation of Example 1.1.4 and Figure 1.1.7 is of particular interest because although it looks utterly simple, there are *no* formulas in terms of elementary functions, or even in terms of integrals of elementary functions, for the solutions. A proof of this surprising fact is certainly not easy, and involves a branch of higher mathematics called Galois theory; we do not provide further details at this level. But what it means is that solving the equation $x' = x^2 - t$ cannot be reduced to computing integrals. As you can clearly see, this does not mean that the solutions themselves do not exist, only that formulas do not exist.

> *An important strength of the qualitative method of sketching solutions on direction fields is that it lets us see these solutions for which there are no formulas, and therefore let us examine their behavior.*

We shall demonstrate in the remainder of this chapter how to use rough graphs of fields of slopes to find more specific information about the solutions. We shall first present definitions and theory, then we shall give examples showing the power of these extended techniques.

1.2 Qualitative Description of Solutions

We can now draw slope fields and solutions upon them, as in Figures 1.2.1. We need a language with which to discuss these pictures. How can we describe them, classify them, distinguish among them?

1. *Funnels and antifunnels.* Usually the main question to ask is "what happens as $t \to \infty$?" In the best circumstances, solutions will come in classes, all tending to infinity in the same way, perhaps asymptotically to a curve which can be explicitly given by an equation. This happens in

the lower halves of each picture in Figure 1.2.1. The solutions that come together behave as if they were in a *funnel,* as shown in Figure 1.2.2.

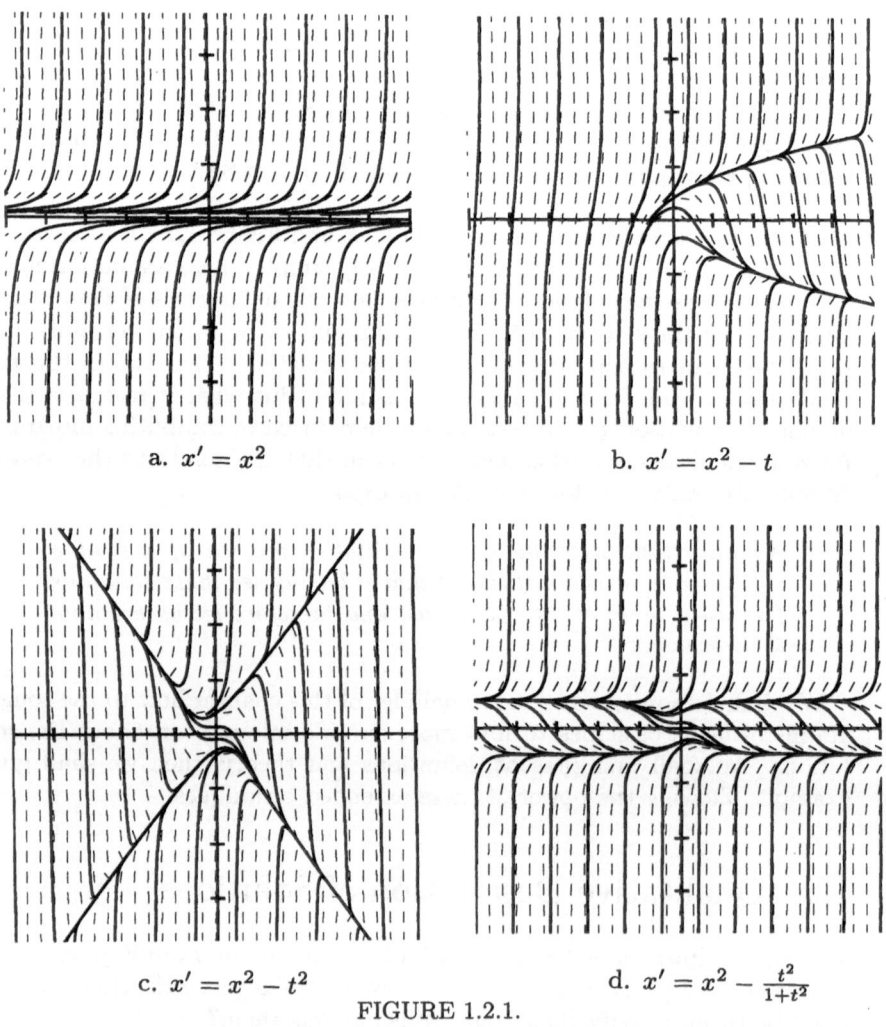

a. $x' = x^2$

b. $x' = x^2 - t$

c. $x' = x^2 - t^2$

d. $x' = x^2 - \frac{t^2}{1+t^2}$

FIGURE 1.2.1.

Classes of solutions which behave in the same way are often separated by solutions with exceptional behavior. The solution $x(t) = 0$ to the equation $x' = 2x - x^2$ (see Figure 1.2.3) is such an exceptional solution, separating the solutions which tend to 2 as $t \to \infty$ from those which tend to $-\infty$. Such solutions often lie in *antifunnels,* parts of the plane in which solutions fly away from exceptional solutions. An antifunnel is just a backwards funnel, but to avoid confusion we adopt the convention of always thinking from left to right along the t-axis. Since in many practical applications t represents time, this means thinking of going forward in time. Figure 1.2.3 shows a funnel on the upper part, an antifunnel on the lower part. Other good antifunnels occur in Figure 1.2.1 (the upper half of each picture).

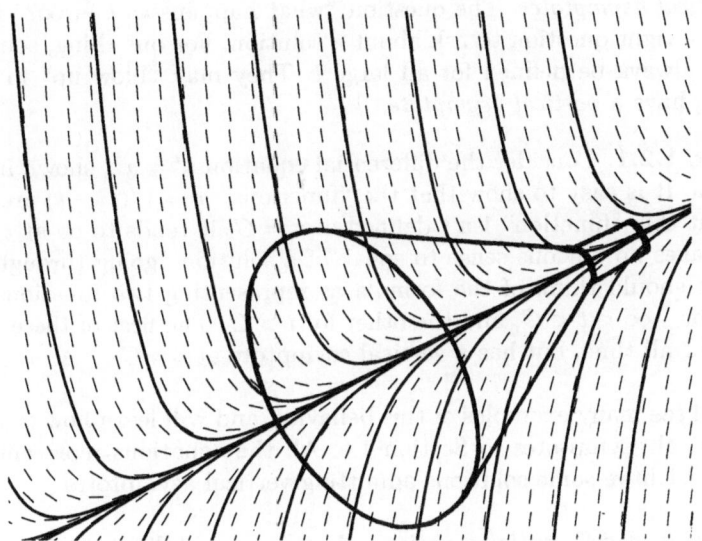

FIGURE 1.2.2. Funnel (of solutions in the t, x-plane).

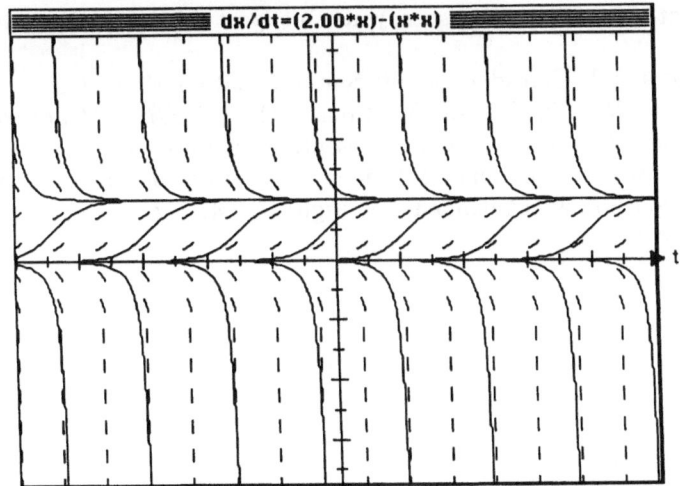

FIGURE 1.2.3. Funnel (in upper part); Antifunnel (in lower part).

Funnels and antifunnels are related to the stability of individual solutions. If the initial condition (the starting point for drawing a solution) is slightly perturbed in the x-direction, what happens? If a solution is *stable,* the perturbed solution is very similar to the original solution, as in a funnel. If a solution is *unstable,* perturbed solutions may fly off in different directions, as in an antifunnel.

2. *Vertical asymptotes.* The question "what happens as $t \to \infty$?" is not always the right question to ask about a solution. For one thing, solutions may not always be defined for all large t. They may "blow up" in finite time, i.e., have a *vertical asymptote.*

Example 1.2.1. Consider the differential equation $x' = x^2$, shown in Figure 1.2.1a. It is easy to show that the "functions" $x = 1/(C - t)$ are solutions. But this "function" isn't defined at $t = C$; it tends to ∞ at $t = C$, and it makes no obvious sense to speak of a solution "going through ∞." Thus one should think of the formula as representing two functions, one defined for $-\infty < t < C$, and the other for $t > C$. The first of these is not defined for all time, but has a vertical asymptote at $t = C$.   ▲

We will see many examples of this behavior, and will learn how to locate such vertical asymptotes in Section 1.6. All the equations represented in Figure 1.2.1 have some solutions admitting vertical asymptotes.

3. *Undefined differential equations.* Another possibility why it may be inappropriate to ask "what happens as $t \to \infty$?" is that a solution may land somewhere where the differential equation is not defined. Anytime a differential equation $x' = f(t, x)$ has f given by a fraction where the denominator sometimes vanishes, you can expect this sort of thing to occur.

Example 1.2.2. Consider the differential equation $x' = -t/x$.
   For every $R \geq 0$, the two functions $x(t) = \pm\sqrt{R^2 - t^2}$ are solutions, defined for $-R < t < R$, and representing an upper or lower half circle depending on the sign. These solutions just end on the t-axis, where $x = 0$ and the equation is not defined. See Figure 1.2.4.   ▲

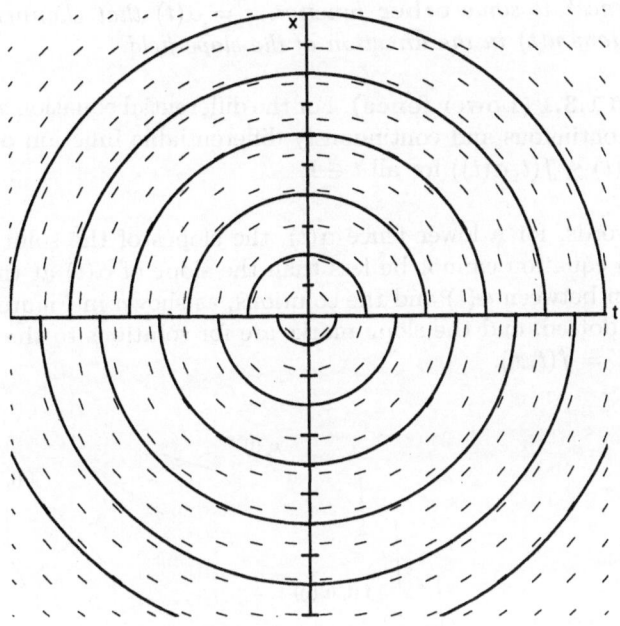

FIGURE 1.2.4. $x' = -t/x$.

Students often think that worrying about domains of definition of functions is just mathematicians splitting hairs; this is not so. Consider the equations of motion under gravity (equation (8) of the Introduction). The right-hand side is a fraction, and the equations are not defined if $\mathbf{x}_i = \mathbf{x}_j$; of course, this corresponds to collision. More generally, the differential equation describing a system will usually be undefined when the system undergoes some sort of catastrophe.

In order to describe quite precisely these various pictorial behaviors, we begin in the next section to formalize these notions.

1.3 Fences

Consider the standard first-order differential equation,

$$x' = f(t, x).$$

On some interval **I** (an open or closed interval from t_0 to t_1, where t_1 might be ∞, t_0 might be $-\infty$), we shall formally define funnels and antifunnels in terms of *fences*.

For a given differential equation $x' = f(t, x)$ with solutions $x = u(t)$, *a "fence" is some **other** function $x = \alpha(t)$ that channels the solutions $u(t)$ in the direction of the slope field.*

Definition 1.3.1 (Lower fence). For the differential equation $x' = f(t, x)$, we call a continuous and continuously differentiable function $\alpha(t)$ a *lower fence* if $\alpha'(t) \leq f(t, \alpha(t))$ for all $t \in \mathbf{I}$.

In other words, for a lower fence $\alpha(t)$, the slopes of the solutions to the differential equation cannot be less than the slope of $\alpha(t)$ at the points of intersection between $\alpha(t)$ and the solutions, as shown in Figure 1.3.1. The fences are dotted, and the slope marks are for solutions to the differential equation $x' = f(t, x)$.

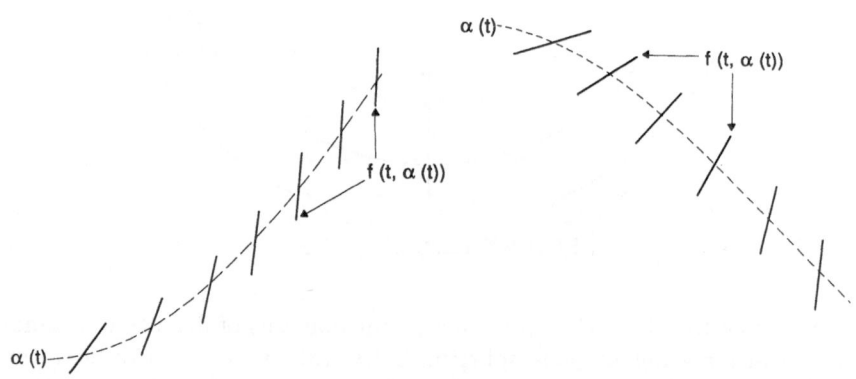

FIGURE 1.3.1. Lower fences $\alpha'(t) \leq f(t, \alpha(t))$.

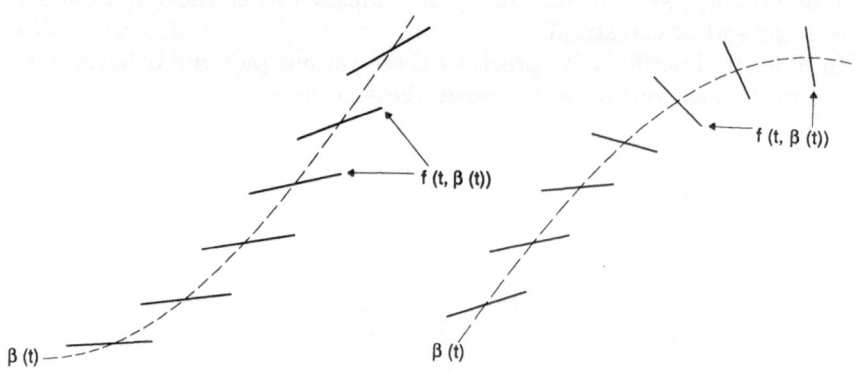

FIGURE 1.3.2. Upper fences $f(t, \beta(t)) \leq \beta'(t)$.

Definition 1.3.2 (Upper fence). For the differential equation $x' = f(t, x)$, we call a continuous and continuously differentiable function $\beta(t)$ an *upper fence* if $f(t, \beta(t)) \leq \beta'(t)$ for all $t \in \mathbf{I}$.

An intuitive idea is that *a lower fence pushes solutions up, an upper fence pushes solutions down.*

Example 1.3.3. Consider $x' = x^2 - t$ as in Example 1.1.4. For the entire direction field for this differential equation, refer to Figures 1.1.7 and 1.2.1b; in our examples (Figure 1.3.3a,b,c) we shall just draw the relevant pieces. ▲

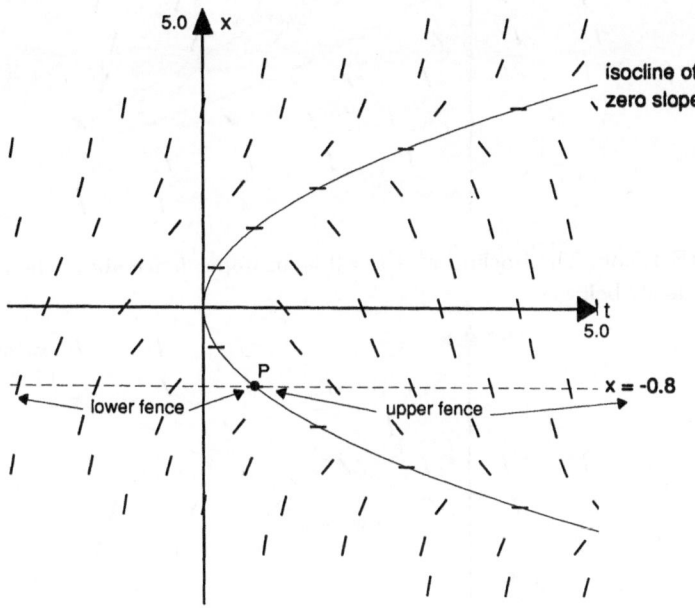

FIGURE 1.3.3a. The line $x = -0.8$ is a lower fence to the left of the point P on the isocline of zero slope and an upper fence to the right.

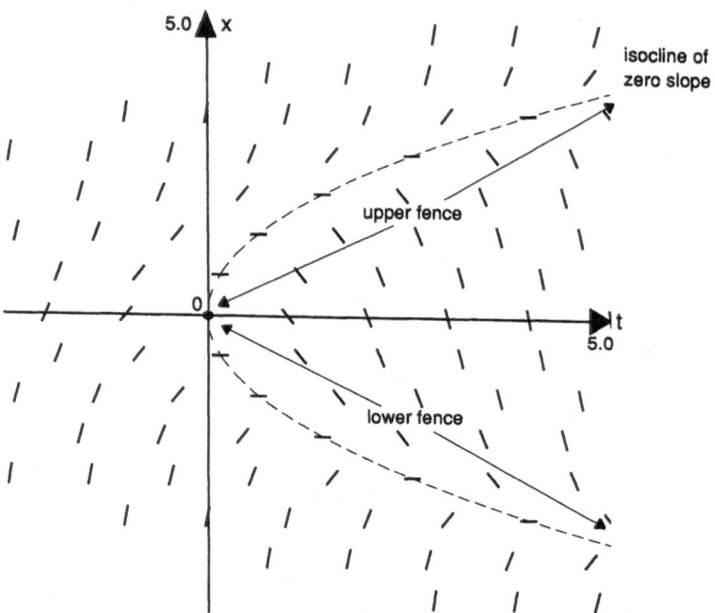

FIGURE 1.3.3b. The isocline $x^2 - t = 0$ is an upper fence above the t-axis and a lower fence below.

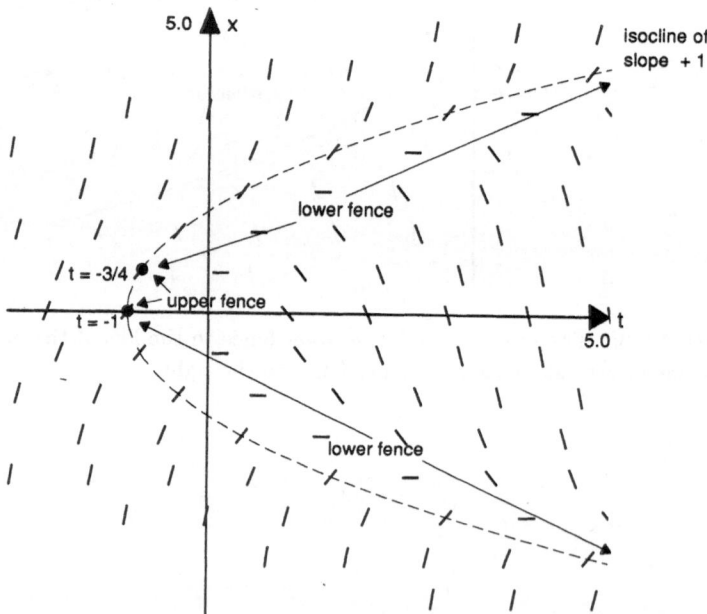

FIGURE 1.3.3c. The isocline $x^2 - t = 1$ is comprised of *three* fences, divided at $t = -1$ (where the isocline has infinite slope) and at $t = -\frac{3}{4}$ (where the isocline has slope = 1).

A fence can be *strong*, when the inequalities are strict:

$$\alpha'(t) < f(t, \alpha(t)) \quad \text{or} \quad f(t, \beta(t)) < \beta'(t),$$

or *weak*, when for some values of t equality is realized, so that the best statements you can make are

$$\alpha'(t) \leq f(t, \alpha(t)) \quad \text{or} \quad f(t, \beta(t)) \leq \beta'(t).$$

Fences can also be *porous* or *nonporous*, depending on whether or not solutions can sneak through them:

Definition 1.3.4. For a differential equation $x' = f(t, x)$ with solution $x = u(t)$,

a *lower* fence $\alpha(t)$ is *nonporous* if whenever $\alpha(t) \leq u(t)$, then $\alpha(t) < u(t)$ for *all* $t > t_0$ in **I** where $u(t)$ is defined;

an *upper* fence $\beta(t)$ is *nonporous* if whenever $u(t) \leq \beta(t)$, then $u(t) < \beta(t)$ for all $t > t_0$ in **I** where $u(t)$ is defined.

The reason for the fence terminology is the following result of the above definitions: *if the fence is nonporous, a solution that gets to the far side of a fence will **stay** on that side.* That is, a solution that gets above a nonporous lower fence will stay above it, and a solution that gets below a nonporous upper fence will stay below. A nonporous fence is like a *semipermeable membrane* in chemistry or biology: a solution to the differential equation may cross it in only one above-below direction from left to right, never in the opposite direction.

Under reasonable circumstances, all fences will turn out to be nonporous. This statement is rather hard to prove, but we shall do so in Chapter 4, where exceptions will also be discussed. Meanwhile, with a *strong* fence we can get almost for free the following theorem:

Theorem 1.3.5 (Fence Theorem for strong fences). *A strong fence for the differential equation $x' = f(t, x)$ is **nonporous.***

Proof. We shall prove the theorem for a *lower* fence.

The hypothesis $\alpha(t_0) \leq u(t_0)$ means that at t_0, the solution $u(t)$ of the differential equation is at or above the fence $\alpha(t)$. The fact that $\alpha(t)$ is a strong lower fence means that $\alpha'(t) < f(t, \alpha(t))$.

The conclusion, $\alpha(t) < u(t)$ for all $t > t_0$, means that $u(t)$ stays above $\alpha(t)$.

i) Suppose first that $\alpha(t_0) < u(t_0)$. Then suppose the opposite of the conclusion, that for some $t > t_0$, $\alpha(t) \geq u(t)$. Let t_1 be the first $t > t_0$ such that $\alpha(t) = u(t)$, as in Figure 1.3.4. At t_1,

$$u'(t_1) = f(t_1, u(t_1)) = f(t_1, \alpha(t_1)) > \alpha'(t_1),$$

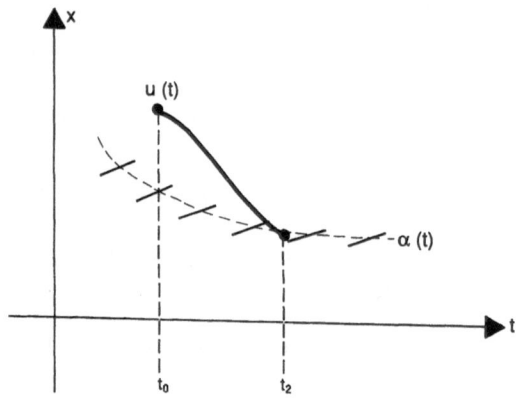

FIGURE 1.3.4.

by definition of lower fence. If $u'(t_1) > \alpha'(t_1)$, then $u(t) - \alpha(t)$ is increasing at t_1. However this contradicts the fact that $u(t) - \alpha(t)$ is positive to the left of t_1 but supposed to be zero at t_1.

ii) If $\alpha(t_0) = u(t_0)$ and

$$u'(t_0) = f(t_0, u(t_0)) = f(t_0, \alpha(t_0)) > \alpha'(t_0),$$

then $u(t) - \alpha(t)$ is increasing at t_0. Therefore the solution first moves above the fence for $t > t_0$, and after that the first case will apply.

The case for an upper fence is proved analogously.   □

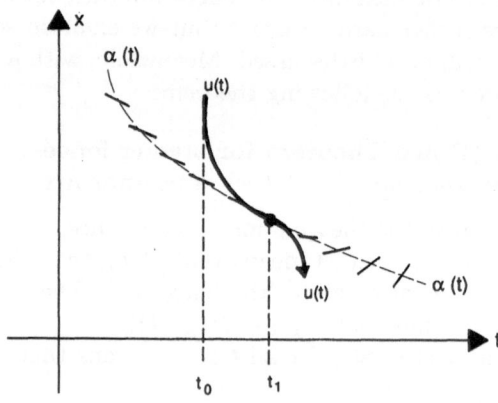

FIGURE 1.3.5.

Note how part i) of the proof of Theorem 1.3.5 requires a strong fence. We would not have reached the contradiction if $\alpha'(t_1) = u'(t_1)$. Figure 1.3.5 shows how a solution might "sneak through" a weak fence. In Chapter 4 we shall show how to close this hole (in the fence and in the argument!).

As you shall soon see, we shall often need to construct fences out of pieces, so a more inclusive definition requires $\alpha(t)$ and $\beta(t)$ to be continuous and *piecewise* differentiable.

> A *piecewise differentiable* function is composed of pieces on which the function is continuous and continuously differentiable except at finitely many points, and at those points left-hand and right-hand derivatives exist though they differ; examples are shown in Figure 1.3.6. (See Exercises 1.2–1.4#6 for discussion of continuous functions that are *not* piecewise differentiable.)

FIGURE 1.3.6. Piecewise differentiable functions.

If $\alpha(t)$ or $\beta(t)$ are piecewise differentiable, the fence definitions also require that the inequalities hold for both the left-hand and right-hand derivatives at the points joining the pieces.

You may well ask what happens to the proof of Theorem 1.3.5 at discontinuities in the derivative of either fence, but the answer is that our argument is a perfectly good one-sided one in each case, left-hand for i) and right-hand for ii), so angles in the graph of the fence will not matter.

Example 1.3.6. Consider again $x' = x^2 - t$.

A lower fence across the whole plane can be constructed from the half-line $x = -1$ and the lower portion of the parabola shown in part b) of Example 1.3.3. This is done in Figure 1.3.7.   ▲

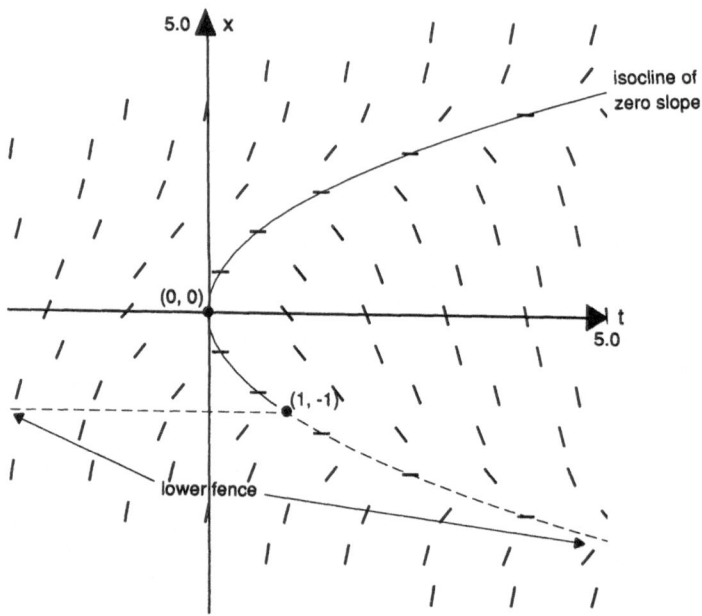

FIGURE 1.3.7.

But what makes a fence "interesting"? What is the qualitative *use* of fences? In Section 1.4 we shall discuss why we want fences, so you will know where to look for them. Then we shall return in Section 1.5 to details of how in practice you *find* fences.

1.4 Funnels and Antifunnels

Fences are especially valuable when there is one of each kind, close together (as often happens with isoclines for a slope field), stretching from some t_0 to infinity. These determine funnels and antifunnels, and the formal definitions provide some extremely useful theorems.

We will want nonporous fences. Recall that the Fence Theorem 1.3.5 assures us that strong fences are nonporous; weak fences may not be. We shall see in Chapter 4 that an additional condition will assure nonporosity for a weak fence. For now we shall not worry about it further; you can rest assured that all the weak fences in the examples and exercises of Chapter 1 will turn out to be nonporous. We shall proceed to show what nonporous fences can do.

Definition 1.4.1 (Funnel). If for the differential equation $x' = f(t, x)$, over some t-inverval **I**, $\alpha(t)$ is a *nonporous lower fence* and $\beta(t)$ a *nonporous* *upper fence,* and if $\alpha(t) < \beta(t)$, then the set of points (t, x) for $t \in \mathbf{I}$ with $\alpha(t) \leq x \leq \beta(t)$ is called a *funnel.*

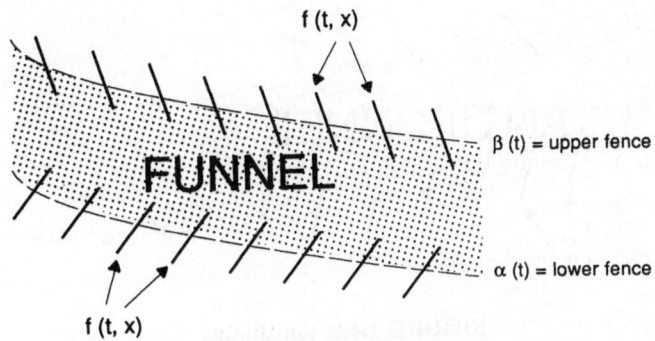

FIGURE 1.4.1. Funnel.

Once a solution enters a funnel, it stays there. This is the message of Theorem 1.4.2, which is especially significant if the funnel narrows as $t \to \infty$.

Theorem 1.4.2 (Funnel Theorem). *If $\alpha(t)$ and $\beta(t)$ determine a funnel for $t \in \mathbf{I}$, and if $u(t)$ is a solution to $x' = f(t, x)$ with $u(t^*)$ in the funnel for some $t^* \in \mathbf{I}$, then $u(t)$ is in the funnel for all $t > t^*$ in $\mathbf{I}$ for which $u(t)$ is defined.*

Proof. The theorem is an immediate consequence of Definition 1.4.1 and the proof of the Fence Theorem 1.3.5, since a nonporous upper fence prevents the solution from escaping from the top of the funnel, and a nonporous lower fence prevents escape at the bottom. □

Definition 1.4.3 (Antifunnel). If for the differential equation $x' = f(t, x)$, over some t-interval $\mathbf{I}$, $\alpha(t)$ is a *nonporous lower fence* and $\beta(t)$ a *nonporous upper fence,* and if $\alpha(t) > \beta(t)$, then the set of points (t, x) for $t \in \mathbf{I}$ with $\alpha(t) \geq x \geq \beta(t)$ is called an *antifunnel.*

Solutions are, in general, *leaving* an antifunnel. But *at least one solution is trapped inside the antifunnel,* as is guaranteed by the following theorem:

Theorem 1.4.4 (Antifunnel Theorem: Existence). *If $\alpha(t)$ and $\beta(t)$ determine an antifunnel for $t \in \mathbf{I}$, then there exists a solution $u(t)$ to the differential equation $x' = f(t, x)$ with $\beta(t) \leq u(t) \leq \alpha(t)$ for all $t \in \mathbf{I}$.*

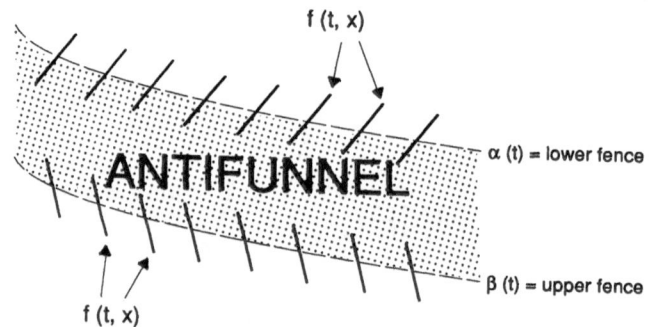

FIGURE 1.4.2. Antifunnel.

To *prove* Theorem 1.4.4, the *existence* of at least one solution in an antifunnel, we must wait until Chapter 4. For certain antifunnels, however, there is another part of the Antifunnel Theorem, *uniqueness* of an exceptional solution that stays in the antifunnel, which is perhaps more surprising and which we *can* prove in a moment as Theorem 1.4.5.

In order to discuss uniqueness of a solution in an antifunnel, we need first to examine the quantity $\partial f/\partial x$ which measures "dispersion" of solutions to the differential equation $x' = f(t, x)$, for which f gives the *slope*.

The *dispersion* $\partial f/\partial x$, if it exists, measures how fast solutions of the differential equation $x' = f(t, x)$ "pull apart," or how quickly the slope f is changing in a vertical direction in the t, x-plane, as shown in Figure 1.4.3. (Recall from multivariable calculus that

$$\frac{\partial f}{\partial x} \equiv \lim_{h \to 0} \frac{f(t, x + h) - f(t, x)}{h},$$

so for a *fixed* t, $\partial f/\partial x$ measures the rate of change with respect to x, that is, in a vertical direction in the t, x-plane.)

> *If the dispersion $\partial f/\partial x$ is large and positive, solutions tend to fly apart in positive time: If $\partial f/\partial x$ is large and negative, solutions tend to fly apart in negative time: if $\partial f/\partial x$ is close to zero, solutions tend to stay together.*

Thus dispersion $\partial f/\partial x$ measures the *stability* of the solutions. If solutions are pulling apart slowly, then an error in initial conditions is less crucial.

We also have the following result: The distance between two solutions $x = u_1(t)$ and $x = u_2(t)$ is *nondecreasing* in a region where $\partial f/\partial x \geq 0$, and *nonincreasing* in a region where $\partial f/\partial x \leq 0$, as we shall show in the next proof.

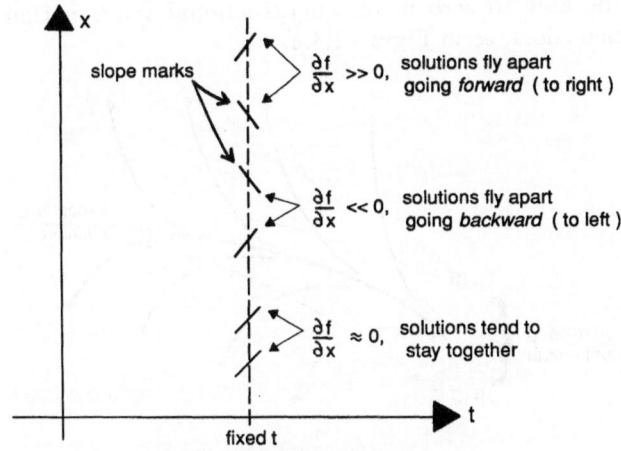

FIGURE 1.4.3. Dispersion.

Theorem 1.4.5 (Antifunnel Theorem: Uniqueness). *If for the differential equation $x' = f(t, x)$, the functions $\alpha(t)$ and $\beta(t)$ determine an antifunnel for $t \in \mathbf{I}$, and if the antifunnel is* **narrowing**, *with*

$$\lim_{t \to t_1} |\alpha(t) - \beta(t)| = 0,$$

and if $\partial f / \partial x \geq 0$ in the antifunnel, then there exists **one and only one** *solution $u(t)$ that stays in the antifunnel.*

Proof (of the "only one" part: the existence was proved in Theorem 1.4.4). The assumption $\partial f / \partial x \geq 0$ implies that the solutions cannot come together as t increases. Indeed, let u_1 and u_2 be two solutions in the antifunnel, with $u_1(t) > u_2(t)$. Then

$$(u_1 - u_2)'(t) = f(t, u_1(t)) - f(t, u_2(t)) = \int_{u_2(t)}^{u_1(t)} \frac{\partial f}{\partial x} (t, u) du \geq 0$$

so that the distance between them can never decrease. This is incompatible with staying between the graphs of α and β which are squeezing together, so there cannot be more than one solution that stays in the antifunnel. $\square$

Antifunnels are sometimes more important than funnels. Although antifunnels correspond to instability, where small changes in initial conditions lead to drastic changes in solutions, you may be able to use them to get some very specific information:

For example, if the functions $\alpha(t)$ and $\beta(t)$ determine a narrowing antifunnel such that

$$\lim_{t \to \infty} |\alpha(t) - \beta(t)| = 0,$$

you may be able to zero in on an *exceptional solution* that divides the
different behaviors, as in Figure 1.4.4.

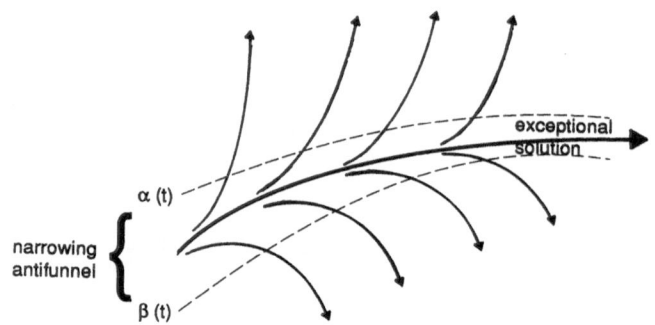

FIGURE 1.4.4. Exceptional solution in narrowing antifunnel.

*The importance of fences, funnels, and antifunnels lies in the
fact that it is **not** necessary to **solve** the differential equation
to recognize them and use them to give vital information about
solutions.*

We shall now in Section 1.5 go on to examples of their use.

1.5   The Use of Fences, Funnels, and Antifunnels

There are many ways to find fences. Almost any curve you can draw is a
fence (or a piecewise combination of upper and lower fences), as shown in
Examples 1.3.3 and 1.3.6. Some *isoclines,* or parts thereof, make obvious
fences; some fences come from *bounds on $f(t,x)$;* other fences, and *these are
the most important,* come from *solutions to simpler but similar differential
equations.*

   In the following examples we shall illustrate the use of each of these
methods in finding fences to construct funnels and antifunnels, and we
shall show the sort of quantitative information that can result.

   1. *Isoclines.* One of the easiest ways to obtain useful fences is to examine
the *isoclines* for a given differential equation.

Example 1.5.1. Consider $x' = x^2 - t$, as in Examples 1.1.4, 1.3.3, 1.3.6
and Exercises 1.2–1.4#1,2.

   The isocline $x^2 - t = 0$, for $t > 0$, $x < 0$, is a lower fence for this
differential equation, and the isocline $x^2 - t = -1$, for $t > 5/4$, $x < 0$, is an
upper fence (Exercises 1.5#2), as shown in Figure 1.5.1.

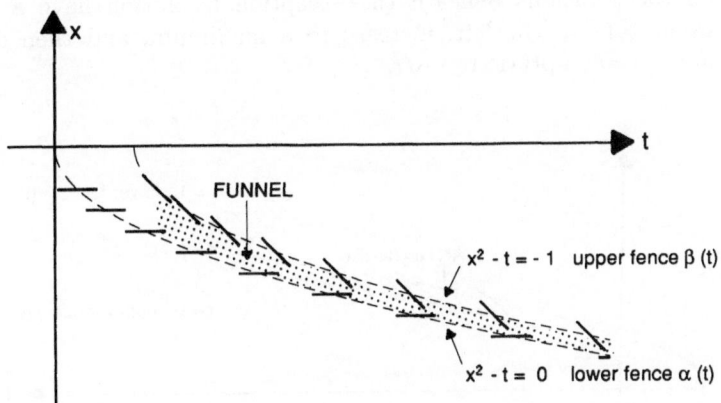

FIGURE 1.5.1.

We label the lower fence $\alpha(t) = -\sqrt{t}$ and the upper fence $\beta(t) = -\sqrt{t-1}$. Then the functions $\alpha(t)$ and $\beta(t)$ determine a *funnel* that swallows all solutions for $t \geq 5/4$, $-\sqrt{t} \leq x \leq \sqrt{t}$. (The upper limit for x becomes clear in Figure 1.5.2.) As you have shown in Exercise 1.2–1.4#5,

$$\lim_{t \to \infty} |\alpha(t) - \beta(t)| = 0.$$

Therefore, for $t \to \infty$, the funnelled solutions to the differential equation behave like $\alpha(t) = -\sqrt{t}$.

Furthermore, the isoclines $x^2 - t = 0$ and $x^2 - t = 1$, for $t > 0$, can be written as $\gamma(t) = +\sqrt{t}$, $\delta(t) = +\sqrt{t+1}$. The curves $\gamma(t)$ and $\delta(t)$ determine a *narrowing antifunnel* for $t > 0$. Moreover, since $\partial f/\partial x = 2x > 0$ in the antifunnel, by Theorem 1.4.5 the solution $\hat{u}(t)$ that stays in the antifunnel is unique (Figure 1.5.2, on the next page).

This is a particularly illustrative case of the information obtainable from direction fields, funnels and antifunnels:

inside the parabola $x^2 - t = 0$, the slopes are negative so the solutions are all decreasing in this region; outside this parabola the slopes are positive so all solutions must be increasing there;

there is *one* exceptional solution $\hat{u}(t)$, $\sqrt{t} < \hat{u}(t) < \sqrt{t+1}$, uniquely specified by requiring it to be asymptotic to $+\sqrt{t}$ for $t \to \infty$;

all the solutions above the exceptional solution are monotone increasing and have vertical asymptotes both to the left and to the right. (The vertical asymptotes need to be proven, which is not so simple, but we shall encounter typical arguments in Example 1.6.1 and Exercises 1.6.);

all the solutions beneath the exceptional solution have a vertical asymptote to the left, increase to a maximum, and then decrease and are asymptotic to $-\sqrt{t}$.

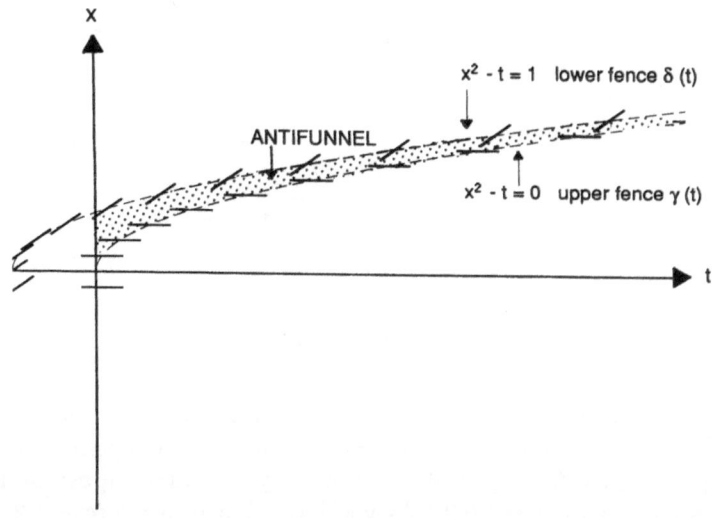

FIGURE 1.5.2.

Let us elaborate on this last point as follows: When a solution *leaves the antifunnel* given by $\gamma(t)$, $\delta(t)$ at a point (t_1, x_1), it is forced to *enter the funnel* given by $\alpha(t)$, $\beta(t)$, since the piecewise linear function $\eta(t)$ defined

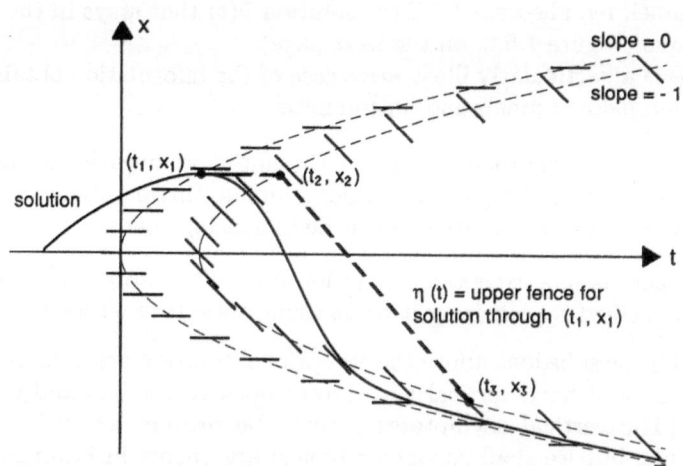

FIGURE 1.5.3. $\eta(t) =$ upper fence for solution through (t_1, x_1).

and graphed in Figure 1.5.3 is an *upper fence*. The fence $\eta(t)$ is the line segment of slope 0 from (t_1, x_1) to (t_2, x_2) where it first intersects the isocline $x_2 - t = -1$, and the line segment of slope -1 from (t_2, x_2) to (t_3, x_3), where it meets the isocline again at the top of the funnel.   ▲

2. *Bounds on $f(t, x)$.* A second source of useful fences is from *bounds on $f(t, x)$*, which we shall add to isocline fences in the following example:

Example 1.5.2. Consider $x' = \sin tx$.

The curves $tx = \eta(\pi/2)$ for integer n are the isoclines corresponding to slopes 0, 1, and -1. These isoclines are hyperbolas, all with the axes as asymptotes, for $n \neq 0$. Because of symmetry, we shall consider only the first quadrant, as shown in Figure 1.5.4.

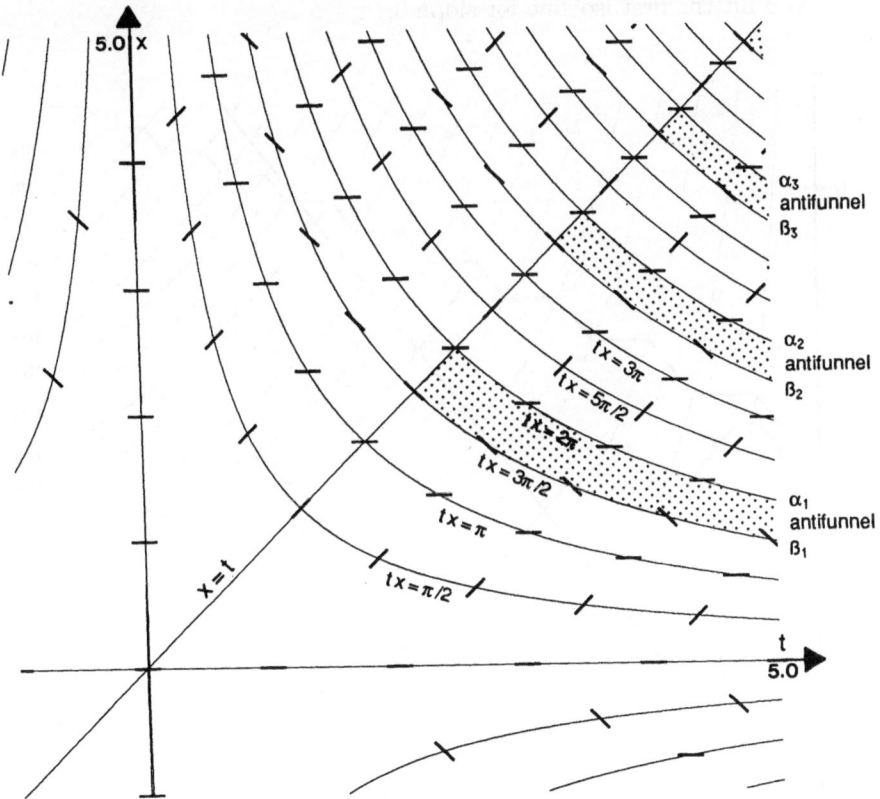

FIGURE 1.5.4. $x' = \sin tx$.

For positive k, you should confirm that the curves

$$\alpha_k(t) = 2k\pi/t \quad \text{and} \quad \beta_k(t) = (2k - (1/2))\pi/t$$

determine an antifunnel, which we shall call the "k^{th} antifunnel," in the
region $t > x$. There are an infinite number of these antifunnels, and in fact,
the regions between the antifunnels are funnels, so we get the following
description of the solutions:

> there are exceptional solutions $u_1, u_2, u_3, \ldots$, one in each of the anti-
> funnels above;

> all the other solutions stay between u_k and u_{k+1} for some k.

We can also derive information about the solutions for $t \leq x$, by building
another fence from the fact that $f(t, x) = \sin tx$ is bounded. This fence is
a piecewise differentiable one, constructed as follows, for any point on the
x-axis:

(i) start at $t = 0$ with a slope 1 and go straight in that direction until
you hit the first isocline for slope 0;

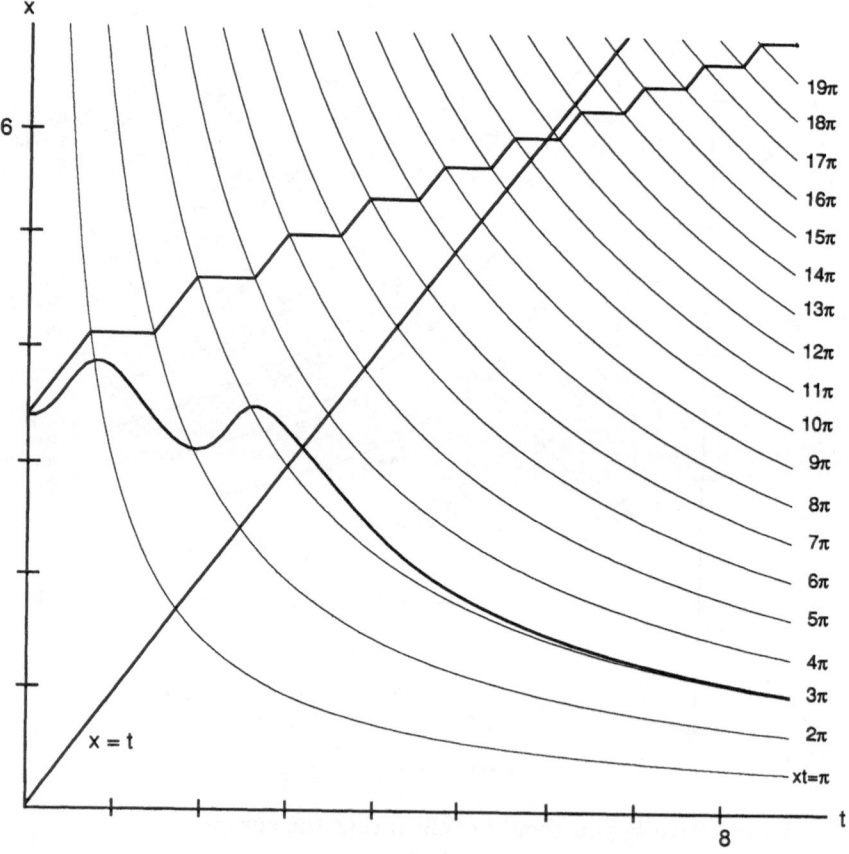

FIGURE 1.5.5. Upper fence for $x' = \sin tx$.

(ii) then go horizontally with slope 0 until you hit the next isocline for slope 0;

(iii) then go up again with slope 1 until you hit the next isocline for slope 0;

(iv) continue in this manner, as shown in Figure 1.5.5.

You can show in Exercises 1.5#3b that this "curve" is a weak upper fence that meets the line $x = t$. A solution starting at a point on the x-axis will stay below this fence until it reaches the line $x = t$, after which it will stay below one of the hyperbola fences described above. Therefore no solutions can escape the funnels and antifunnels described above.

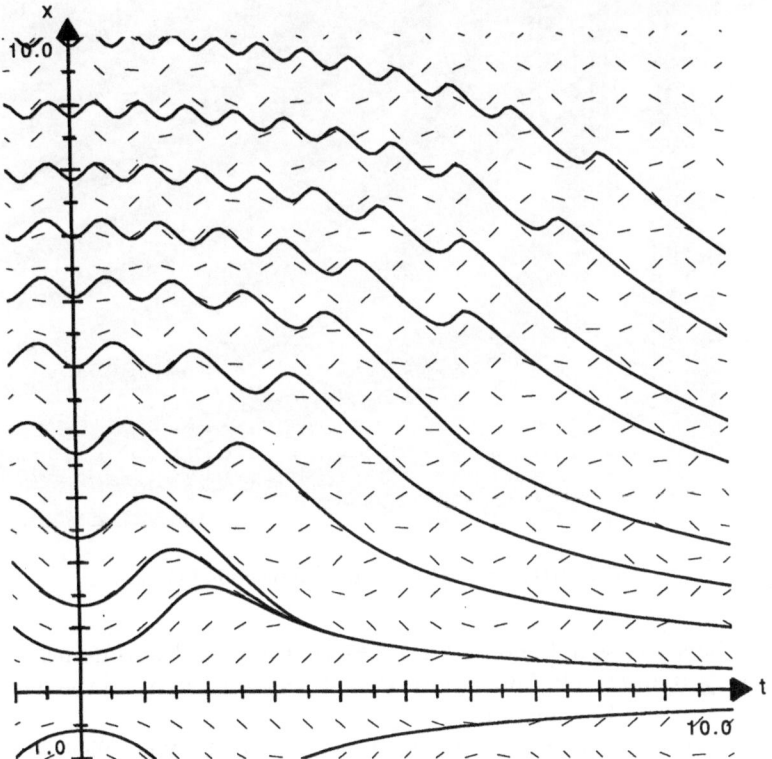

FIGURE 1.5.6. $x' = \sin tx$. Slopes marked on a grid.

It can also be shown that for positive k, the solutions to this differential equation have $2k$ maxima (k on each side of the x-axis) and that nonexceptional solutions in the first quadrant lie in a funnel, which we shall call the "k^{th} funnel," described by the curves

$$\alpha_k^*(t) = (2k-1)(\pi/t) \quad \text{and} \quad \beta_k(t) = (2k-(1/2))(\pi/t)$$

for sufficiently large t (Exercise 1.5#3c). The results of all this analysis are shown in Figures 1.5.6 and 1.5.7.

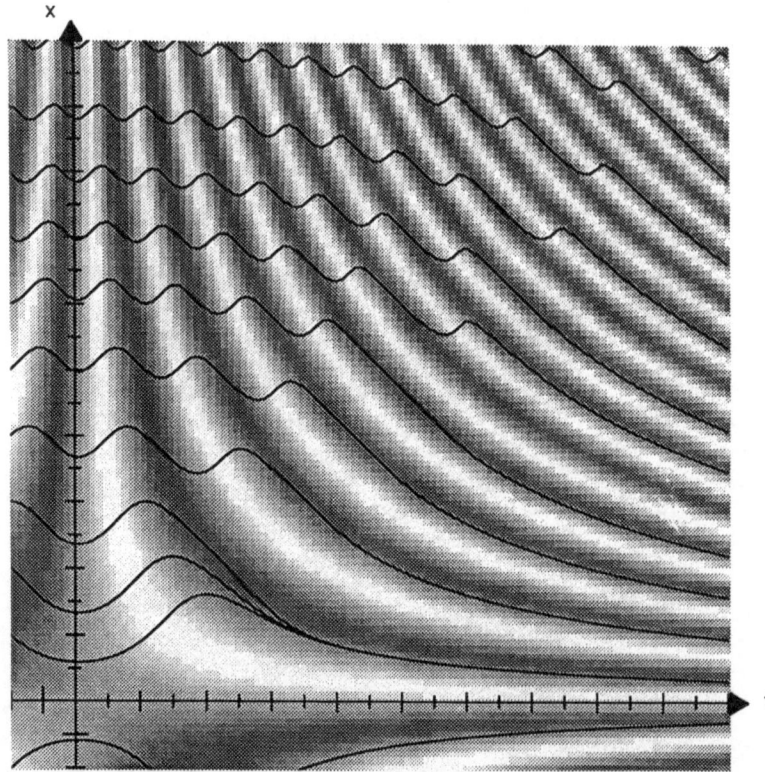

FIGURE 1.5.7. $x' = \sin tx$. Color-coded slopes.

You have shown in Exercise 1.2–1.4#8 that in the kth antifunnels, $\partial f/\partial x = t\cos tx \geq 0$; in the kth funnels, $\partial f/\partial x < 0$, corresponding to the discussion of *dispersion* in Section 1.4.

For further discussion of this example, see Mills, Weisfeller and Krall, *Mathematical Monthly*, November, 1979. ▲

3. *Solutions to similar equations.* A third and *most important* source of fences, often more useful than isoclines, is *solutions to similar differential equations,* which we shall illustrate with Examples 1.5.3 and 1.6.3. These are the fences which help us zero in on quantitative results for solutions of differential equations, even when we cannot explicitly find those solutions.

Example 1.5.3. Consider

$$x' = 1 + A\frac{\cos^2 x}{t^2} = f(t, x), \quad \text{for} \quad A > 0,$$

an equation we will meet in Volume III, in the analysis of Bessel functions for a vibrating membrane. What can you say about the solutions as $t \to \infty$? There are two good equations (because you can solve them and use them for fences) with which to compare the given differential equation. Because for all x

$$1 \leq 1 + A\frac{\cos^2 x}{t^2} \leq 1 + \frac{A}{t^2},$$

one similar equation is

$$x' = 1, \quad \text{with solution} \quad \alpha(t) = t + c_1,$$

and the *other similar equation* is

$$x' = 1 + \frac{A}{t^2}, \quad \text{with solution} \quad \beta(t) = t + c_2 - \frac{A}{t}.$$

If we choose $c_1 = c_2 = c$, then $\alpha(t)$ and $\beta(t)$ determine a narrowing antifunnel for $t > 0$, as graphed in Figure 1.5.8, with the slope marks for solutions to the differential equation as indicated according to the inequality.

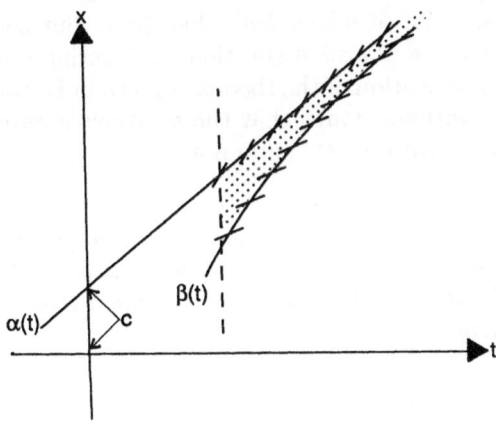

FIGURE 1.5.8. Funnel for $x' = 1 + A(\cos^2 x)/t^2$.

(Actually, according to the inequality, $\alpha(t)$ and $\beta(t)$ are weak fences, so we are a bit ahead of ourselves with this example, but this is a very good example of where we will want to use solutions to similar equations as fences. The Antifunnel Theorem of Chapter 4 will in fact be satisfied in this case.)

The antifunnel is narrowing, with

$$\lim_{t \to \infty} |\alpha(t) - \beta(t)| = 0,$$

so we can hope to have one unique solution that remains in the antifunnel. However

$$\frac{\partial f}{\partial x} = -(A/t^2)2\cos x \sin x = -(A/t^2)\sin 2x$$

is both positive and negative in this antifunnel, so our present criterion for uniqueness is not satisfied. Nevertheless, we will show in Example 4.7 that there indeed is for any given c a unique solution in the antifunnel, behaving as $t \to \infty$ like $(t + c)$.   ▲

See Example 1.6.3 as additional illustration of using similar equations for finding fences.

Remark. Pictures *can* be misleading. This is one reason why we so badly need our fence, funnel, and antifunnel theorems.

For instance, look ahead just for a minute to Figure 5.4.6 of Section 5.4. There you see three pictures of "solutions" to $x' = x^2 - t$, our equation of Examples 1.1.4, 1.3.3, 1.3.6, 1.5.1. You can ignore the fact that the Chapter 5 pictures are made by three different methods (which will be thoroughly discussed in Chapter 3). Just notice that a lot of spurious "junk" appears in these pictures, some of which *looks* like junk, but some of which (in the middle) *looks* like a plausible additional attracting solution in another funnel. With the application of the theorems given in Example 1.5.1, we can say precisely and without doubt that the solutions approaching $x = -\sqrt{t}$ really belong there, and the others do not.

So, what we are saying is:

> You can (and should) use the pictures as a **guide** to proving where the solutions go, but you must apply some method of proof (that is, the theorems) to justify a statement that such a guess is in fact true.

1.6   Vertical Asymptotes

If $|f(t, x)|$ grows very quickly with respect to x, the solutions to $x' = f(t, x)$ will have *vertical asymptotes*.

Example 1.6.1. Consider $x' = kx^2$, for $k > 0$.

A computer printout of the solutions looks like Figure 1.6.1.

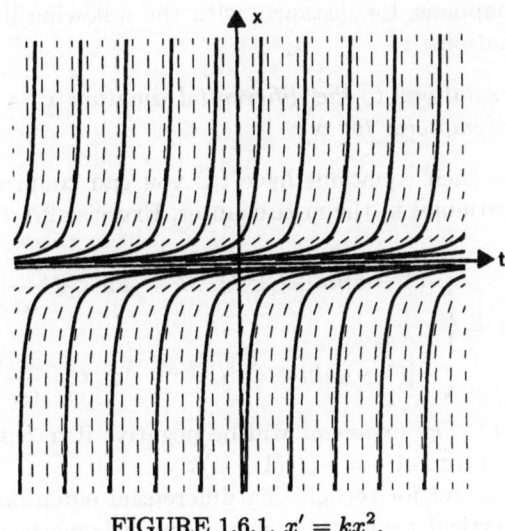

FIGURE 1.6.1. $x' = kx^2$.

Remark. Because the slope $x' = f(t, x)$ has no explicit dependence on t, the solutions are *horizontal translates* of one another.

As you can verify, individual nonzero solutions are $u(t) = 1/(C - kt)$ with a vertical asymptote at $t = C/k$, as in Figure 1.6.2.   ▲

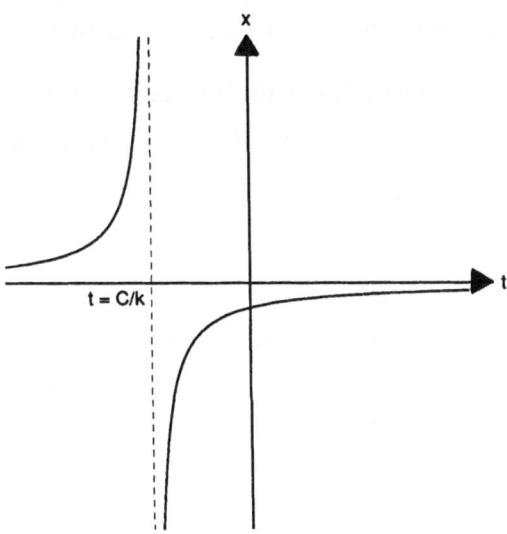

FIGURE 1.6.2. Two solutions to $x' = kx^2$.

Example 1.6.1 illustrates the fact that *a solution need not be defined for every t.* This happens, for instance, with the following important class of differential equations:

> *Nonzero solutions to the differential equation $x' = k|x|^\alpha$ have vertical asymptotes for $\alpha > 1$.*

This is because such equations have (as you can confirm analytically by separation of variables in the next chapter, Exercise 2.1#1j) nonzero solutions

$$
u(t) = \begin{cases} [(1-\alpha)kt + C]^{1/(1-\alpha)} & \text{for } t < \dfrac{C}{(\alpha-1)k} \\[2ex] -[(1-\alpha)kt + C]^{1/(1-\alpha)} & \text{for } t > \dfrac{C}{(\alpha-1)k} \end{cases}
$$

with constant C. The exponent will be negative if $\alpha > 1$, and then there will be an asymptote at $t = -C/(1-\alpha)k$.

This fact is useful for recognizing differential equations that may have asymptotes. Vertical asymptotes may occur whenever $|f(t,x)|$ grows at least as fast as $k|x|^\alpha$ for $\alpha > 1$, such as, for instance, functions with terms like x^3 or e^x as dominant terms. Furthermore, we can use functions of the form $|x|^\alpha$ for finding *fences* with vertical asymptotes, as will be shown in our next two examples.

The following example proves the *existence* of some of the vertical asymptotes for our favorite example:

Example 1.6.2. Consider again $x' = x^2 - t$, as in Examples 1.1.4, 1.3.3, 1.3.6, 1.5.1.

There is an *antifunnel* (Example 1.5.1) determined by

$$
x^2 - t = 1 \quad \text{and} \quad x^2 - t = 0 \quad \text{for } t > 0,
$$

with a unique solution $\hat{u}(t)$ that remains in the antifunnel for all time.

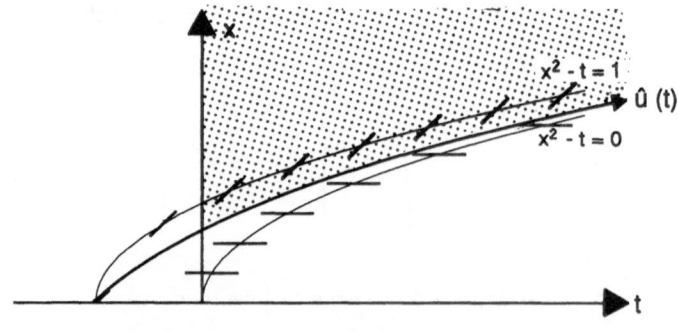

FIGURE 1.6.3. Antifunnel for $x' = x^2 - t$.

We shall show that any solution to the differential equation $x' = x^2 - t$ that lies above $\hat{u}(t)$ at some $t \geq 0$ (i.e., within the shaded region of Figure 1.6.3 at some point) has a vertical asymptote.

Since x' grows as x^2, the basic plan is to try as a lower fence solutions to $x' = x^{3/2}$, because the exponent $3/2$ is less than 2 but still greater than 1. Solutions to $x' = x^{3/2}$ are of the form

$$v(t) = \frac{4}{(C - t)^2}$$

with a vertical asymptote at $t = C$.

These solutions $v(t)$ to $x' = x^{3/2}$ will be *strong lower fences* for $x' = x^2 - t$ only where $x^2 - t > x^{3/2}$. The boundary of that region is the curve $x^2 - t = x^{3/2}$, as shown in Figure 1.6.4.

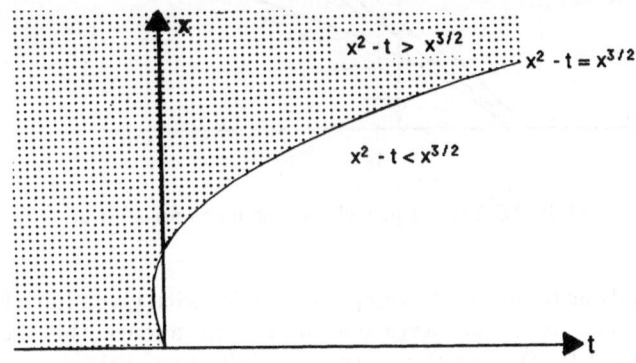

FIGURE 1.6.4. Strong lower fence for $x' = x^2 - t$.

We combine Figures 1.6.3 and 1.6.4 in Figure 1.6.5, where we will construct a lower fence in two pieces as follows: Every solution $u(t)$ lying above $\hat{u}(t)$ must first leave the antifunnel by crossing the isocline $x^2 - t = 1$ with slope 1 at some point $(t_1, u(t_1))$. For $t > t_1$, the slopes for $u(t)$ continue to increase, so the lower fence begins for $t > t_1$ with the line segment from $(t_1, u(t_1))$ having slope 1. The fence then meets the curve $x^2 - t = x^{3/2}$, at some point (t_2, x_2). (For $t > 0$, the curve $x^2 - t = x^{3/2}$ lies above $x^2 - t = 1$ and has slope approaching zero as $t \to \infty$; therefore a segment of slope 1 must cross it.)

Thus $u(t)$ enters the region where the solutions $v(t)$ to $x' = x^{3/2}$ are strong lower fences. By the Fence Theorem 1.3.5, each solution $u(t)$ must now remain above the $v(t)$ with $v(t_2) = x_2$. Since the $v(t)$ all have vertical asymptotes, so must the $u(t)$. ▲

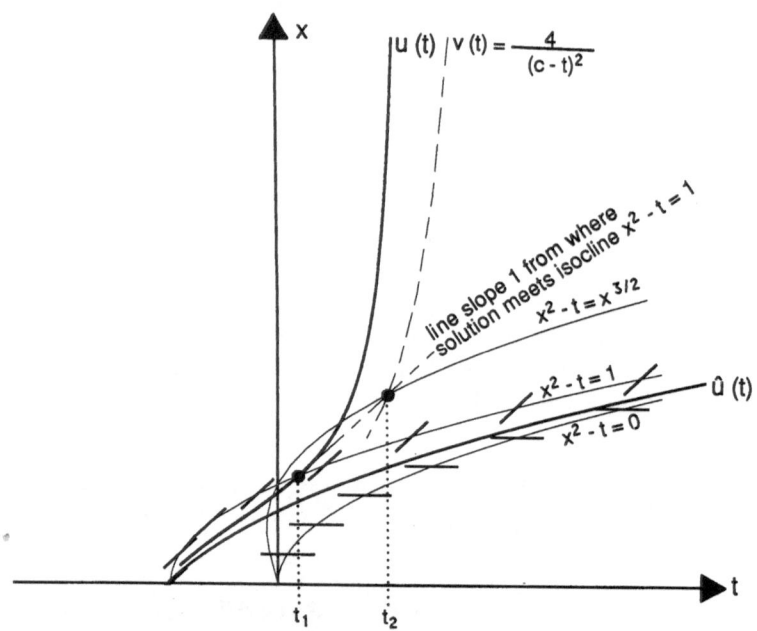

FIGURE 1.6.5. Piecewise lower fence for $x' = x^2 - t$.

The analysis of our next example allows us, with given initial conditions, to zero in on a *particular* vertical asymptote for a fence; this therefore gives a bound on t for the solution with those initial conditions.

Example 1.6.3. Consider

$$x' = x^2 + t^2. \tag{1}$$

This equation (1) can be compared with its simpler (solvable) relatives, $x' = t^2$ and $x' = x^2$. Solutions to either of these simpler equations are lower fences, because the slope of a solution to equation (1) is greater or equal at every point.

To show the powerful implications of this observation, let us see what it tells us about the solution to equation (1) *through the origin* (i.e., the solution $x = u(t)$ for which $u(0) = 0$.) The first simple relation

$$x' = t^2 \quad \text{has solutions} \quad \alpha_k(t) = (t^3/3) + k,$$

so $\alpha(t) = (t^3/3)$ is a lower fence for $t \geq 0$ (a weak lower fence at $t = 0$), as shown in Figure 1.6.6. All solutions to $x' = x^2 + t^2$ that cross the x-axis above or at 0 lie above this curve $\alpha(t)$.

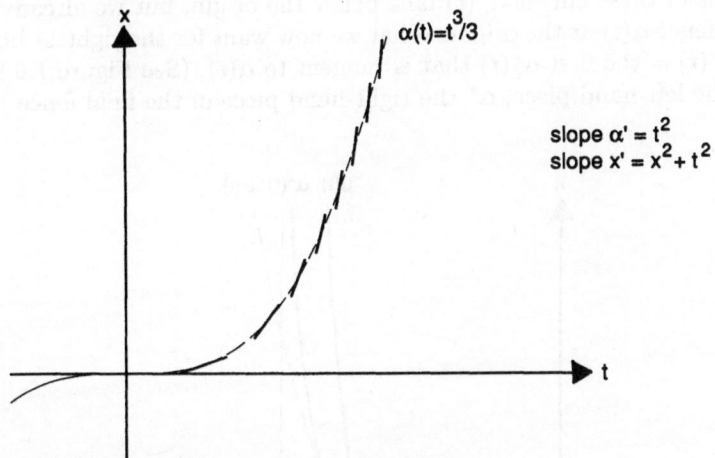

FIGURE 1.6.6. Lower fence for $x' = x^2 + t$.

But we can do better than this, because the solutions to equation (1) will also lie above another fence, $\alpha_c^*(t)$, from the second simple relation

$$x' = x^2, \quad \text{which has solutions} \quad \alpha_c^*(t) = 1/(c-t).$$

The important thing about $\alpha_c^*(t)$ is that it has a *vertical asymptote, so we shall be able to put an upper bound on t.* The family of curves $\alpha_c^*(t)$ are shown in Figure 1.6.7.

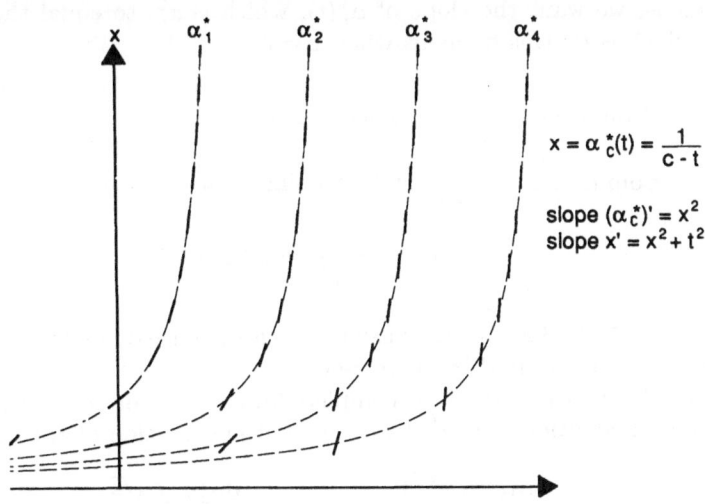

FIGURE 1.6.7. Other lower fences for $x' = x^2 + t^2$.

None of these curves $\alpha_c^*(t)$ falls below the origin, but we already have a lower fence $\alpha(t)$ at the origin. What we now want for the tightest bound on t is $\alpha^*(t) =$ the first $\alpha_c^*(t)$ that is tangent to $\alpha(t)$. (See Figure 1.6.8 where α is the left-hand piece, α^* the right-hand piece of the final fence.)

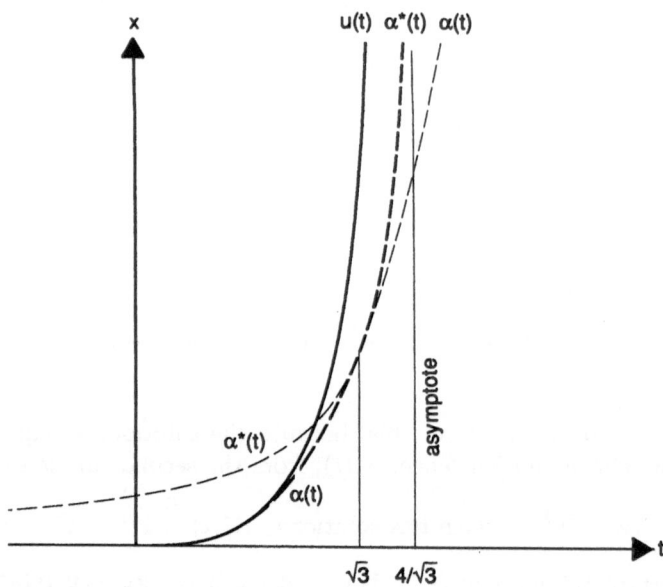

FIGURE 1.6.8. Piecewise lower fence, in bold dashes, with vertical asymptote for $x' = x^2 + t^2$.

That is, we want the slope of $\alpha_c^*(t)$, which is x^2, to equal the slope of $\alpha(t)$, which is t^2; this happens when $x = t$.

from $\alpha : x = \dfrac{t^3}{3} = t$ when $t = \sqrt{3} = x.$

from $\alpha_c^* : x = \dfrac{1}{c-t}$ at that point means $\sqrt{3} = \dfrac{1}{(c - \sqrt{3})}$

or $c = \dfrac{4}{\sqrt{3}} \approx 2.3094.$

Hence $c = 4/\sqrt{3}$ is a vertical asymptote for a fence α^* on the right, which is the best you can find by these fences.

With this method, the best complete lower fence for the solution to the differential equation $x' = x^2 + t^2$, with initial condition at the origin, is

$$\alpha(t) = t^3/3 \qquad\qquad 0 \le t \le \sqrt{3}$$

$$\alpha^*(t) = 1/((4/\sqrt{3}) - t) \quad t > \sqrt{3}$$

as shown in Figure 1.6.8. Since the fence has a vertical asymptote, we have shown that the solution through the origin is only defined for $t < t_0$, where t_0 is a number satisfying

$$t_0 < \frac{4}{\sqrt{3}} \approx 2.3094. \quad \blacktriangle$$

Exercises 1.1 Slope Fields, Sketching Solutions

1.1#1. Consider the following differential equations:

(a) $x' = x$ (b) $x' = x - t$

(i) Using isoclines, sketch by hand a slope field for each equation. Draw the isoclines using different colors, to eliminate confusion with slopes. Then draw in some solutions by drawing curves that follow the slopes.

(ii) Confirm by differentiation and substitution that the following functions are actually solutions to the differential equations

for (a): $x = Ce^t$ for (b): $x = t + 1 + Ce^t$

1.1#2. Consider the following differential equations:

(a) $x' = x^2$ (b)° $x' = x^2 - 1$

(i) Using isoclines, sketch by hand a slope field for each equation. Draw the isoclines using different colors, to eliminate confusion with slopes. Then draw in some solutions by drawing curves that follow the slopes.

(ii) Tell in each case which solutions seem to become vertical.

(iii) Confirm by differentiation and substitution that the following functions are actually solutions to the differential equations

for (a): $x = 1/(C - t)$ for (b): $x = (1 - Ce^{2t})/(1 + Ce^{2t})$

(iv) Use the computer programs *Analyzer* and *DiffEq* to confirm these solutions graphically. Show which values of C lead to which members of the family of solutions.

(v) In each case, find another solution that is not given by the formula in part (iii). Hint: Look at your drawings of solutions on the direction field. This provides a good warning that analytic methods may not yield *all* solutions to a differential equation. It also emphasizes that the drawing contains valuable information that might otherwise be missed.

1.1#3. Consider the following differential equations:

(a) $x' = -x/t$ (c) $x' = x/(2t)$

(b) $x' = x/t$ (d) $x' = (1-t)/(1+x)$

(i) Using isoclines, sketch by hand a slope field for each equation. Indicate clearly where the differential equation is undefined. Draw the isoclines using different colors, to eliminate confusion with slopes. Then draw in some solutions by drawing curves that follow the slopes.

(ii) Confirm by differentiation (implicit if necessary) and substitution that the following equations actually define parts of solutions to these differential equations. State carefully which parts of the curves give solutions, referring to where the differential equations are undefined.

for (a): $x = C/t$ for (c): $x = C|t|^{1/2}$

for (b): $x = Ct$ for (d): $(t-1)^2 + (x+1)^2 = C^2$

1.1#4. (Courtesy Michèle Artigue and Véronique Gautheron, Université de Paris VII.) On each of the following eight direction fields,

(i) Sketch the isocline(s) of zero slope.

(ii) Mark the region(s) where the slopes are undefined.

(iii) Sketch in a few solutions.

(iv)° Match the direction fields with the following equations. Each numbered equation corresponds to a different one of the lettered graphs.

1. $x' = 2$	5. $x' = x/t$
2. $x' = t$	6. $x' = -t/x$
3. $x' = x - t$	7. $x' = (x-2)/(t-1)$
4. $x' = x$	8. $x' = tx^2 + t^2$

a) 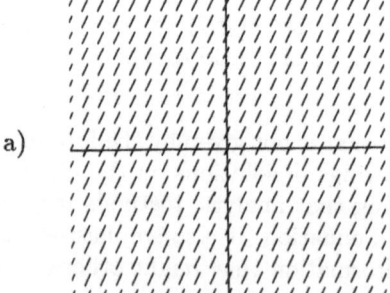 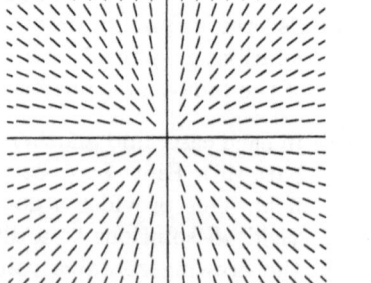 b)

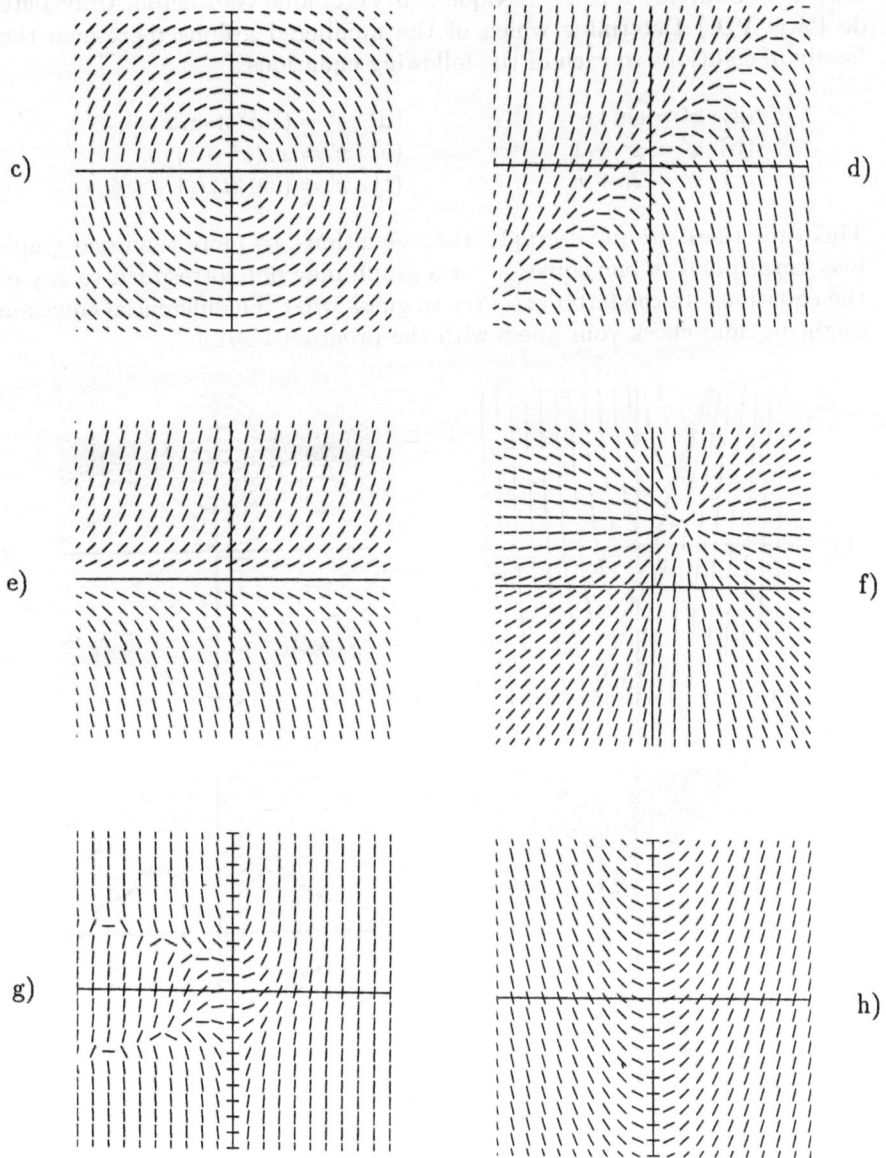

c)

d)

e)

f)

g)

h)

1.1#5°. (Courtesy Michèle Artique and Véronique Gautheron, Université de Paris VII.) Determine which of the numbered graphs represents the family of solutions to each of the following equations:

(a) $x' = \sin tx$ (d) $x' = (\sin t)(\sin x)$

(b) $x' = x^2 - 1$ (e) $x' = x/(t^2 - 1)$

(c) $x' = 2t + x$ (f) $x' = (\sin 3t)/(1 - t^2)$

This time there are more graphs than equations, so more than one graph may correspond to one equation, or a graph may not correspond to any of the equations. In the latter case, try to guess what the differential equation might be, and check your guess with the program *DiffEq*.

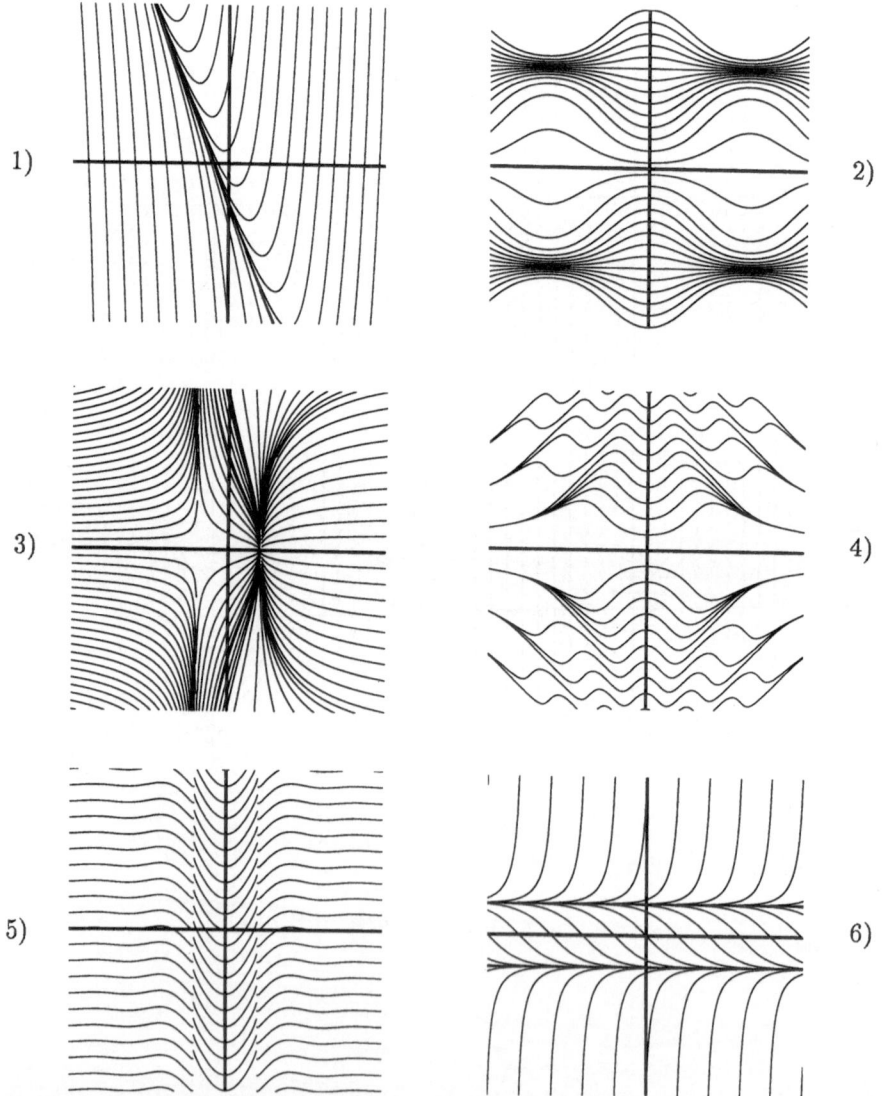

1) 2)

3) 4)

5) 6)

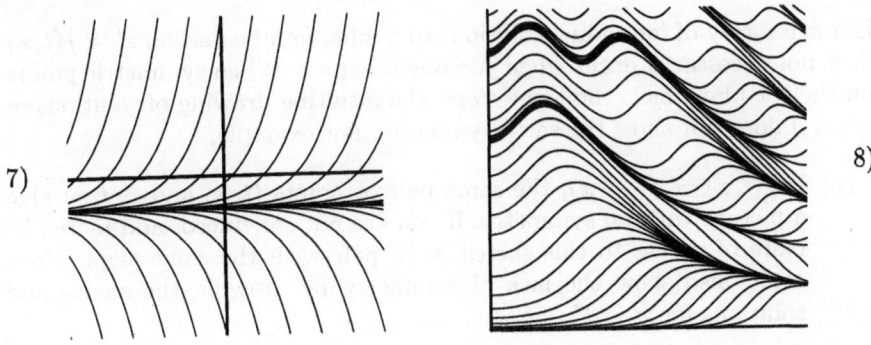

7)

8)

1.1#6. To gain insight into how solutions fit into the direction fields, consider the following two equations:

$$\text{(a)} \ \ x' = x^2 - t^2 \qquad\qquad \text{(b)} \ \ x' = x^2 - \frac{t^2}{1 + t^2}$$

With the help of the computer program *DiffEq*, which makes quick work of the more complicated $f(t, x)$'s, print out direction fields only and try by hand to draw in solutions. For part (a) you can (and should) clarify the situation by calculating the *isoclines* directly from the differential equation and graphing them (perhaps on a separate piece of paper or in a different color) with the proper slope marks for the solutions. You can check your final results by allowing the computer to draw in solutions on slope fields. Describe the crucial differences in the drawings for (a) and (b).

1.1#7. Symmetry of individual solutions to a differential equation $x' = f(t, x)$ occurs when certain conditions are met by $f(t, x)$. For instance,

(a) If $f(t, x) = -f(-t, x)$, every pair of symmetric points (t, x) and $(-t, x)$ have slopes that are symmetric with respect to the vertical x-axis, so solutions are symmetric about this axis. Illustrate this fact by drawing a direction field and solutions to $x' = tx$, and to $x' = tx^2$. See Figure (a) below.

But symmetry of individual solutions to a differential equation $x' = f(t, x)$ does not necessarily occur when you might expect. When symmetric points on the t, x-plane have the *same slope*, the resulting drawing of solutions in general does *not* show the same symmetry. For example,

(b) If $f(t, x) = f(-t, x)$, the same pair of points (t, x) and $(-t, x)$ give a picture with no symmetry. If you are not convinced, add more like pairs of points to this sketch, each pair with the same slopes. You could also show the lack of symmetry by drawing the slopes and solutions for $x' = t^2 - x$.

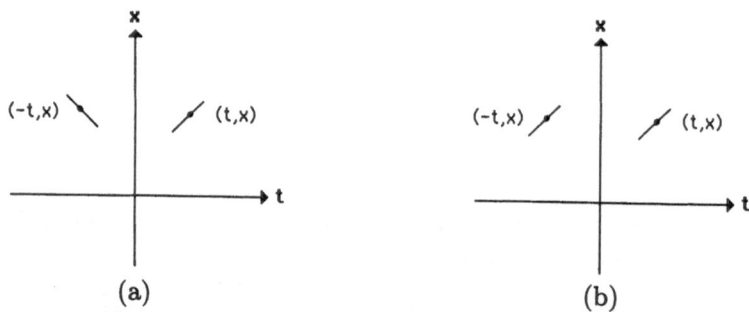

(a)   (b)

(c) Of the other possible permutations of plus and minus signs, show which cases lead to symmetric solutions and state with respect to what they are symmetric. That is, consider

$$f(t, x) = f(t, -x); \qquad f(t, x) = f(-t, -x);$$
$$f(t, x) = -f(t, -x); \qquad f(t, x) = -f(-t, -x).$$

1.1#8. A quite different question of symmetry arises in the matter of whether the picture of solutions for one differential equation $x' = f(t, x)$ corresponds to that for another with a related f, such as $x' = -f(t, x)$. The answer is quite likely "no"; for example

(a) Compare the drawings of solutions for $x' = tx$ with those for $x' = -tx$.

But there are times when the answer is "yes"; for example

(b) Compare the drawings of solutions for $x' = x^2 - t$ with those for $x' = t - x^2$.

This problem serves as a warning that symmetry of solutions to one differential equation $x' = f(t, x)$ with respect to another differential equation with a related f is *not* a simple question. You should *not* assume there will be a relation. If you wish to explore this matter further, see the next exercise.

1.1#9°. Since you have a computer graphics program *DiffEq*, you have a headstart if you wish to explore some of the symmetry questions raised by the previous problem. Consider the differential equation

$$x' = x^2 + \sin x + t + \cos t + 3,$$

carefully chosen to include both an odd function and an even function in each variable (as well as avoiding extra symmetries that could creep in with a strictly polynomial function). Print out slope fields with some solutions for each of the various sign variants:

(a) $x' = f(t, x)$ (e) $x' = -f(t, x)$
(b) $x' = f(-t, -x)$ (f) $x' = -f(-t, -x)$
(c) $x' = -f(t, -x)$ (g) $x' = f(t, -x)$
(d) $x' = -f(-t, x)$ (h) $x' = f(-t, x)$.

The graphs should all be the same size, default is fine. It will be helpful if at least some of your initial conditions are the same from graph to graph. Then compare the results and conjecture when there might be a relation between the solutions of $x' = f(t, x)$ and the graph of one of the sign variants. Test your conjecture(s) by holding pairs of computer drawings up to the light together. Be careful to match the axes exactly, and see if the solutions are the same—you may find that some cases you expected to match in fact do not under this close inspection. You can test for the various types of symmetry as follows:

with respect to the x-axis: (flip one drawing upside down)
with respect to the t-axis: (flip one drawing left to right)
with respect to the origin: (flip upside down *and* left to right).

Now that you have honed your list of conjectures, prove which will be true in general. This is the sort of procedure to follow for proper analysis of pictures: *experiment, conjecture, seek proof* of the results. The pictures are of extreme help in knowing what can be proved, but they do not stand alone without proof.

1.1#10. Given a differential equation $x' = f(t, x)$ and some constant c, find what can be said when about solutions to $x' = f(t, x) + c$. Use the "experiment, conjecture, prove" method of the previous exercise.

1.1#11. To become aware of the limitations of the computer program *DiffEq*, use it to graph direction fields and some sample solutions for the following equations, for the default window $-10 \le t \le 10$, $-7.5 \le x \le 7.5$, setting the stepsize $h = 0.3$. Explain what is wrong with each of the computer drawings, and tell what you could do to "fix" it.

(a) $x' = \sin(t^3 - x^3)$ (b) $x' = (t^2/x) - 1$ (c) $x' = t(2 - x)/(t + 1)$.

1.1#12. Again using the computer program *DiffEq*, sketch direction fields and then some isoclines and solutions for

(a) $x' = (2 + \cos t)x - (1/2)x^2 - 1$

(b) $x' = (2 + \cos t)x - (1/2)x^2 - 2$.

1.1#13°. Notice that many of the solutions sketched in the previous exercises have *inflection points*. These can be precisely described for a solution $x = u(t)$ in the following manner: obtain x'' by differentiating $f(t, x)$ with respect to t (and substituting for x'). This expression for x'' gives information at every point (t, x) on concavity and inflection points, which can be particularly valuable in hand-sketching. Carry out this process for the equation $x' = x^2 - 1$ (direction field and solutions were drawn in Exercise 1.1#2b).

1.1#14. Use the method described in the previous exercise to derive the equation for the locus of inflection points for solutions of $x' = x^2 - t$. Hint: it is easier to plot $t = g(x)$; label your axes accordingly. Use the computer program *Analyzer* to plot and print this locus.

Confirm that the inflection points of the solutions indeed fall on this locus by superimposing your *Analyzer* picture on a same scale *DiffEq* picture of the direction field and solutions (i.e., holding them up to the light together with the same axis orientation—this will mean turning one of the pages over and rotating it).

You will get to *use* the results of this problem in Exercise 1.2–1.4#17.

1.1#15. Use the method described in Exercise 1.1#13 to derive the equation for the locus of inflection points for solutions of $x' = \sin tx$. This time part of the locus is not so easy to plot. For various values of t, use the computer program *Analyzer* to find corresponding x values satisfying that locus equation. Then mark these (t, x) pairs on a *DiffEq* picture of the direction field and solutions, and confirm that they indeed mark the inflection points for those values of t.

1.1#16. Sketch by hand solutions to $x' = ax - bx^2$ by finding where the slope is zero, positive, and negative. Use the second derivative (as in Exercise 1.1#13) to locate the inflection points for the solutions. This equation is called the *logistic equation;* it occurs frequently in applications—we shall discuss it further in Section 2.5.

Exercises 1.2–1.4 Qualitative Description of Solutions: Fences, Funnels, and Antifunnels

1.2–1.4#1. Consider the differential equation

$$\frac{dx}{dt} = x^2 - t.$$

For each of the lines $x = t$, $x = 2$ and each of the isoclines with $c = -2$, 0, 2, show on a drawing which parts are upper fences and which parts are lower fences. Furthermore, show the combination of these curves (or parts thereof) that form funnels and antifunnels.

1.2–1.4#2. Identify funnels and antifunnels in these slope fields, drawn by isoclines:

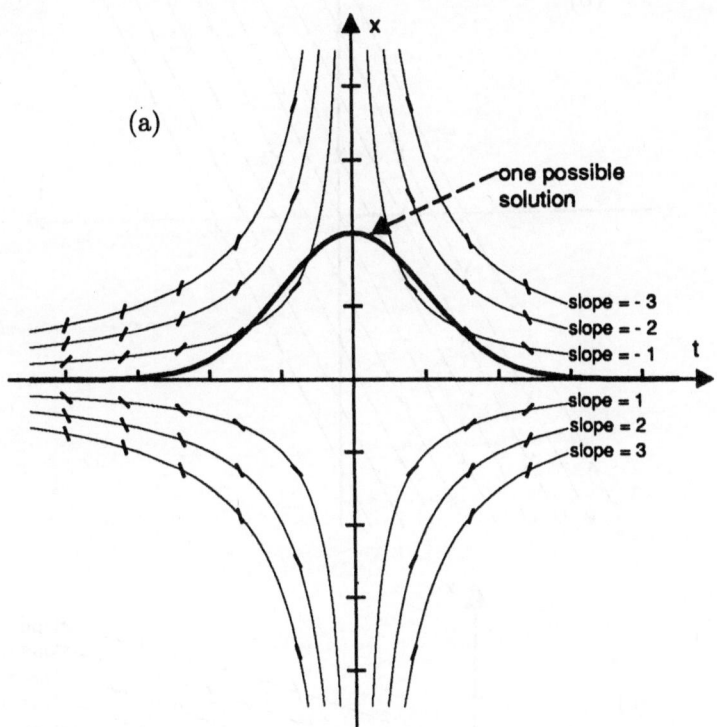

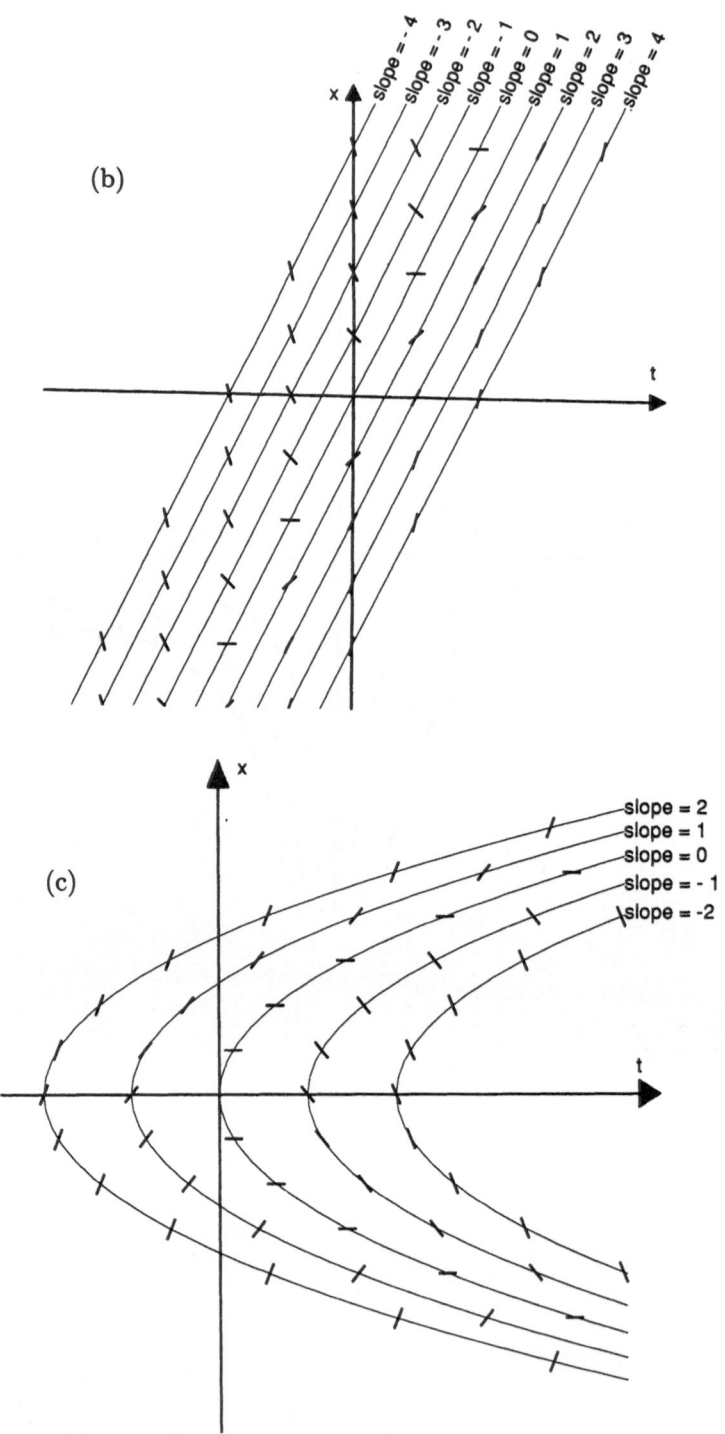

Then look back at the solutions drawn in these fields in Figures 1.1.4, 1.1.6, 1.1.7.

1.2–1.4#3. For each of the following equations, use *Analyzer* to draw a few relevant isoclines on the same graph and print it out. For each isocline, draw on your printout the appropriate slope marks for the solution to the differential equation. Shade and identify funnels and antifunnels. You may find it helpful to refer back to Exercises 1.1#1–3 where you drew these slope fields from isoclines. (You can use such handsketches for this exercise if you prefer.)

(a) $x' = x$ (e) $x' = -\dfrac{x}{t}$

(b) $x' = x^2$ (f) $x' = \dfrac{x}{t}$

(c)° $x' = x^2 - 1$ (g) $x' = \dfrac{x}{2t}$

(d) $x' - x - t$ (h) $x' = \dfrac{1-t}{1+x}$

In order to partly check your work, you can use the computer program *DiffEq* to print out a slope field with some solutions. Then hold it up to the light with your *Analyzer* printout showing isoclines and some hand-drawn solutions. The computer-drawn slope field should "fit" the isoclines; the extent to which the hand-drawn solutions match the computer-drawn solutions is an indication of your skill at following the direction field.

1.2–1.4#4°. Why do we *need* fences and funnels if the computer pictures *show* us the solutions? The following will show you why a picture alone is not enough: Use the computer program *DiffEq* on our favorite differential equation, $x' = x^2 - t$, for $0 \le t \le 15$, $-7.5 \le x \le 7.5$, with stepsize $h = 0.4$. You should see spurious solutions that don't go into the funnel, sometimes even appearing to go into a different funnel; yet the analysis of Example 1.5.1 proves that in fact there *is* a funnel, but only one. When we get to Section 5.4, you will see an explanation of why the erroneous pictures of this exercise happen.

1.2–1.4#5. Show that the isoclines for $x' = x^2 - t$ (Example 1.5.1) form *narrowing* funnels and antifunnels, that is, that

$$\lim_{t \to \infty} |\sqrt{t+c} - \sqrt{t}| = 0.$$

1.2–1.4#6. Our fence definitions require functions that are continuous and piecewise differentiable, as in Figure 1.2.6. Determine which of the following continuous functions are piecewise differentiable. State exactly why the others fail, and where they could not be used as a fence:

(a) $x = |\sin t|$ (c) $x = t \sin(1/t)$
(b) $x = \sqrt[3]{t}$ (d) $x = \sqrt{|t|}$

1.2–1.4#7. (a) Can there exist some curve $x = \psi(t)$ such that $\psi(t)$ is both an upper and a lower fence for $x' = f(t, x)$ on some interval $t \in [a, b]$? If so, what can we say about $\psi(t)$?

(b) Can there exist some curve $x = \psi(t)$ such that $\psi(t)$ is both a strong upper fence and a strong lower fence for $x' = f(t, x)$ on some interval $[a, b]$? If so, what can we say about $\psi(t)$?

1.2–1.4#8. For the following differential equations $x' = f(t, x)$, make drawings of the regions in the t, x-plane where the dispersion $\partial f/\partial x$ is positive and where it is negative. Then with the computer program *DiffEq* sketch some solutions to the differential equation and see that they behave as the dispersion predicts.

(a) $x' = \sin tx$ (b) $x' = x^3 - x^2 t$

1.2–1.4#9°. (a) Show that for the differential equation $x' = x - x^2$, the region $|x| \leq 1/2$ is an antifunnel, and the region $1/2 \leq x \leq 3/2$ is a funnel.

(b) The funnel and antifunnel found in part (a) are not *narrowing*. Find a narrowing antifunnel containing the solution $x(t) = 0$, and a narrowing funnel containing the solution $x(t) = 1$. (The first part is very easy, but the second is a bit more subtle.) Hint: Consider functions like $1 + (a/t)$.

1.2–1.4#10. Consider the differential equation $x' = x - x^2 + 1/t$.

(a) Show that the region $-2/t \leq x \leq 0$, $t \geq 1$ is an antifunnel.

(b) Show that the solution in (a) is the unique solution that stays in the antifunnel.

1.2–1.4#11. Consider further the differential equation $x' = x - x^2 + 1/t$ of the previous exercise. It is no longer as obvious as in Exercise 1.2–1.4#9 that there are solutions asymptotic to $x = 1$, although it looks likely that there is a funnel here for t sufficiently large.

(a) Sketch the slope field around $x = 1$ for $x' = x - x^2$, and then in a different color show how the extra term $1/t$ affects the slope field. It should appear that the solutions above $x = 1$ will now approach more slowly. Where do the solutions go that cross $x = 1$? Can they go to ∞?

(b) Show that $x = 1$ is now a lower fence.

(c) Find an upper fence that *is* asymptotic to $x = 1$.

(d) Find a lower fence *between* the upper fence of (c) and $x = 1$, showing that the solutions are bounded a little bit *away* from $x = 1$.

(e) Use the computer program *DiffEq* with its blowup feature to verify these results.

1.2–1.4#12. Consider the differential equation $x' = x - x^2 + 2/t$.

(a) Use the computer program *DiffEq* to sketch the slope field and a few solutions.

(b) Show that there *is* a solution asymptotic to $x = 0$, although we seem not to be able to easily state it. Hint: consider the same antifunnel as in Exercises 1.2–1.4#10.

1.2–1.4#13. Consider the differential equation $x' = x - x^2 + 2/t$ as in the previous problem, but now consider its behavior near $x = 1$.

(a) What sorts of fences are the curves $\alpha(t) = 1 + a/t$ for $0 < a < \infty$?

(b) Show that any solution $u(t)$ which at some t_0 satisfies

$$1 \leq u(t_0) \leq 1 + 2/t_0$$

is defined for all t and satisfies $\lim_{t \to \infty} u(t) = 1$.

1.2–1.4#14. Consider the following differential equations, and show how the different perturbations affect the behavior compared to $x' = x - x^2$. Use the program *DiffEq* to draw direction fields with some solutions. Then analyze what you can say about what you see. Will there be funnels and antifunnels near $x = 0$ and $x = 1$ if t gets big enough? (In the default window size, the behavior of $x' = x - x^2$ hasn't "won out" yet; extend the t-axis until solutions level out horizontally.)

Explain the behavior for $0 < t < 2$. (Try $1/(t - 1)$; is that an upper fence? Do *all* solutions turn up before $t = 1$?)

(a) $x' = x - x^2 + 1/\sqrt{t}$ (b) $x' = x - x^2 + 1/\ln t$.

1.2–1.4#15°. Consider the differential equation $x' = -x/t$.

(a) Show that there are solutions tending to 0.

(b) Does this behavior change when we add a perturbation? Find out what happens near $x = 0$ for $x' = -(x/t) + x^2$. Start from a computer drawing and proceed to look for fences and funnels.

1.2–1.4#16. Consider the differential equation $x' = -x + (1 + t)x^2$. The first term on the right indicates a decrease, but the second indicates an increase. The factor $(1 + t) \ldots$

(a) Show that for any $t_0 > 0$, there exists a number α such that if $0 < \alpha < \varepsilon$ the solution with $x(t_0) = \alpha$ tends to 0 at ∞.

(b) Show that there is an antifunnel above the t-axis, but that with a fence of $1/t$ you cannot get uniqueness for a solution in the antifunnel. Can you invent a better fence?

1.2–1.4#17. Consider the differential equation $x' = x^2 - t$. Use the results of Exercise 1.1#14 to show that there is a *higher* lower fence for the funnel than $x = -\sqrt{t}$.

Exercises 1.5   Uses of Fences, Funnels, and Antifunnels

1.5#1. Analyze the following differential equations, using the computer program *DiffEq* to get a general picture, then doing what you can to make fences, funnels, and antifunnels that give more precise information about the solutions. (The exercises of the previous section give some ideas of what to try.)

(a) $x' = e^t - h$

(b) $x' = \dfrac{t^2}{x} - 1$

(c) $\theta' = -1 + \dfrac{\cos^2 \theta}{4t^2}$

(d) $x' = x^2 \sqrt{1 + t^2} - 1$

(e) $x' = \cos x - t$

(f) $x' = (t + 1) \cos x$

(g) $x' = x^2 + t^2$

(h)° $x' = \dfrac{t(2 - x)}{t + 1}$

1.5#2. Fill in the steps in Example 1.5.1 to show that for $x' = x^2 - t$,

(a) the isocline $x^2 - t = 0$ is a lower fence for $t > 0$,

(b) the isocline $x^2 - t = -1$ is an upper fence for $t > 5/4$.

1.5#3. Consider $x' = \sin tx$, the equation of Example 1.5.2.

(a) (i) Show that the piecewise function constructed in Example 1.5.2, Figure 1.5.5 is an upper fence.

   (ii) Furthermore, show that for any $x_0 = b$, the line $x = t + b$ is a weak upper fence, and the line $x = -t + b$ is a weak lower fence.

(b) Show why part (a)(i) means that *every* solution to this differential equation gets trapped below one of the hyperbola isoclines. This means to show that the fence constructed in Figure 1.5.5 indeed intersects the diagonal. Hint: show that the slopes of the inclined pieces is $< 1/2$; show also that for fixed x the hyperbolas are equidistant horizontally.

(c) Show that the nonexceptional solutions (that is, those not in the antifunnels) collect in *funnels* between $\alpha_k^*(t) = (2k-1)(\pi/t)$ and $\beta_k(t) = (2k-1/2)(\pi/t)$. Then show that each of these solutions have $2k$ maxima (k on each side of the x-axis).

(d) Show that symmetry gives for every k^{th} antifunnel another for $k = -1, -2, -3, \ldots$ with symmetrical solutions $u_{-k}(t) = -u_k(t)$, and that by this notation, α_k is symmetrical to β_{-k}, β_k to α_{-k}.

(e) Show that α_k and $\beta_{-\ell}$ (from the last part, (c)) for $k > 0, \ell > 0$ form a *narrowing* antifunnel. Show that $\partial f/\partial x$ exists everywhere but is both positive and negative in this antifunnel. Then show that there are infinitely many solutions that *stay* in the antifunnel. Why is this not a counterexample to the Antifunnel Uniqueness Theorem 1.4.5?

1.5#4. Analyze the following differential equations, using isoclines to sketch the direction field and then identifying fences, funnels, and antifunnels to give precise information about the solutions:

$$\text{(a)} \ \ x' = -\sin tx \qquad\qquad\qquad \text{(b)} \ \ x' = \cos tx$$

1.5#5. Sketch some isoclines, slopes, and solutions for the following functions. Locate funnels and antifunnels. Compare the behavior of solutions for these equations:

$$\text{(a)} \ \ x' = t\cos x - 1 \qquad\qquad \text{(c)} \ \ x' = t\cos x + 2$$
$$\text{(b)} \ \ x' = t\cos x + 1$$

1.5#6. Consider the differential equation $x' = x^2 - t^2$.

(a) Find the isoclines and draw them for $c = 0, \pm 1, \pm 2$. In each quadrant, show how some of these isoclines can be used to form funnels and antifunnels.

(b) In the first quadrant find an antifunnel A satisfying the Antifunnel Uniqueness Theorem 1.4.5. Find in the fourth quadrant a narrowing funnel F. Show that every solution leaving the antifunnel A *below* the exceptional solution is forced to enter the funnel F.

(c) Which kinds of symmetry are exhibited by this differential equation, considering such facts as $f(-t, -x) = f(t, x)$?

1.5#7. For the equation $x' = 1 - \dfrac{\cos^2 x}{t^2}$, show fences and funnels (as in Example 1.5.3).

1.5#8. Consider $\dfrac{dx}{dt} = \sin(t^3 - x^3)$.

(a) As in Exercise 1.1#11a, use the computer to draw solutions with stepsize $h = 0.3$. The picture is *not* convincing. Change the stepsize to get a more reliable picture of the solutions; tell which step you have used.

(b) This differential equation can be analyzed using isoclines. You can use *Analyzer* as a "precise" drawing aid for the isoclines. Sketch and label some isoclines for slopes of $0, \pm 1, \ldots$.

(c) Explain how isoclines can be used to define antifunnels and funnels in the first quadrant similar to ones defined for $x' = \sin tx$ in Example 1.5.2. Especially choose antifunnels, so that Theorem 1.4.5 can be applied.

1.5#9. For the differential equation $x' = (x^2 - t)/(t^2 + 1)$,

(a) Plot the isoclines for slopes $0, -1, 1$.

(b) Sketch several solutions.

(c) Find a funnel and an antifunnel.

(d) Show that the solution in the antifunnel is unique.

1.5#10°. Consider the differential equation $x' = x^2/(t^2 + 1) - 1$.

(a) What kinds of curves are the isoclines? Sketch the isoclines of slope 0 and -1. Describe the regions where the slope is positive and negative.

(b) Show that the solutions $u(t)$ to the equation with $|u(0)| < 1$ are defined for all t.

(c) What are the slopes of the asymptotes for the isoclines for slope $m = (1+\sqrt{5})/2$? Show that the isocline of slope m and its asymptote define an antifunnel. How many solutions are there in this antifunnel?

(d) Sketch the solutions to the equation.

Exercises 1.6 Vertical Asymptotes

1.6#1. Prove there exist vertical asymptotes for

(a) $x' = x^3 - t$

(b) $x' = e^x + t$

(c) $x' = x^2 - xt$

(d) $x' = -2x^3 + t$

(e) $x' = ax - bx^2$

1.6#2. Consider the differential equation $x' = x^2 - t^2$.

(a) You found in Exercises 1.5#6b an antifunnel A in the first quadrant, with an exceptional solution trapped inside. Show that every solution leaving the antifunnel A *above* this exceptional solution has a vertical asymptote.

(b) Use fences to show that the solution of $x' = x^2 - t^2$ through the point $(0, 2)$ has a vertical asymptote at $t = t_0$ and find t_0 to one significant digit. Confirm your results with a computer drawing, blowing up near the asymptote.

1.6#3°. Consider the differential equation $x' = x^3 - t^2$.

(a) Show that there is a narrowing antifunnel along $x = t^{2/3}$.

(b) Show that all solutions above the exceptional solution in the antifunnel have vertical asymptotes.

1.6#4. Consider again $x' = x^2 - t$ from Example 1.6.2. Show that the solutions *below* the funnel bounded below by $x^2 - t = 0$ for $t > 0$ have vertical asymptotes to the left (by working with negative time).

1.6#5. Consider $x' = x^2 - t$ from Example 1.6.2. Find a bound on the vertical asymptotes for the solutions, and confirm your results with a computer drawing.

$\qquad$ (a) through $(0, 1)$ $\qquad\qquad$ (b) through $(-1, 0)$.

1.6#6°. By comparing $x' = x^2 - t$ with $x' = (1/2)x^2$, show that the solution to $x' = x^2 - t$ with $x(0) = 1$ has a vertical asymptote $t = t_0$ for some $t_0 < 2$. Confirm your results with a computer drawing.

1.6#7. Consider $x' = x^2 + t^2$ as in Example 1.6.1. Find a bound on the vertical asymptotes for the solutions, and confirm your results with a computer drawing.

$\qquad$ (a) through $(1, 0)$ $\qquad\qquad$ (b) through $(0, 1)$.

2

Analytic Methods

In this chapter we actually have two themes: methods of solution for differential equations that can be solved analytically, and some discussion of how such equations and their solutions are used in real-world applications.

The first task is to present traditional methods for *analytic solutions* to differential equations—that is, the non-graphic, non-computer approach that can yield *actual formulas for solutions to differential equations* falling into certain convenient classes.

For first order differential equations, $x' = f(t, x)$, there are two classes that suffice for solving most of those equations that actually have solutions which can be written in elementary terms:

$$\text{\textit{separable} equations:} \qquad x' = g(t)h(x)$$

$$\text{\textit{linear} equations:} \qquad x' = p(t)x + q(t).$$

If $x' = f(t, x)$ does not depend explicitly on x, then $x' = g(t)$, and to find the solutions means simply to find the antiderivatives of $g(t)$.

If, on the other hand, $x' = f(t, x)$ does not depend explicitly on t, then $x' = h(x)$, and the equation is called *autonomous*. Every first order autonomous equation is separable, which means it is possibly solvable by a single integration.

In Section 2.1 we shall discuss the separable equations, traditionally the easiest to attack, but they yield implicit solutions. In Sections 2.2 and 2.3 are the linear equations, which have the advantage of giving an explicit formula for solutions to equations in this class. Later in Section 2.6 we shall discuss another class, the *exact* equations, which actually include the separable equations, and which also produce implicit solutions. Finally, Section 2.7 introduces *power series* solutions. The Exercises (such as 2.2–2.3#8,9 and Miscellaneous Problems #2) explore yet other traditional classes, such as *Bernoulli* equations and *homogeneous* equations.

These presentations will probably review methods you have seen in previous courses, but you should find here new perspectives on these methods, particularly for the linear equations and exact equations.

Partly because, however, this chapter may be to a great extent a review, we also attend here to a second theme of *applicability*—how differential equations arise from modeling real-world situations. In Sections 2.4 and 2.5 we shall discuss important applications of differential equations that also shed a lot of light on the meaning of the linear equations and the methods for their solution.

2.1 Separation of Variables

The method of separation of variables applies to the case where $x' = f(t, x)$ can be written as

$$x' = g(t)h(x).$$

If $h(x_0) = 0$, then $u(t) = x_0$ is a solution.

If $h(x) \neq 0$, then the variables can be *separated* by rewriting the equation as

$$\frac{dx}{h(x)} = g(t)dt, \quad \text{and} \quad \int \frac{dx}{h(x)} = \int g(t)dt. \tag{1}$$

If you can compute the integrals,

$$\int \frac{dx}{h(x)} = H(x) + C_H; \qquad \int g(t)dt = G(t) + C_G,$$

then the solution is

$$H(x) = G(t) + C, \tag{2}$$

which explicitly or implicitly will define solutions $x = u(t)$.

Remark. This derivation illustrates the beauty of Leibniz' notation dx/dt for a derivative; the same result can be achieved by writing

$$\int \frac{1}{h(x)} \frac{dx}{dt} dt = \int g(t)dt.$$

Example 2.1.1. Let us consider the particularly useful general class of equations of the form $x' = g(t)x$.

For $x = 0$, $u(t) = 0$ is a solution.

For $x \neq 0$,

$$\int (1/x)dx = \int g(t)dt$$

$$\ln|x| = \int g(t)dt$$

$$|x| = e^{\int g(t)dt}.$$

If we are given an initial condition $u(t_0) = x_0$, then the solution is

$$x = u(t) = x_0 e^{\int_{t_0}^{t} g(\tau)d\tau}. \quad \blacktriangle$$

Example 2.1.2. Consider $x' = ax - bx^2$ $a > 0, b > 0$, an autonomous equation that arises frequently in population problems, to be discussed in Section 2.5.

Since $x' = (a - bx)x$, two solutions are $u(t) = 0$ and $u(t) = a/b$.

For $x \neq 0$ and $x \neq a/b$, we get by separation of variables

$$\int \frac{dx}{ax - bx^2} = \int dt,$$

which by partial fractions gives for the left side

$$\int \frac{dx}{ax} + \int \frac{b\,dx}{a(a - bx)} = \frac{1}{a} \ln |x| - \frac{1}{a} \ln |a - bx|$$

$$= \frac{1}{a} \ln \left| \frac{x}{a - bx} \right|,$$

and for the right side

$$\int dt = t + C.$$

Setting these two expressions equal and exponentiating leads to

$$\left| \frac{x}{a - bx} \right| = Q e^{at},$$

where we write Q as constant instead of a more complicated expression involving C, the original constant of integration. Solving for x we get

$$u(t) = \begin{cases} \dfrac{aQe^{at}}{bQe^{at} - 1} & \text{for } x < 0 \\[2ex] 0 & \text{for } x = 0 \\[2ex] \dfrac{aQe^{at}}{bQe^{at} + 1} & \text{for } 0 < x < a/b \\[2ex] a/b & \text{for } x = a/b \\[2ex] \dfrac{aQe^{at}}{bQe^{at} - 1} & \text{for } x > a/b. \end{cases}$$

This information is confirmed on the next page by Figure 2.1.1, the picture of solutions in the direction field (as constructed in Exercise 1.1#16). For $x < 0$ or $x > a/b$, the independent variable t is restricted, and there are *vertical asymptotes* (as proved in Exercise 1.6#1e).   ▲

Remark. In Example 2.1.2, as in Example 1.6.1, the slope does not depend explicitly on t, so the solutions from left to right are horizontal translates of one another. This is an important aspect of autonomous systems.

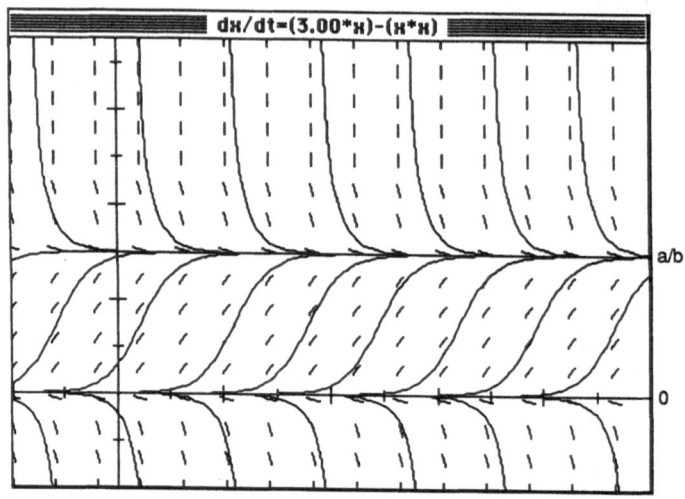

FIGURE 2.1.1. $x' = ax - bx^2$.

Example 2.1.3. Consider $x' = (t+1)\cos x$.

Since $\cos x = 0$ for x an odd multiple of $\pi/2$, some solutions are

$$u_k(t) = (2k+1)(\pi/2).$$

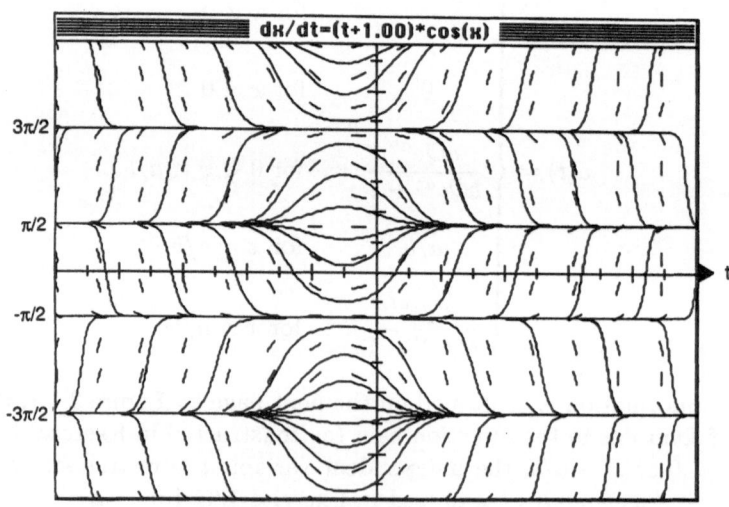

FIGURE 2.1.2. $x' = (t+1)\cos x$.

For other values of x, the differential equation leads to

$$\int \frac{dx}{\cos x} = \int (t+1)dt$$

$$\frac{1}{2} \ln \frac{1+\sin x}{1-\sin x} = \frac{t^2}{2} + t + C$$

$$\frac{1+\sin x}{1-\sin x} = e^{2((t^2/2)+t+C)} \tag{3}$$

Note again that formula (3) states the solution *implicitly*. The task of algebraically solving such a function for $x = u(t)$ or of graphing it is often impossible, but qualitative solution sketching from the differential equation itself is quite simple and enlightening, as illustrated in Figure 2.1.2. ▲

Notice that, as shown by Example 2.1.3, it is often far more difficult to graph the functions $H(x) = G(t) + C$ produced by separation of variables than to treat graphically the original differential equation $x' = g(t)h(x)$. The same could even be said for Example 2.1.2, for which the qualitative picture of Figure 2.1.1 "says it all" at a glance. Since Example 2.1.2 arises from a population problem, as we shall show in Section 2.5, the usual question is simply "what happens, for $x \geq 0$, as $t \to \infty$?" The answer to that is immediately deducible, with minimum effort, from the qualitative picture.

Summary. For separation of variables, if $x' = h(x)g(t)$, then

1. For all x_0 such that $h(x_0) = 0$, $u(t) = x_0$ is a solution.

2. For all x_0 such that $h(x_0) \neq 0$, separate variables and solve by

$$\int \frac{dx}{h(x)} = \int g(t)dt.$$

2.2   Linear Differential Equations of First Order

If $p(t)$ and $q(t)$ are two continuous functions defined for $a < t < b$, then the differential equation

$$x' = p(t)x + q(t) \tag{4}$$

is the general linear first order equation. It is called *homogeneous* if $q(t) \equiv 0$, *nonhomogeneous* otherwise. For a given equation of the form (4), the equation

$$x' = p(t)x \tag{5}$$

is called its *associated* homogeneous equation, or sometimes its *complementary* equation.

Note that *if* a linear first order differential equation is *homogeneous*, then it is separable; as we have seen in Example 2.1.1, the solution $x = u(t)$ of equation (5) with $u(t_0) = x_0$ is

$$u(t) = x_0 e^{\int_{t_0}^t p(\tau)d\tau}. \tag{6}$$

We can use the homogeneous case to help solve the *nonhomogeneous* case, with the following theory which is basic to *linear* equations of all sorts, differential or otherwise. (Here we prove the result for linear differential equations.)

Theorem 2.2.1 (Solving nonhomogeneous linear equations). *The solution to $x' = p(t)x + q(t)$ with initial condition $u(t_0) = x_0$ is*

$$u(t) = u_c(t) + u_p(t),$$

the **sum** *of a solution $u_p(t)$ of the original equation, $x' = p(t)x + q(t)$,* **and** *the solution $u_c(t)$ of the associated homogeneous equation, $x' = p(t)x$, with initial condition $u_c(t_0) = x_0 - u_p(t_0)$.*

Proof. Substitute $u(t)$ into the original differential equation. $\qquad\square$

Since equation (6) has given us $u_c(t)$, the only other necessity for solving nonhomogeneous linear equations is to find a $u_p(t)$. There are two famous methods for doing this: *undetermined coefficients,* which is often the easiest *when* it can be applied, and *variation of parameters,* which has the immense advantage that it will *always* work, although it frequently leads to rather unpleasant computations. We shall describe the first method here, and the second in the following section.

METHOD OF UNDETERMINED COEFFICIENTS

For some differential equations $x' = p(t)x + q(t)$, a particular solution $u_p(t)$ can be found by guessing a solution of the appropriate form to work with $q(t)$. Let us start with three examples, for which we already know from Example 2.1.1 that

$$u_c = x_0 e^{\int p(t)dt} = Ce^t.$$

Example 2.2.2. $x' = x + \sin t$.

Let us try for a solution of the form $u_p(t) = \alpha \sin t + \beta \cos t$. Substituting this "guess" into the differential equation says that we must have

$$\alpha \cos t - \beta \sin t = \alpha \sin t + \beta \cos t + \sin t.$$

Setting equal coefficients of like terms on the left and on the right implies that $\alpha = \beta$ (from the $\cos t$ terms) and $-\beta = \alpha + 1$ (from the $\sin t$ terms). This system of two nondifferential equations requires that

$$\alpha = \beta = -1/2.$$

Therefore we know $u_p(t) = -\frac{1}{2}\sin t - \frac{1}{2}\cos t$, and the general solution to the original differential equation is

$$u(t) = u_c(t) + u_p(t) = Ce^t - \frac{1}{2}\sin t - \frac{1}{2}\cos t. \quad \blacktriangle$$

Example 2.2.3. $x' = x + t^2$.

Let us try for a particular solution of the form

$$u_p(t) = \alpha + \beta t + \gamma t^2.$$

To satisfy the differential equation, we must have (by substitution)

$$\beta + 2\gamma t = \alpha + \beta t + \gamma t^2 + t^2,$$

and the coefficients of the constants, the linear terms, and the quadratic terms in turn each produce an equation; the algebraic solution of that system of three nondifferential equations requires that

$$\gamma = -1, \quad \beta = -2, \quad \alpha = -2.$$

Thus $u_p(t) = -2 - 2t - t^2$, and the general solution to the differential equation is

$$u(t) = u_c(t) + u_p(t) = Ce^t - 2 - 2t - t^2. \quad \blacktriangle$$

Example 2.2.4. $x' = x + e^{2t}$.

If you guess $u_p(t) = \alpha e^{2t}$ and try substituting that into the differential equation, you will get

$$2\alpha e^{2t} = \alpha e^{2t} + e^{2t},$$

which implies by the coefficients of e^{2t} that $2\alpha = \alpha + 1$, so $\alpha = 1$.

This gives a general solution to the differential equation of

$$u(t) = u_c(t) + u_p(t) = Ce^t + e^{2t}. \quad \blacktriangle$$

Why did these examples work? Why was it a good idea to "try" the functions we did? If we refer to linear algebra (Appendix L in Volume II), we can explain in terms of *vector spaces*.

Proposition 2.2.5. *Consider the differential equation* $x' = p(t)x + q(t)$. *Suppose that* $q(t)$ *is an element of a finite-dimensional vector space* V *of functions* f, **closed** *under the operation*

$$f(t) \to f'(t) - p(t)f(t) = \left[\frac{d}{dt} - p(t)\right]f(t).$$

Let $f_1, \ldots, f_n$ be a basis of the vector space. Set

$$x(t) = a_1 f_1(t) + \cdots + a_n f_n(t).$$

If this expression is substituted into the equation, the result is a system of n ordinary linear equations for the coefficients a_i, which may or may not admit a solution if $d/dt - p(t)$ is singular.

The fact that the vector space V must be *closed* under the operation $f' - pf$ means that $f \in V \Rightarrow f' - pf \in V$.

In Example 2.2.2, the vector space was the space of functions $\alpha \sin t + \beta \cos t$, obviously closed under derivatives $f \to f' - f$.

In Example 2.2.3, the vector space was the space of polynomials of degree at most 2, closed under $f \to f' - f$.

In Example 2.2.4, the vector space was the space of functions αe^{2t}, also closed under derivatives $f \to f' - f$.

These examples all happened to have $p(t) = 1$; the method proceeds in similar fashion according to Proposition 2.2.5 for other $p(t)$. The following example illustrates how to apply the same proposition in another case.

Example 2.2.6. $tx' = x + t^2$.

If we rewrite this equation as $x' = (1/t)x + t$, and try a particular solution of the form $u_p(t) = \alpha + \beta t$, we run into trouble because that won't give a vector space closed under $f \to f' - pf$.

However, we can avoid such difficulty in this case if we stick with the equation in its original form, $tx' = x + t^2$, then focus on t^2 instead of a strict $q(t)$ and try for a particular solution of the form

$$u_p(t) = \alpha + \beta t + \gamma t^2.$$

The natural operator here is $f \to tf' - f$ for a vector space to which t^2 must belong, and we otherwise proceed as above.

To solve the equation, we must have

$$\beta t + 2\gamma t^2 = \alpha + \beta t + \gamma t^2 + t^2,$$

which implies, by the equations for the coefficients of each type of term, that

$$\alpha = 0, \quad \beta = \text{anything}, \quad \gamma = 1.$$

Thus any function of the form $t^2 + \beta t$ is a particular solution, and

$$u(t) = u_c(t) + u_p(t) = Ct + t^2 + \beta t = t^2 + (C + \beta)t. \quad \blacktriangle$$

Thus you see that the method of undetermined coefficients is somewhat adaptable, and further extensions will be discussed in Exercises 2.2#7 and 8.

However, life isn't always so easy. Even when you can find a seemingly appropriate vector space of functions to try the method of undetermined coefficients, the system of linear equations that you get for the coefficients may be *singular,* and have no solutions. This does not mean that the *differential* equation has no solution, it just means that your guess was bad.

Example 2.2.7. $x' = x + e^t$.

A reasonable vector space is simply the 1-dimensional vector space of functions αe^t. But substituting $u_p(t) = \alpha e^t$ gives:

$$\alpha e^t = \alpha e^t + e^t, \quad \text{i.e.} \quad \alpha = \alpha + 1,$$

which has *no* solution for α, so $u_p(t) = \alpha e^t$ was a bad guess.   ▲

In such a case as Example 2.2.7, however, there exists another, *bigger* vector space of the form $P(t)e^t$, for $P(t)$ a polynomial.

Example 2.2.8. For $x' = x + e^t$, try substituting

$$u_p(t) = \alpha e^t + \beta t e^t.$$

A vector space of these functions also contains $q(t)$ and is closed under $f \to f' - pf$. You can confirm by this method that $u_p(t) = te^t$ is a particular solution to the differential equation, and

$$u(t) = u_c(t) + u_p(t) = (C + t)e^t. \quad ▲$$

Actually, it is only exceptionally that such a large vector space as shown in Example 2.2.8 is needed; it is when $q(t)$ include solutions to the associated homogeneous equation. But those are exactly the cases that are important, as we shall discuss further in Volume II, Section 7.7.

The method of undetermined coefficients therefore works, by Proposition 2.2.5, whenever you can *find* a sufficiently large *finite*-dimensional vector space closed under $f \to f' - pf$. The examples illustrate the possibilities. For functions $q(t)$ that include terms like $\tan t$ or $\ln t$, another method is necessary for finding a particular solution $u_p(t)$. In the next section we present a method that works for *any* linear equation, and furthermore gives some real insight into linear equations.

Summary. For a first order linear differential equation $x' = p(t)x + q(t)$, the general solution is

$$x = u(t) = u_c(t) + u_p(t) = Ce^{\int_{t_0}^{t} p(\tau)d\tau} + u_p(t),$$

where $u_p(t)$ is a *particular* solution to the entire nonhomogeneous equation, and $u_c(t)$ is the solution to the associated homogeneous equation.

One way to find $u_p(t)$ is by the *method of undetermined coefficients*, which uses an educated guess when $q(t)$ is amenable. See Examples 2.2.2, 2.2.3, 2.2.4, and 2.2.6.

Another way to find $u_p(t)$, which *always* works, is the *method of variation of parameters*, to be presented in the next section.

2.3   Variation of Parameters

Variation of parameters is a technique that *always* produces a particular solution $u_p(t)$ to a nonhomogeneous linear equation, at least if you can compute the necessary integrals. The idea is to

1. *Assume a solution* of the form of the solution (6) to the associated homogeneous equation (5) with the *constant* x_0 replaced by a *variable* $v(t)$:

$$u(t) = v(t)e^{\int_{t_0}^{t} p(\tau)d\tau},\qquad(7)$$

with $u(t_0) = v(t_0) = x_0$.

2. *Substitute in the non-homogeneous differential equation* (4) the assumed solution (7) prepared in the last step:

$$u'(t) = v'(t)e^{\int_{t_0}^{t} p(\tau)d\tau} + v(t)p(t)e^{\int_{t_0}^{t} p(\tau)d\tau} \quad (\text{differentiating (7)})$$

$$= p(t)\underbrace{v(t)e^{\int_{t_0}^{t} p(\tau)d\tau}}_{u(t)} + q(t), \quad (\text{from differential equation (4)})$$

which, amazingly enough, we *can* simplify and solve for $v'(t)$:

$$v'(t) = q(t)e^{-\int_{t_0}^{t} p(\tau)d\tau} = r(t).\qquad(8)$$

3. *Solve for the varying parameter* $v(t)$ *using equation (8) and substitute in the assumed solution* (7).

As a result of step (2), $v(t)$ is a solution to the differential equation $x' = r(t)$, with initial condition $v(t_0) = x_0$. Therefore

$$v(t) = \int_{t_0}^{t} r(s)ds + x_0 = \int_{t_0}^{t}\left[q(s)e^{-\int_{t_0}^{s} p(\tau)d\tau}\right]ds + x_0;$$

so, from step (1),

$$\boxed{u(t) = e^{\int_{t_0}^{t} p(\tau)d\tau}\left\{\int_{t_0}^{t}\left[e^{-\int_{t_0}^{s} p(\tau)d\tau}q(s)\right]ds + x_0\right\}}\qquad(9)$$

where $x_0 = u(t_0)$, and the entire $\{...\}$ is $v(t)$.

This procedure does indeed find solutions for the nonhomogeneous differential equation (4), as we shall illustrate in the following example.

Example 2.3.1. Consider $x' = ax + b$, with $u(t_0) = x_0$ and $a \neq 0$.

This is not just an arbitrary example; it turns out to be an important differential equation that we shall use in proving theorems, as we shall see in Chapter 4. (Notice that this equation is in fact separable as well as linear, so in Exercise 2.1#1c you solve the same equation by separation of variables.)

The homogeneous equation is again the equation of Example 2.1.1, so:

1). Assume $u(t) = v(t)e^{\int_{t_0}^{t} a\, d\tau} = v(t)e^{a(t-t_0)}$.

2). Substitution in the differential equation gives

$$v'(t)e^{a(t-t_0)} + av(t)e^{a(t-t_0)} = av(t)e^{a(t-t_0)} + b$$

$$v'(t) = be^{-a(t-t_0)}, \quad \text{with } v(t_0) = x_0.$$

3). Solving for $v(t)$ gives

$$v(t) = -\left(\frac{b}{a}\right)e^{-a(t-t_0)} + c, \quad \text{with } c = x_0 + \frac{b}{a},$$

so from step 1),

$$u(t) = \left[-\left(\frac{b}{a}\right)e^{-a(t-t_0)} + c\right]e^{a(t-t_0)}$$

$$= x_0 e^{a(t-t_0)} + \left(\frac{b}{a}\right)[e^{a(t-t_0)} - 1]. \quad \blacktriangle$$

The method of variation of parameters always works in the same way, so you can bypass the steps and simply write the results directly as equation (9). We shall revisit the equation of Example 2.3.1 and illustrate this method.

Example 2.3.2. Consider again $x' = ax + b$, with $u(t_0) = x_0$. In this case,

$$e^{\int_{t_0}^{t} a\, d\tau} = e^{a(t-t_0)},$$

so

$$u(t) = e^{a(t-t_0)}\left\{\int_{t_0}^{t} be^{-a(s-t_0)}\, ds + x_0\right\}$$

$$= e^{a(t-t_0)}\left[\frac{-b}{a}[e^{-a(t-t_0)} - 1] + x_0\right]$$

$$= x_0 e^{a(t-t_0)} + \left(\frac{b}{a}\right)[e^{a(t-t_0)} - 1]. \quad \blacktriangle$$

So far we have obtained equation (9) as *one* solution to the differential equation (4) with initial condition $x(t_0) = x_0$. In Exercise 2.2–2.3#4 you can verify that substitution of (9) in (4) confirms its status as a solution.

We have therefore *existence* of solutions for every initial condition (t_0, x_0). In Section 4.5 we shall prove *uniqueness* of this solution, so formula (9) is in fact the *complete* solution to (4).

It is well worth your effort to *learn* this equation (9), or at least to understand it thoroughly. The summary at the end of this section and Section 2.4 should both give additional insight.

But first let us proceed with another example.

Example 2.3.3. Consider $x' = \dfrac{-x + \sin t}{t}$.

First,

$$p(t) = -\left(\frac{1}{t}\right),$$

so

$$\int_{t_0}^t p(\tau)d\tau = \ln\left|\frac{t_0}{t}\right|,$$

and

$$q(t) = \left(\frac{1}{t}\right)\sin t.$$

So now we are ready to use equation (9):

$$x(t) = e^{\ln\left|\frac{t_0}{t}\right|}\left\{e^{-\ln\left|\frac{t_0}{s}\right|}\left(\frac{1}{s}\right)\sin s\, ds + x_0\right\}$$

$$= \frac{t_0}{t}\left\{\int\left(\frac{s}{t_0}\right)\left(\frac{1}{s}\right)\sin s\, ds + x_0\right\}$$

$$= \frac{t_0}{t}\left\{\left(\frac{1}{t_0}\right)(\cos t_0 - \cos t) + x_0\right\}$$

$$= \frac{1}{t}\{-\cos t + \underbrace{\cos t_0 + x_0 t_0}_{C}\}.$$

The constant terms can be grouped together, so there is indeed only the one constant overall that is expected in the solution to a first order differential equation. ▲

We close this section with a summary of the equations.

Summary. For a first order linear equation $x' = p(t)x + q(t)$, the solution with $u(t_0) = x_0$ can be written, by *variation of parameters*, as

$$\boxed{u(t) = e^{\int_{t_0}^t p(\tau)d\tau}\left\{\int_{t_0}^t [e^{-\int_{t_0}^s p(\tau)d\tau}q(s)]ds + x_0\right\}.}$$

(9, again)

If you first compute

$$w(s,t) = e^{\int_s^t p(\tau)d\tau}$$ (10)

then the solution with $u(t_0) = x_0$ can be written more simply as

$$u(t) = w(t_0, t)\left\{\int_{t_0}^t w(s, t_0)q(s)ds + x_0\right\}.$$ (11)

The quantity $w(s,t)$ defined by equation (10) is sufficiently important to the solutions to linear differential equations that it is called the *fundamental solution*. The next section give a real world interpretation for the fundamental solution; the concept will later be generalized to higher dimensions in Volume II, Chapter 12.

2.4 Bank Accounts and Linear Differential Equations

One "universal" way of "understanding" the linear differential equation (4)

$$x' = p(t)x + q(t)$$

is to consider the solutions as the value of a bank account, having a variable interest rate $p(t)$; $q(t)$ represents rate of deposits (positive values) and withdrawals (negative values). Note that the equation representing the value of such an account will only be a linear equation if:

1. the interest rate depends only on time, not on the amount deposited, and

2. the deposits and withdrawals are also functions of time only, and do not depend on the value of the account.

Despite their complication, the equations (9) and (11) for solution of a linear first order equation can be understood rather well in these terms. Observe that these formulas write the solution as *the sum of two terms*.

In terms of bank accounts, this is simply the following rather obvious fact: *the value of your deposit* is the *sum* of *what is due to your initial deposit, and what is due to subsequent deposits and withdrawals.*

We next ask what is represented by the fundamental solution

$$w(s,t) \equiv e^{\int_s^t p(\tau)d\tau}$$ (10, again)

defined at the end of the last section?

Recall (Example 2.1.1) that the solutions to the associated homogeneous equation $x' = p(t)x$ are

$$u_c(t) = x_0 e^{\int_{t_0}^t p(\tau)d\tau} = x_0 w(t_0, t).$$

Thus $w(s, t)$ has the property that $w(s, t)q(s)$ is the solution, at time t of $x' = p(t)x$ with initial condition $q(s)$ at time s.

In terms of bank accounts, $w(s, t)$ measures the value at time t of a unit deposit at time s. This is meaningful for all (s, t): If $s < t$, $w(s, t)$ measures the value at time t of a unit deposit at time s. If $s > t$, $w(s, t)$ measures the deposit at time t which would have the value of one unit at time s.

We are now ready to interpret the "bank account equation" in two ways:

1. One possibility is to update each deposit, x_0 and $q(s)$, *forward* to time t. Then

$$u(t) = \underbrace{w(t_0, t)x_0}_{u_c} + \int_{t_0}^t w(s, t)q(s)ds. \tag{12}$$

To see that equation (12) is another possible form of (11), notice that

$$w(s, t)w(t_0, s) = w(t_0, t),$$

from equation (10). Therefore

$$w(s, t) = w(t_0, t)(w(t_0, s))^{-1},$$

and we can factor $w(t_0, t)$ out of the integral in the last term of (12).

2. Another possibility is to date every deposit *backward* to time t_0, such that the total deposit at time t_0 should have been

$$x_0 + \int_{t_0}^t w(s, t_0)q(s)ds$$

in order to give the same amount at time t as before, i.e.,

$$u(t) = w(t_0, t)\left[x_0 + \int_{t_0}^t w(s, t_0)q(s)ds\right],$$

which is exactly the equation (11). Thus the idea of equations (9) and (11) resulting from variation of parameters is to take all the deposits and withdrawals that are described by the function $q(t)$, and for each one to figure out how large a deposit (or overdraft) at t_0 would have been required to duplicate the transaction when it occurs at time s.

You have to be careful before you add sums of money payable at different times. The idea is to avoid adding apples and oranges. Lottery winners are well aware of this problem. The states running the lotteries say that the grand prize is \$2 million, but what they mean is that the winner gets \$100,000 a year for 20 years. They are adding apples and oranges, or rather money payable today to money payable 10 or 15 years hence, neglecting the factor $w(s,t)$ by which it must be multiplied.

Example 2.4.1. If interest rates are to be constant at 10% during the next 20 years, how much did a winner of a \$2 million dollar lottery prize win, in terms of today's dollars, assuming that the money is paid out continuously? (If the payoff was not continuous, you would use a sum at the end instead of an integral; however the continuous assumption gives results close to that for a weekly or monthly payoff, which is reasonable.)

Solution. Use one year as the unit of time. The relevant function $w(s,t)$ is

$$w(s,t) = e^{0.1\,(t-s)}.$$

Let x_0 be the desired initial sum equivalent to the prize. Then an account started with x_0 dollars, from which \$100,000 is withdrawn every year for twenty years will have nothing left, i.e.

$$e^{(.1)(20)}\left(\int_0^{20}(-100,000)\,e^{-0.1\,s}ds + x_0\right) = 0.$$

This gives, after integrating,

$$x_0 = (100,000)/(0.1)\,(1 - e^{-2}) \approx 864,664.$$

So, although the winner will still get \$100,000 a year for 20 years, the organizers only had to put rather less than half of their \$2 million claim into their bank account.   ▲

2.5   Population Models

The *growth of a population* under various conditions is a never-ending source of differential equations. Let $N(t)$ denote the size of the population at time t (a common notation among biologists). Then the very simplest possible model for growth is that of a constant per capita growth rate:

Example 2.5.1. If a population $N(t)$ grows at a constant per capita rate, then

$$\frac{1}{N}\frac{dN}{dt} = r = \text{constant}. \tag{13}$$

The constant r is called a *fertility coefficient*. This equation can be rewritten as $N' = rN$; the techniques of Chapter 1 and Section 2.1 give the following picture, Figure 2.5.1, and the solution $N = N_0 e^{rt}$.

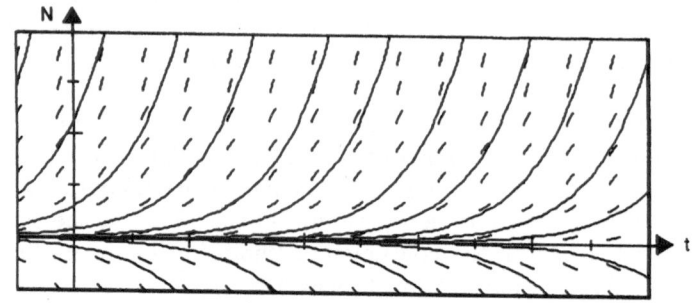

FIGURE 2.5.1. $N' = rN$.

The effect of r (for $r > 0$) is to increase the slope of solutions as r increases; for a fixed r the equation is autonomous and the solutions are horizontal translates. For this problem only the top half-plane is relevant, since a negative population has no meaning. ▲

Obviously the model of Example 2.5.1 is too simple to describe for long the growth of any real population; other factors will come into play.

Example 2.5.2. Consider what happens when the basic assumption of Example 2.1.1, as formulated in (13), is modified to show a decrease in per capita rate of growth due to *crowding* or *competition* that is directly proportional to N:

$$\frac{1}{N}\frac{dN}{dt} = r - bN. \tag{14}$$

Then we can write equation (14) as $N' = rN - bN^2$. This is the *logistic equation*, which we have already solved in Example 2.1.2; the solutions are shown again in Figure 2.5.2. (Note that for a real world population problem we are only interested in positive values of N.) See Exercise 2.4–2.5#3 for support of this model from U.S. Census data. ▲

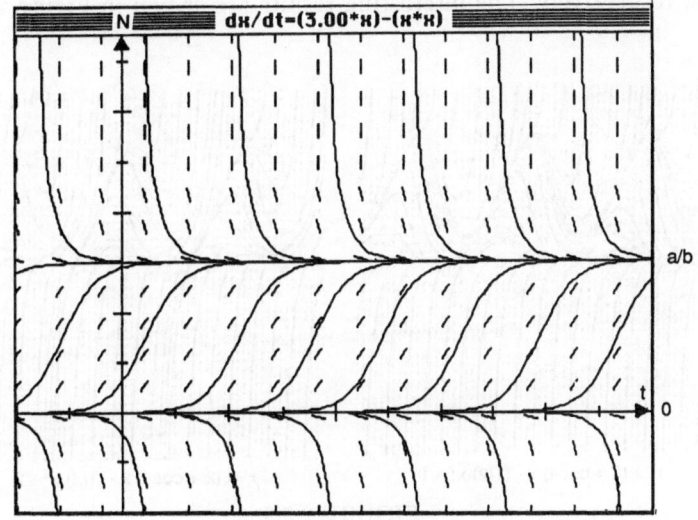

FIGURE 2.5.2. $N' = aN - bN^2$.

There are many variations on the theme of population models. One of these describes the equations of Exercise 1.1#12. The form of those equations is discussed as follows:

Example 2.5.3. Consider

$$x' = (2 + \cos t)x - \left(\frac{1}{2}\right)x^2 + \alpha(t). \tag{15}$$

This equation does not fall under the classes amenable to our analytic methods, but it represents a more complicated population model that we would like to discuss and show what you can do with it by the qualitative methods of Chapter 1.

A reasonable interpretation of equation (15) is that x represents the size of a population that varies with time. Therefore on the right-hand side, the first term represents unfettered growth as in Example 2.5.1, with a fertility coefficient $(2+\cos t)$ that varies seasonally but always remains positive. The second term represents competition as in Example 2.5.2, with competition factor $(\frac{1}{2})$. The third term $\alpha(t)$ is a function with respect to time alone, representing changes to the population that do not depend directly on its present size; e.g., positive values could represent stocking, as in stocking a lake with salmon; negative values might represent hunting, in a situation where there is plenty of prey for all hunters who come, or some fixed number of (presumably successful) hunting licenses are issued each year.

In Exercise 1.1#12, the values of the function $\alpha(t)$ were constant, at -1 and -2, respectively. The interesting results are shown in Figure 2.5.3.

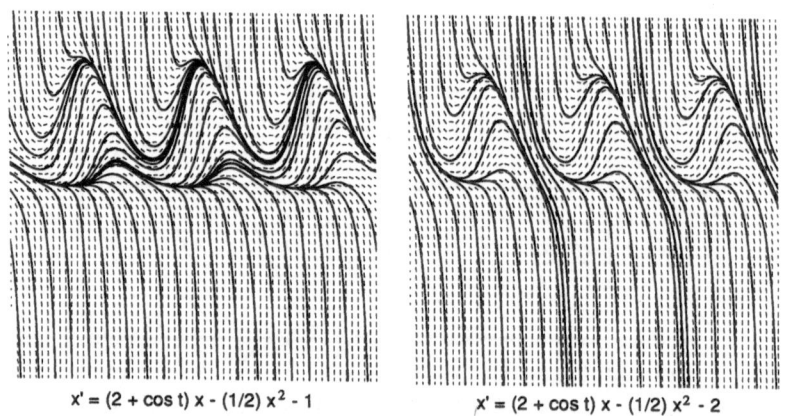

$$x' = (2 + \cos t)\, x - (1/2)\, x^2 - 1 \qquad\qquad x' = (2 + \cos t)\, x - (1/2)\, x^2 - 2$$

FIGURE 2.5.3.

In the first case there are exactly two periodic solutions: there exists a funnel around the upper one, which is stable, and an antifunnel around the lower one, which is unstable. In the second case, there are no periodic solutions: *every* solution dives off to negative infinity, so no stability is possible. Furthermore, as a population size, $x = u(t) < 0$ makes no sense, so note that once $u(t)$ reaches zero, a population is extinct; harvesting was too severe for the population to survive.

A good question for further exploration is "what value of the constant α will separate these two behaviors?" You can use the computer program *DiffEq* to explore this question; the accuracy of your results is limited only by your patience and perseverance (to the limits of the accuracy of the computer!).   ▲

A population model could be helpful in understanding the linear first order differential equation (4) of Section 2.2. In that case, $p(t)$ is a fertility rate, and $q(t)$ represents stocking (positive values) or hunting (negative values). Again, the equation is linear only if there is to be no decrease in fertility due to competition, for instance, and the number killed by hunting must depend on factors other than the actual population. If you are more interested in populations than in bank accounts, you could reread Section 2.4 from this point of view.

The word "population" may be interpreted rather broadly to include other growth models, and decay models as well. The case of *radioactive decay* is a particularly famous one; the equation is basically the same as (13) in the simplest growth model, with a negative rather than positive coefficient.

> The rate of disintegration of a radioactive substance is propor-
> tional at any instant to the amount of the substance present.

Example 2.5.4. *Carbon dating.* The carbon in living matter contains a
minute proportion of the radioactive isotope C^{14}. This radiocarbon arises
from cosmic ray bombardment in the upper atmosphere and enters living
systems by exchange processes, reaching an equilibrium concentration in
these organisms. This means that in living matter, the amount of C^{14} is in
constant ratio to the amount of the stable isotope C^{12}. After the death of
an organism, exchange stops, and the radiocarbon decreases at the rate of
one part in 8000 per year.

Therefore carbon dating enables calculation of the moment when an
organism died, and we set t = the number of years after death. Then
$x(t)$ = ratio of C^{14} to C^{12} satisfies the differential equation

$$x' = -\frac{1}{8000}x,$$

so, from Example 2.1.1,

$$x(t) = x_0\, e^{-t/8000}. \tag{16}$$

A sample use of equation (16) is the following:

> Human skeletal fragments showing ancient Neanderthal char-
> acteristics are found in a Palestinian cave and are brought to a
> laboratory for carbon dating. Analysis shows that the propor-
> tion of C^{14} to C^{12} is only 6.24% of the value in living tissue.
> How long ago did this person live?

We are asked to find t when $x = 0.0624x_0$. From equation (16),

$$0.0624x_0 = x_0 e^{-t/8000}, \quad \text{so}$$

$$t = -8000\, \ln 0.0624 \approx 22,400 \text{ years},$$

the number of years before the analysis that death occurred. ▲

Example 2.5.4 discusses a differential equation with a unique solution
through any set of initial conditions; it illustrates just one of the many
ways in which this might be done. We shall go on in Chapter 4, in Example
4.2.3, to study the question of uniqueness and will contrast the behavior of
this example with another situation.

2.6 Exact Differential Equations

Consider $F(t,x)$, an arbitrary function in two variables, which describes a
surface in 3-space if you set $z = F(t,x)$. The *level curves* of this surface

are the functions given implicitly by

$$F(t, x) = C. \tag{17}$$

Differentiating equation (17) implicitly gives the following differential equation satisfied by these level curves:

$$\frac{\partial F}{\partial x}(t, x)x' + \frac{\partial F}{\partial t}(t, x) = 0, \tag{18}$$

and we can rewrite equation (18) as

$$M(t, x)x' + L(t, x) = 0. \tag{19}$$

We now observe that we might use equation (19) to work the above argument *backwards*. This will be possible for *some* differential equations, which are then called *exact* differential equations. In order to do this, we need the following theorem.

Theorem 2.6.1 (Condition for exactness). *Consider a differential equation written in the form*

$$M(t, x)x' + L(t, x) = 0,$$

with $M(t, x)$ and $L(t, x)$ two suitably differentiable functions defined in a rectangle $R = [a, b] \times [c, d]$. Then there exists a function $F(t, x)$ such that

$$\frac{\partial F}{\partial x} = M(t, x) \quad and \quad \frac{\partial F}{\partial t} = L(t, x)$$

if and only if

$$\frac{\partial M}{\partial t} = \frac{\partial L}{\partial x}. \tag{20}$$

Proof. The *necessity* of condition (20) is due, from multivariable calculus, to the equality of crossed partials (if these functions are continuous):

$$\frac{\partial^2 F}{\partial x \partial t} = \frac{\partial^2 F}{\partial t \partial x}. \tag{21}$$

To prove the *sufficiency* of condition (20), pick some point (t_0, x_0) in R and set

$$F(t, x) = \int_{t_0}^{t} L(\tau, x_0)d\tau + \int_{x_0}^{x} M(t, \sigma)d\sigma. \tag{22}$$

Then

$$\frac{\partial F}{\partial x} = M(t, x),$$

and

$$\frac{\partial F}{\partial t} = L(t, x_0) + \int_{x_0}^{x} \frac{\partial M}{\partial t}(t, \sigma) d\sigma$$

$$= L(t, x_0) + \int_{x_0}^{x} \frac{\partial L}{\partial x}(t, \sigma) d\sigma$$

$$= L(t, x_0) + L(t, x) - L(t, x_0)$$

$$= L(t, x). \quad \square$$

Remark. The sufficiency derivation in this proof may seem a bit miraculous and the choice of F unmotivated. It is closely related to the statement that a force field is conservative, in a simply connected region, if its curl vanishes there. A detailed explanation will appear in a related volume on differential forms, as a special case of Poincaré's Lemma.

The theoretical procedure for *solving* an exact differential equation is very simple: If Theorem 2.6.1 is satisfied, then the solutions are the functions $x = u(t)$ defined implicitly by $F(t, x) = C$ for various constants C. The process of actually finding $F(t, x)$ is illustrated in the following three examples.

Example 2.6.2. Consider $x' = \dfrac{ct - ax}{at + bx}$ (which is not defined along the line at $at + bx = 0$), which can be written as $(at + bx)x' + (ax - ct) = 0$.

This is an exact equation because $\partial M / \partial t = \partial L / \partial x = a$. The fact that

$$M(t, x) = at + bx = \partial F / \partial x$$

implies that

$$F_M(t, x) = atx + \left(\frac{1}{2}\right) bx^2 + \psi(t), \tag{23}$$

and the fact that

$$L(t, x) = ax - ct = \frac{\partial F}{\partial t}$$

implies that

$$F_L(t, x) = axt - \left(\frac{1}{2}\right) ct^2 + \varphi(x). \tag{24}$$

Equations (23) and (24) can be reconciled to give

$$F(t, x) = axt + \left(\frac{1}{2}\right) bx^2 - \left(\frac{1}{2}\right) ct^2.$$

Therefore the solution to this exact differential equation is

$$F(t, x) = axt + \left(\frac{1}{2}\right) bx^2 - \left(\frac{1}{2}\right) ct^2 = C,$$

which is a description for

$$\begin{aligned} \textit{hyperbolas,} &\quad \text{if } a^2 + bc > 0; \\ \textit{ellipses,} &\quad \text{if } a^2 + bc < 0. \end{aligned}$$

These observations are confirmed by Figure 2.6.1.

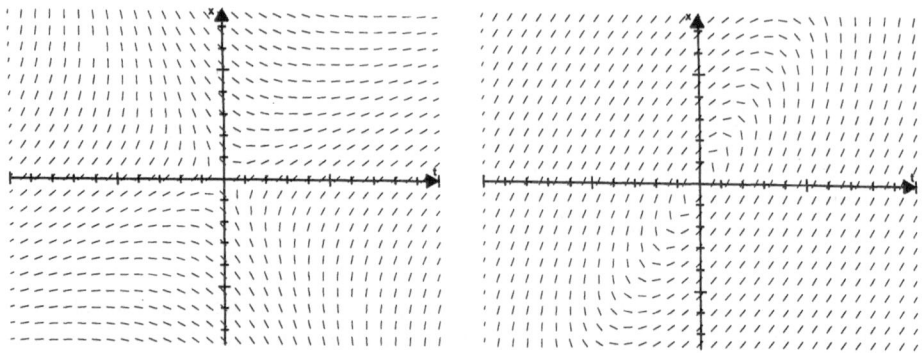

FIGURE 2.6.1. $x' = \frac{t-x}{t+x}$ on left, $x' = \frac{2t-x}{t-x}$ on right.

You will find if you try to draw solutions in these direction fields using the computer program *DiffEq* that there may be trouble, as shown in Figure 2.6.2, wherever the slope gets vertical—precisely along the lines where the differential equation is not defined.

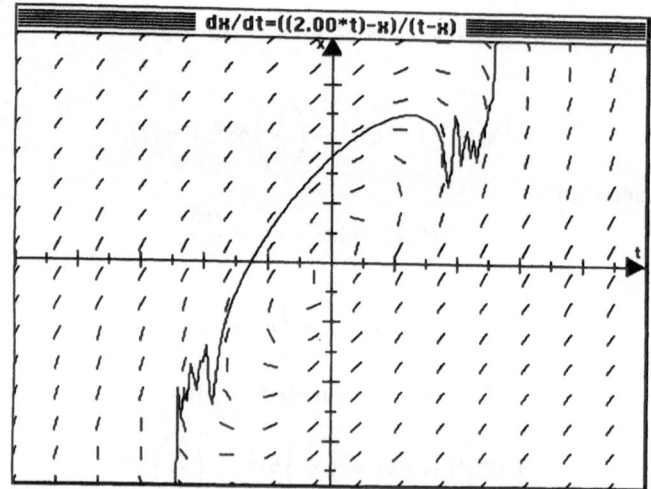

dx/dt=((2.00*t)-x)/(t-x)

FIGURE 2.6.2. $x' = \frac{2t-x}{t-x}$ (undefined along $t - x = 0$).

This is to be expected, because the solutions to a differential equation are *functions*. What appears to be a hyperbola or an ellipse in the slope field is, as shown in Figure 2.6.3, actually two completely different solutions; one above and one below the line where the differential equation is not defined.

▲

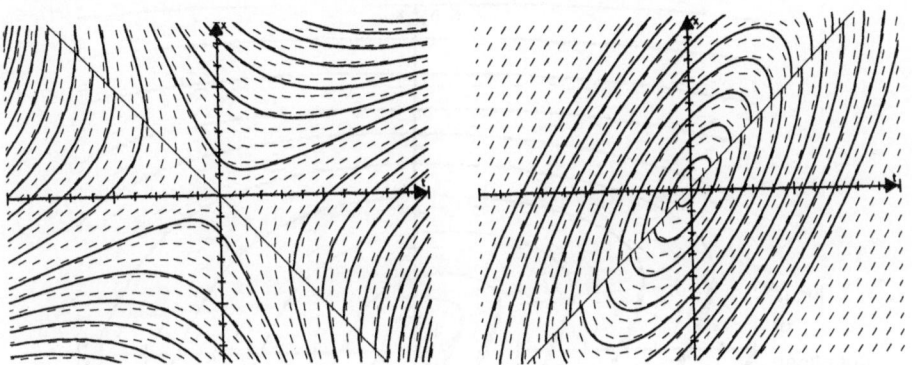

FIGURE 2.6.3. Solutions do not cross lines where slopes are undefined.

Example 2.6.3. Consider

$$\underbrace{(t\cos x + 3x^2)}_{M(t,x)}x' + \underbrace{\sin x + 2t}_{L(t,x)} = 0,$$

(undefined along $t\cos x + 3x^2 = 0$).

You can confirm that $\partial M/\partial t = \partial L/\partial x$. Then the fact that

$$M(t,x) = t\cos x + 3x^2 = \frac{\partial F}{\partial x}$$

implies that

$$F_M(t,x) = t\sin x + x^3 + \psi(t), \tag{25}$$

and the fact that

$$L(t,x) = \sin x + 2t = \frac{\partial F}{\partial t}$$

implies that

$$F_L(t,x) = t\sin x + t^2 + \varphi(x). \tag{26}$$

The reconciliation of equations (25) and (26) tells us that

$$F(t,x) = t\sin x + x^3 + t^2, \tag{27}$$

so the solution to this exact differential equation is

$$F(t,x) = t\sin x + x^3 + t^2 = C.$$

It is hard to imagine the surface $z = F(t, x)$ described by equation (27) standing alone, but using the computer we can easily get a lovely picture of its level curves from the direction field and solution curves of the differential equation, illustrated in Figure 2.6.4.

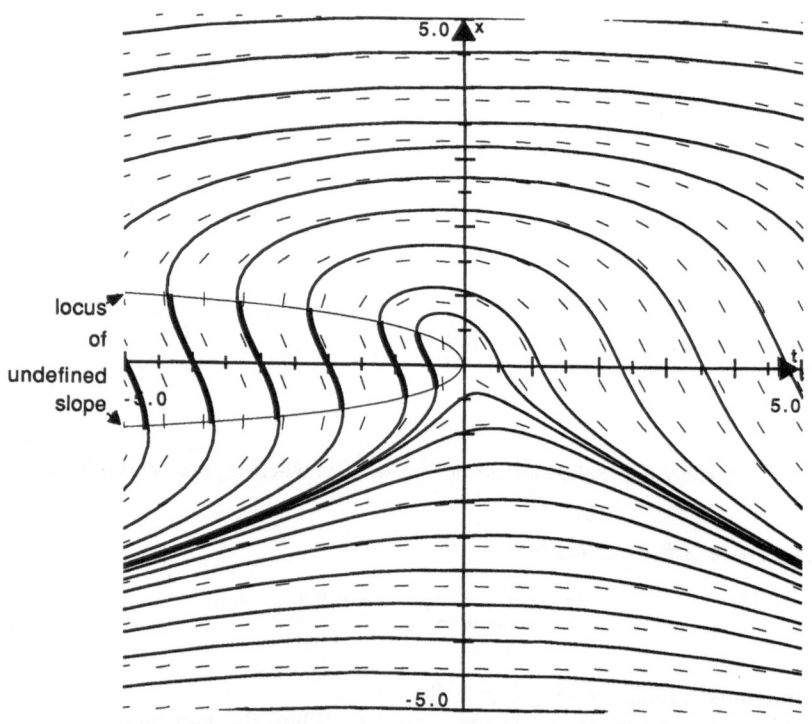

FIGURE 2.6.4. $x' = -\frac{\sin x - 2t}{t \cos x + 3x^2}$ (undefined along $t \cos x + 3x^2 = 0$).

Recall that this differential equation is undefined where $t \cos x + 3x^2 = 0$. The "level curves" drawn in this direction field are sometimes composed of two or three functions, where they cross this locus. In Figure 2.6.4 the third function is drawn thicker. (As suggested in the MacMath Documentation under *DiffEq*, you can enter the differential equation as a system of two equations, to be further discussed in Section 6.1, to get good computer drawings of these level curves.)   ▲

The illustration of Example 2.6.3 and Figure 2.6.4 demonstrates a technique that is useful for helping to visualize the surface described by *any* particular ugly $F(t, x)$: follow the steps of equations (17) through (19), and then use the computer to get the direction field on which the solution curves of the differential equation are the level curves of the surface.

The geometrical interpretation of solutions to exact equations as level curves of a function $F(t, x)$ will be very useful in other ways, as you will begin to see in Volume II, Chapter 6.

Example 2.6.4. All *separable* equations, $x' = g(t)h(x)$, can be written in exact form. They can be written as

$$\frac{1}{h(x)}x' - g(t) = 0,$$

and we see that

$$\frac{\partial}{\partial t}\left(\frac{1}{h(x)}\right) = \frac{\partial}{\partial x}(-g(t)) = 0.$$

Therefore, from (22), the solutions are along the level curves of

$$F(t, x) = \int_{t_0}^{t} g(\tau)d\tau - \int_{x_0}^{x} \frac{1}{h(\sigma)}d\sigma,$$

or $G(t) - H(x) = -C$ according to the notation of Section 2.1. ▲

There is also a relationship between *linear* equations and exact equations, which you will explore in Exercise 2.6#2, but perhaps this analogy is misleading. Linear and exact equations generalize to higher dimensions in completely different ways. The linear equations are the beginning of the entire theory of ordinary differential equations, but the exact equations are historical tricks that work only for one-dimensional ordinary differential equations and don't generalize at all to the higher dimensional theory of ODE's. (Basically the linear equations generalize to the theory studied in Volume II, Chapters 6 and 7 with linear algebra, whereas the exact equations generalize to total partial differential equations, the Poincaré Lemma, and the Frobenius Theorem, which occur in the study of differential forms.)

On a more immediately practical level, you should notice that for both *separable* and *exact* differential equations, the solutions are *implicit*, and more often than not are therefore inconvenient or impossible to solve explicitly for $x = u(t)$. The *linear* equations, on the other hand, have the advantage of producing *explicit* solutions, as given by the famous formula (9).

Summary. An equation of the form $M(t, x)x' + L(t, x) = 0$ is *exact* if

$$\frac{\partial M}{\partial t} = \frac{\partial L}{\partial x}.$$

If the equation is exact, the solutions are expressed implicitly by

$$F(t, x) = C,$$

where

$$\frac{\partial F}{\partial x} = M(t, x) \quad \text{and} \quad \frac{\partial F}{\partial t} = L(t, x).$$

The procedure for finding $F(t, x)$ is as follows:

Holding t constant, integrate M with respect to x, to get F_M (remembering that a "constant" in x might be a function in t);

Holding x constant, integrate L with respect to t, to get F_L (remembering that a "constant" in t might be a function in x);

Finally, reconcile the results of these last two computations to give an $F(t, x)$ that satisfies both F_M and F_L.

2.7   Series Solutions of Differential Equations

The method of undetermined coefficients introduced in Section 2.2, can also be used to compute power series solutions of the differential equation $x' = f(t, x)$, at least if the function $f(t, x)$ is itself given by a power series.

More precisely, suppose we are looking for a solution $u(t)$ such that $u(t_0) = x_0$. Then set

$$u(t) = x_0 + a_1(t - t_0) + a_2(t - t_0)^2 + \ldots$$

and substitute into the equation. You will then be able to calculate the a_i recursively.

Example 2.7.1. Find the power series of the solution to $x' = x$ with $x(0) = 1$.

Set

$$x(t) = 1 + a_1 t + a_2 t^2 + a_3 t^3 + \ldots;$$

then substitution gives

$$a_1 + 2a_2 t + 3a_3 t^2 + \ldots = 1 + a_1 t + a_2 t^2 + a_3 t^3 + \ldots$$

and identifying the coefficients of successive powers of t gives

$$\begin{aligned}
a_1 &= 1, \\
2a_2 &= a_1, \quad \text{i.e. } a_2 = 1/2 \\
3a_3 &= a_2, \quad \text{i.e. } a_3 = 1/6, \quad \text{etc.}
\end{aligned}$$

Actually, we see that $na_n = a_{n-1}$, so since $a_1 = 1$ we get $a_n = 1/(n!)$, and

$$x(t) = \sum_{n=0}^{\infty} \frac{t^n}{n!},$$

which you should recognize as the power series defining e^t.   ▲

The fact that made the computation of Example 2.7.1 work is that each coefficient appeared on the left with a lower power of t than it did on the right, and as such could be computed from the lower coefficients, which had already been computed. This fact is always true, but for nonlinear equations it is usually impossible to find a formula for the coefficients. See, for example, Exercise 2.7#4.

Example 2.7.2. Find the power series of the solution of $x' = x^2 - t$, the famous equation of Chapter 1, with $x(0) = 1$. As in the last example, set

$$x(t) = 1 + a_1 t + a_2 t^2 + a_3 t^3 + \dots$$

and substitute. You find

$$a_1 + 2a_2 t + 3a_3 t^2 + \dots = (1 + a_1 t + a_2 t^2 + a_3 t^3 + \dots)^2 - t$$
$$= 1 + (2a_1 - 1)t + (a_1^2 + 2a_2)t^2$$
$$+ (2a_1 a_2 + 2a_3)t^3 + \dots ,$$

leading to

$$
\begin{array}{lll}
a_1 \ = 1 & \Rightarrow & a_1 = 1 \\
2a_2 = 2a_1 - 1 & \Rightarrow & a_2 = 1/2 \\
3a_3 = a_1^2 + 2a_2 & \Rightarrow & a_3 = 2/3 \\
4a_4 = 2a_1 a_2 + 2a_3 & \Rightarrow & a_4 = 7/12 \\
\vdots
\end{array}
$$

Clearly this process could be continued ad infinitum with sufficient effort; on the other hand, no simple pattern seems to be emerging for the coefficients. In particular, *it is not clear what the radius of convergence of the series is,* or even whether the series converges at all.   ▲

In the sense of not knowing a radius of convergence, power series solutions are usually pretty useless: unless you know something about how they converge, they convey next to no information, and it is usually hard to know anything about the convergence unless a pattern for the coefficients emerges. A recognizable pattern will usually exist only for linear equations.

In Example 2.7.2, the radius of convergence of the power series cannot be infinite, and in fact must be smaller than 2, because we know from Exercise 1.6#5 that the solution blows up before $t = 2$. This means that the series

$$\Sigma \, 2^n a_n$$

diverges, but we do not know how to show this from looking at the recursive definition of the coefficients; we can show it only from fence techniques.

The following theorem, which we will not prove (the proof is rather devious, requiring Picard iteration and complex analytic functions) says that in fact the power series does converge.

Theorem 2.7.3. *If the function $f(t, x)$ is the sum of a convergent power series in a neighborhood of (t_0, x_0), then substituting the series*

$$x(t) = x_0 + \Sigma a_n (t - t_0)^n$$

into the differential equation $x' = f(t, x)$ and equating equal powers of t leads to a unique power series, which does converge for t in some neighborhood of t_0, and whose sum is the solution to $x' = f(t, x)$ with $x(t_0) = x_0$.

Notice the condition that f be given by a power series. If this condition is not satisfied, *it may be quite possible to determine a power series solution of $x' = f(t, x)$ with radius of convergence 0.* This will be shown by the next example; it is rather lengthy, but carries an important explanation.

Example 2.7.4. Consider the differential equation $x' = t^2/x - 1$, which is singular (i.e., not well defined) if $x = 0$. Determine the power series solutions with $x(0) = 0$.

As above, set

$$x(t) = a_1 t + a_2 t^2 + a_3 t^3 + a_4 t^4 + \dots$$

and substitute in the equation, which we will write $x(x' + 1) = t^2$. We find

$$(a_1 t + a_2 t^2 + a_3 t^3 + a_4 t^4 + \dots)((1 + a_1) + 2a_2 t + 3a_3 t^2 + \dots) = t^2. \quad (28)$$

In particular, $a_1(1 + a_1) = 0$, leading to two solutions, one with $a_1 = 0$ and the other with $a_1 = -1$. It turns out that these behave very differently, so we will discuss them separately.

(i) *The solution with $a_1 = 0$.* In this case, equation (28) becomes

$$(a_2 t^2 + a_3 t^3 + a_4 t^4 + \dots)(1 + 2a_2 t + 3a_3 t^2 + 4a_4 t^3 + \dots) = t^2$$

and identifying coefficients on both sides of the equation leads to

$$
\begin{aligned}
a_2 &= 1 & &\Rightarrow & a_2 &= 1 \\
a_3 + 2a_2^2 &= 0 & &\Rightarrow & a_3 &= -2 \\
a_4 + 2a_3 a_2 + 3a_2 a_3 &= 0 & &\Rightarrow & a_4 &= 10 \\
a_5 + 2a_4 a_2 + 3a_3^2 + 4a_4 a_2 &= 0 & &\Rightarrow & a_5 &= -72 \\
&\vdots
\end{aligned}
$$

The coefficients appear to alternate in sign, and to grow rather rapidly. The following result makes this precise:

The coefficient a_n has the sign of $(-1)^n$, and $|a_n| > (n - 1)!$.

The proof is by induction, and this induction is started above. The recursion formula can be written

$$a_n = -(n+1) \sum_{i=2}^{n/2} a_i a_{n+1-i} \qquad \text{if } n \text{ is even,}$$

$$a_n = -(n+1) \sum_{i=2}^{(n-1)/2} a_i a_{n+1-i} - \frac{n+1}{2}(a_{(n+1)/2})^2 \quad \text{if } n \text{ is odd.}$$

In both cases, the induction hypothesis implies that all the terms on the righthand side have the same sign, negative if n is even and positive if n is odd, and the factor $-(n+1)$ then shows that a_n is positive if n is even and negative if n is odd, as claimed. Moreover, since there is no cancelling, we have

$$|a_n| > (n+1)|a_2 a_{n-1}| > (n+1)(n-2)! > (n-1)! \,.$$

In particular, the power series under discussion has radius of convergence equal to 0. The line above gives

$$\left| \frac{a_n}{a_{n-1}} \right| > n+1,$$

and the result follows from the ratio test.

> *Just because a power series solves a differential equation does not mean that it represents a solution!*

(ii) *The solution with $a_1 = -1$.* In this case, equation (28) becomes

$$(-t + a_2 t^2 + a_3 t^3 + a_4 t^4 + \ldots)(2a_2 t + 3a_3 t^2 + 4a_4 t^3 + \ldots) = t^2,$$

which leads to

$$
\begin{array}{llll}
-2a_2 = 1 & & \Rightarrow & a_2 = -1/2 \\
-3a_3 + 2a_2^2 = 0 & & \Rightarrow & a_3 = 1/6 \\
-4a_4 + 3a_3 a_2 + 2a_2 a_3 = 0 & & \Rightarrow & a_4 = -5/48 \\
-5a_5 + 4a_4 a_2 + 3a_3^2 + 2a_2 a_4 = 0 & & \Rightarrow & a_5 = 19/240 \\
\vdots & & &
\end{array}
$$

This time the coefficients still alternate, but appear to be decreasing in absolute value. This is misleading, because the coefficients eventually start growing, and the 100th is about 1300. Still, they don't grow too fast, but the proof is pretty tricky. We show that the coefficient a_n has the sign of $(-1)^{n+1}$, and, for $n > 2$, $|a_n| < 2^{n-1}/n^2$, proceeding again by induction.

To start, notice that it is true for a_2. Next, observe that the recurrence relation for computing the coefficients can be written

$$a_n = \frac{n+1}{n} \sum_{i=2}^{n/2} a_i a_{n+1-i} \qquad\qquad \text{if } n \text{ is even,}$$

$$a_n = \frac{n+1}{n} \sum_{i=2}^{(n-1)/2} a_i a_{n+1-i} + \frac{n+1}{2n}(a_{(n+1)/2})^2 \quad \text{if } n \text{ is odd.}$$

It follows as above that if the signs of the a_i alternate for $i < n$, then all terms on the right-hand side have the same sign, which is positive if n is odd and negative if n is even. So far, so good. Now, by induction, we find if n is even that

$$|a_n| = \frac{n+1}{n} \sum_{i=2}^{n/2} |a_i a_{n+1-i}| < \frac{4(n+1)}{n} \sum_{i=2}^{n/2} \frac{A^{i-2}}{i^2} \frac{A^{n-i-1}}{(n+1-i)^2}$$

$$= 4A^{n-3} \frac{n+1}{n} \sum_{i=2}^{n/2} \frac{1}{i^2} \frac{1}{(n+1-i)^2}.$$

We need to bound the sum, and this is done by comparison with an integral, as follows:

$$\sum_{i=2}^{n/2} \frac{1}{i^2} \frac{1}{(n+1-i)^2} < \int_1^{n/2} \frac{dx}{x^2(n-x)^2} = \frac{1}{n^2} \int_1^{n/2} \left(\frac{1}{x} + \frac{1}{(n-x)} \right)^2 dx$$

$$= \frac{1}{n^2} \left(\int_1^{n/2} \frac{dx}{x^2} + \int_1^{n/2} \frac{dx}{(n-x)^2} + \int_1^{n/2} \frac{2\,dx}{x(n-x)} \right)$$

$$= \frac{1}{n^2} \int_1^{n-1} \frac{dx}{x^2} + \frac{2}{n^3} \int_1^{n-1} \frac{dx}{x}$$

$$= \frac{1}{n^2} \left(1 - \frac{1}{n-1} + \frac{2}{n} \ln(n-1) \right).$$

The first inequality is the one that needs verifying, but we leave it to the reader. This bound leads to

$$|a_n| < \frac{2A^{n-2}}{n^2} \left(\frac{2}{A} \frac{n+1}{n} \left(1 - \frac{1}{n-1} + \frac{2\ln(n-1)}{n} \right) \right)$$

and we see that in order for the induction to work, it is necessary that A be chosen sufficiently large so that the quantity in the outer parentheses will have absolute value smaller than 1. This is certainly possible, since the sequence in the inner parentheses tends to 1 as $n \to \infty$, and hence has a finite upper bound. It is not too hard to show, using a bit of calculus, that

$A = 4$ actually works. Using the computer program *Analyzer,* for instance, you can show that the greatest value for

$$y = \left(\frac{x+1}{x} \left(1 - \frac{1}{x-1} + \frac{2 \ln(x-1)}{x} \right) \right)$$

occurs for $5 < x < 6$, and that the maximum y is less than 1.57. So the smallest A for which the argument will work is approximately 3.14.

The argument for n odd is exactly similar: we find

$$\sum_{i=2}^{n/2} \frac{1}{i^2} \frac{1}{(n+1-i)^2} + \frac{1}{2} \frac{1}{((n+1)/2)^2} < \int_1^{n/2} \frac{dx}{x^2(n-x)^2},$$

with the last part of the integral, from $(n-1)/2$ to $n/2$, bounding the extra term in the sum.

In particular, the radius of convergence of this power series is at least $1/A$, and this power series does represent a solution to the differential equation.

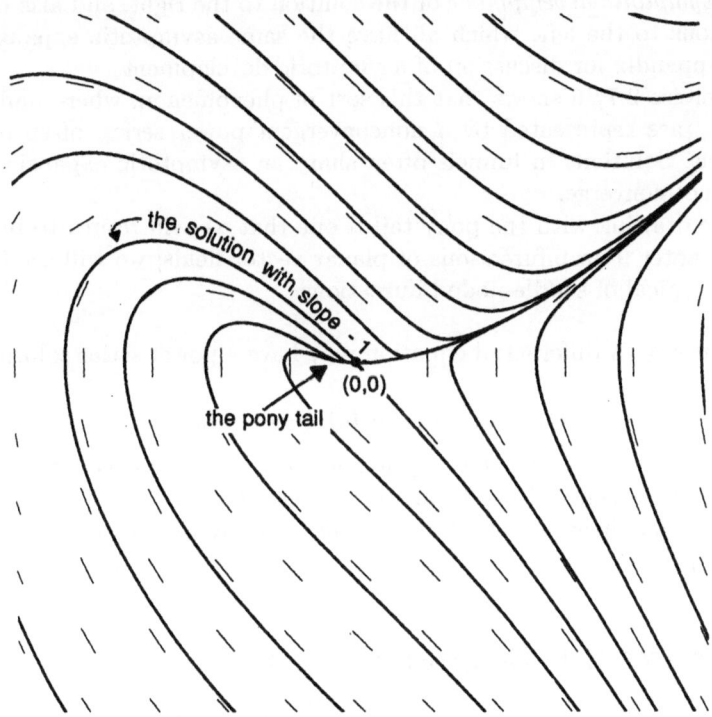

the solution with slope -1

(0,0)

the pony tail

FIGURE 2.7.1. $x' = \frac{t^2}{x} - 1$.

This drastic difference in the behavior of the power series for the two different values of a_1 requires a bit of explanation. The solutions to the differential equation appear in Figure 2.7.1.

In particular, there actually is a unique solution to the differential equation through $(0,0)$ with slope -1, but with slope 0 there is a unique solution to the right of 0, which is continued on the left by a whole "pony tail" of solutions.   ▲

PONY TAIL BEHAVIOR AND POWER SERIES

If the first power series of Example 2.7.4 had converged, it would have picked out one hair of this pony tail as prettier than the others, namely the sum of the power series. So a reasonable philosophy is that unless there is a good reason to think that a power series represents something particular, there is good reason to worry about its convergence. In this case, there is no reason for one solution in the pony tail to be distinguished.

You might wonder, considering that the first power series of Example 2.7.4 does not converge, whether it means anything at all, and it does. It is the *asymptotic development* of the solution to the right, and also of all the solutions to the left, which all have the same asymptotic expansion. See the Appendix for discussion of asymptotic development.

Exercise 2.7#6 shows that this sort of phenomenon, where multiple solutions are represented by a nonconvergent power series, often occurs in funnels. Solutions in funnels often share an asymptotic expansion, which does not converge.

The drawing with the pony tail is one that we will return to in Volume II, Chapter 9 on bifurcations of planar vector fields; we will see that it is quite typical of saddle-node bifurcations.

Summary. A differential equation may have a power series solution of the form

$$u(t) = x_0 + a_1(t - t_0) + a_2(t - t_0)^2 + \dots,$$

if you can substitute this expression into the differential equation and equate coefficients of like terms to evaluate the a_i's.

You *must,* however, check to see where (if anywhere) the resulting series converges.

Exercises 2.1   Separable Equations

2.1#1. Go back and solve by separation of variables wherever possible:

(a) the equation $x' = -tx$ of Example 1.1.1;

(b) the equation $x' = kx^\alpha$ of Section 1.6;

(c) the equation $x' = ax + b$, a very basic equation that you will meet in Example 2.2.2 and again in Chapters 3 and 4;

(d) the appropriate parts of Exercises 1.1#1,2,3.

2.1#2°. Solve these differential equations; then use either *Analyzer* or *Diff-Eq* to get a picture of the solutions:

(a) $(1 + x)tx' + (1 - t)x = 0$

(b) $(1 + x) - (1 - t)x' = 0$

(c) $(t^2 - xt^2)x' + x^2 + tx^2 = 0$

(d) $(x - \alpha) + t^2 x' = 0$

(e) $x - (t^2 - \alpha^2)x' = 0$

(f) $\dfrac{dx}{dt} = \dfrac{1 + x^2}{1 + t^2}$

(g) $\sec^2 \theta \tan \varphi \, d\theta + \sec^2 \varphi \tan \theta \, d\varphi = 0$

(h) $\sec^2 \theta \tan \varphi \, d\varphi + \sec^2 \varphi \tan \theta \, d\theta = 0$

(i) $3e^t \tan x \, dt + (1 - e^t) \sec^2 x \, dx = 0$

(j) $(t - x^2 t)dt + (x - t^2 x)dx = 0$

See also Exercises 2, Miscellaneous Problems, at the end of the chapter.

Exercises 2.2–2.3 Linear Equations

2.2–2.3#1. Go back and solve as linear first order differential equations wherever possible.

(a) the equation $x' = 2t - x$ of Example 1.1.3

(b) the appropriate parts of Exercises 1.1#1,2,3.

2.2–2.3#2. Go back and solve all those equations that are either separable or linear in the following:

(a)–(h) the parts of Exercises 1.5#1.

2.2–2.3#3. Solve $x' = e^t - x$ and explain, for different possible values of the constant of integration, what happens when $t \to \infty$. Compare your results by matching computer drawings (held up to the light) from *Analyzer* (for the equations of the solutions with different values of c) and *DiffEq* (for the direction field and some sample solutions).

2.2–2.3#4°. Solve the following linear differential equations; then use *Analyzer* or *DiffEq* to get a picture of the solutions:

(a) $x' - \dfrac{2x}{t+1} = (t+1)^2$

(b) $x' - \alpha\dfrac{x}{t} = \dfrac{t+1}{t}$

(c) $(t - t^3)x' + (2t^2 - 1)x - \alpha t^3 = 0$

(d) $\dfrac{dx}{dt} + x\cos t = \dfrac{1}{2}\sin 2t$

(e) $x' - \dfrac{n}{t}x = e^t t^n$

(f) $x' + \dfrac{n}{t}x = \dfrac{\alpha}{t^n}$

(g) $x' + x = \dfrac{1}{e^t}$

(h) $x' + \dfrac{1-2t}{t^2}x - 1 = 0$

2.2–2.3#5. Verify by direct substitution in the linear differential equation $x' = p(t)x + q(t)$ that the boxed formula (9) is a solution.

2.2–2.3#6. Another traditional approach to the linear differential equation and its solution is as follows: rewrite equation (4) as

$$\frac{dx}{dt} - p(t)x = q(t).$$

Multiply both sides of this equation by $e^{-\int p(\tau)d\tau}$, which is called an *integrating factor* because of what happens next.

The left side of the new equation becomes

$$e^{-\int p(\tau)d\tau}\left\{\frac{dx}{dt} - p(t)x\right\} = \frac{d}{dt}\left\{e^{-\int p(\tau)d\tau}x\right\}.$$

Now it is easy to integrate both sides of the new equation with respect to t. Finish this process now to arrive at formula (9). You may work with indefinite integrals, but notice where the constant of integration will come in. This method actually comes under the umbrella of exact equations, the subject of Section 2.6; see Exercises 2.6#2.

2.2–2.3#7. The examples of Section 2.2 are all *linear* equations (which, after all, *is* the subject of this section). Nevertheless, this is a good place to remark that there are certain nonlinear equations to which the method of undetermined coefficients can also be applied to give a particular solution. For instance, consider

$$(t-1)x'' + (x')^2 + x = t^2,$$

for which the natural operator on x is

$$(t-1)\frac{d^2}{dt^2} + \left(\frac{d}{dt}\right)^2 + 1.$$

The quantity t^2 appears on the right, and we see that on the left if we assume a quadratic polynomial, the highest power of t will also be 2. So assume a particular solution of the form

$$u_p(t) = \alpha + \beta t + \gamma t^2,$$

and proceed to substitute in the original differential equation; then set up equations for the coefficients of the powers of x. You will obtain a system of nonlinear equations in α, β, and γ, you will find it can be solved. The interesting thing is that there will be *different* values possible for γ, each giving rise to a *different* particular solution. However, since the differential equation is not linear, you will *not* be able to superpose these *particular* solutions, so the method of undetermined coefficients will not be much help at getting a *general* solution.

2.2–2.3#8°. Try the method of undetermined coefficients, as in the last exercise, on the following differential equation:

$$(t^2 + 1)x'' + (x')^2 + kx = t^2.$$

Discuss the effect of the parameter k.

2.2–2.3#9. *Bernoulli equations* are those of the form

$$x' + P(t)x = Q(t)x^n,$$

where $P(t)$ and $Q(t)$ are continuous functions of t and $n \neq 0$, $n \neq 1$. These are important because they can be reduced to linear equations in $z(t)$ by a substitution $z = x^{-n+1}$. Show that this is true.

2.2–2.3#10°. Use the substitution suggested in the last exercise to solve the following Bernoulli equations by transforming them to linear differential equations. Use *Analyzer* or *DiffEq* to get a picture of the solutions.

(a) $x' + tx = t^3x^3$

(b) $(1 - t^2)x' - tx - \alpha tx^2 = 0$

(c) $3x^2x' - \alpha x^3 - t - 1 = 0$

(d) $x - x'\cos t = x^2\cos t(1 - \sin t)$

See also Exercises 2, Miscellaneous Problems, at the end of the chapter.

Exercises 2.4–2.5 Models Requiring Differential Equations

2.4–2.5#1. What rate of interest payable annually is equivalent to 6% continuously compounded?

2.4–2.5#2°. Consider the differential equation

$$\frac{dx}{dt} = (2 + \cos t)x - \frac{1}{2}x^2 + \alpha(t)$$

in the special case where $\alpha(t)$ is a *constant* (as in Example 2.5.2).

(a) Show that if $x = u(t)$ is a solution then $x = u(t + 2\pi)$ is also a solution.

(b) For $\alpha = -1$ there are exactly *two* periodic solutions,

$$u_i(t + 2\pi) = u_i(t); \qquad i = 1, 2.$$

Draw in a funnel around the upper periodic solution and an antifunnel around the lower one.

(c) There is *no* periodic solution for $\alpha = -2$. By *experimenting* with the computer in the program *DiffEq*, find the value (to two significant digits) of the constant α between -1 and -2, which separates the two behaviors: there exists a periodic solution and there is no periodic solution.

2.4–2.5#3.

(a) Consider the equation $N' = rN(1 - \frac{N}{k})$, which is a form of the logistic equation discussed in Example 2.5.1 and solved in Example 2.1.2. Confirm that one way of writing the solution is

$$N = \frac{k}{1 + \frac{e^{-rt}}{c}}.$$

(b) The formula of part (a) was used successfully by R.L. Pearl and L.J. Read (*Proceedings of the National Academy of Sciences*, 1920, p. 275) to demonstrate a rather good fit with the population data of the United States gathered in the decennial census from 1790 to 1910. Using 1790, 1850, and 1910 as the points by which to evaluate the constants they obtained

$$N = \frac{197,273,000}{1 + e^{-0.0313395\,t}}$$

and then calculated a predicted $N(t)$ for each of the decades between, to compare with the census figures. The results are given in the table, with four more decades added by the Dartmouth College Writing Group in 1967.

Year	Population from Decimal Census	Population from Formula (6)	Error	%Error
1790	3,929,000	3,929,000	0	0.0
1800	5,308,000	5,336,000	28,000	0.5
1810	7,240,000	7,228,000	-12,000	-0.2
1820	9,638,000	9,757,000	119,000	1.2
1830	12,866,000	13,109,000	243,000	1.9
1840	17,069,000	17,506,000	437,000	2.6
1850	23,192,000	23,192,000	0	0.0
1860	31,443,000	30,412,000	-1,031,000	-3.3
1870	38,558,000	39,372,000	814,000	2.1
1880	50,156,000	50,177,000	21,000	0.0
1890	62,948,000	62,769,000	-179,000	-0.3
1900	75,995,000	76,870,000	875,000	1.2
1910	91,972,000	91,972,000	0	0.0
1920	105,711,000	107,559,000	1,848,000	1.7
1930	122,775,000	123,124,000	349,000	0.3
1940	131,669,000	136,653,000	4,984,000	3.8
1950	150,697,000	149,053,000	-1,644,000	-1.1
1960	179,300,000[1]			
1970	204,000,000[1]			
1980	226,500,000[1]			

[1]Rounded to the nearest hundred thousand.

(c) Update and revise the table using the more recent census data. Do you think it advisable to change the three base years used to evaluate the constants? How much difference would it make? Would it be sufficient to change visually computer pictures from *Analyzer* or *DiffEq*?

2.4–2.5#4. What is the half-life of C^{14}? (See Example 2.5.3. *Half-life* is the time required for half the amount of the radioactive isotope to disintegrate.)

2.4–2.5#5. At Cro Magnon, France, human skeletal remains were discovered in 1868 in a cave where a railway was being dug. Philip van Doren Stern, in a book entitled *Prehistoric Europe, from Stone Age Man to the Early Greeks* (New York: W.W. Norton, 1969), asserts that the best estimates of the age of these remains range from 30,000 to 20,000 B.C. What range of laboratory C^{14} to C^{12} ratios would be represented by that range of dates?

2.4–2.5#6°. A population of bugs on a plate tend to live in a circular colony. If N is the number of bugs and r_1 is the per capita growth rate, then the Malthusian growth rule states that $dN/dt = r_1 N$. However, those bugs on the perimeter suffer from cold, and they die at a rate proportional to their number, which means that they die at a rate proportional to $N^{1/2}$. Let this constant of proportionality be r_2. Find the differential equation satisfied by N. Without solving it, sketch some solutions. Is there an equilibrium? If so, is it stable?

2.4–2.5#7. By another of Newton's laws, the rate of cooling of some body in air is proportional to the difference between the temperature of the body and the temperature of the air. If the temperature of the air is 20°C and the body cools for 20 minutes from 100° to 60°C, how long will it take for its temperature to drop to 30°C?

2.4–2.5#8. Nutrients flow into a cell at a constant rate of R molecules per unit time and leave it at a rate proportional to the concentration, with constant of proportionality K. Let N be the concentration at time t. Then the mathematical description of the rate of change of nutrients in the above process is

$$\frac{dN}{dt} = R - KN;$$

that is, the rate of change of N is equal to the rate at which nutrients are entering the cell minus the rate at which they are leaving. Will the concentration of nutrients reach an equilibrium? If so, what is it and is it stable? Explain, using a graph of the solutions to this equation.

2.4–2.5#9. Water flows into a conical tank at a rate of k_1 units of volume per unit time. Water evaporates from the tank at a rate proportional to $V^{2/3}$, where V is the volume of water in the tank. Let the constant of proportionality be k_2. Find the differential equation satisfied by V. Without solving it, sketch some solutions. Is there an equilibrium? Is it stable?

2.4–2.5#10. Water containing 2 oz. of pollutant/gal flows through a treatment tank at a rate of 500 gal/min. In the tank, the treatment removes 2% of the pollutant per minute, and the water is thoroughly stirred. The tank holds 10,000 gal. of water. On the day the treatment plant opens, the tank is filled with pure water. Find the function which gives the concentration of pollutant in the outflow.

2.4–2.5#11. At time $t = 0$, two tanks each contain 100 gallons of brine, the concentration of which then is one half pound of salt per gallon. Pure water is piped into the first tank at 2 gal/min, and the mixture, kept uniform by stirring, is piped into the second tank at 2 gal/min. The mixture in the second tank, again kept uniform by stirring, is piped away at 1 gal/min. How much salt is in the water leaving the second tank at any time $t > 0$?

2.4–2.5#12. The following model predicts glucose concentration in the body after glucose infusion: Infusion is the process of admitting a substance into the veins at a steady rate (this is what happens during intravenous feeding from a hanging bottle by a hospital bed). As glucose is admitted, there is a drop in the concentration of free glucose (brought about mainly by its combination with phosphorous); the concentration will decrease at a rate proportional to the amount of glucose. Denote by G the concentration of glucose, by A the amount of glucose admitted (in mg/min), and by B the volume of liquid in the body (in the blood vessels). Find whether and how the glucose concentration reaches an equilibrium level.

2.4–2.5#13. A criticism of the model of the last exercise is that it assumes a constant volume of liquid in the body. However, since the human body contains about 8 pt of blood, infusion of a pint of glucose solution would change this volume significantly. How would you change this model to account for variable volume? I.e., how would you change the differential equation? Will this affect your answer about an equilbrium level? How? What are the limitations of *this* model? (Aside from the fact you may have a differential equation which is hideous to solve or analyze, what criticisms or limitations do you see physically to the variable volume idea?) What sort of questions might you ask of a doctor or a biologist in order to work further on this problem?

2.4–2.5#14. A spherical raindrop evaporates at a rate proportional to its surface area. Find a formula for its volume V as a function of time, and solve this differential equation.

Exercises 2.6 Exact Equations

2.6#1°. Solve the following differential equations that are exact:

(a) $(t^2 + x)dt + (t - 2x)dx = 0$

(b) $(x - 3t^2)dt - (4x - t)dx = 0$

(c) $(x^3 - t)x' = x$

(d) $\left[\dfrac{x^2}{(t-x)^2} - \dfrac{1}{t}\right]dt + \left[\dfrac{1}{x} - \dfrac{t^2}{(t-x)^2}\right]dx = 0$

(e) $2(3tx^2 + 2t)dt - 3(2t^2x + x^2)dx = 0$

(f) $\dfrac{t\,dt + (2t + x)dx}{(t + x)^2} = 0$

2.6#2. Show that the linear equation $x' = p(t)x + q(t)$, multiplied throughout by $e^{\int p(\tau)d\tau}$ is exact. The quantity $e^{\int p(\tau)d\tau}$ is called an *integrating factor* because it makes the equation exact and able to be "integrated." (This idea was presented in different format as Exercises 2.2–2.3#6.)

2.6#3. Using the introductory idea of Section 2.6, use the computer program *DiffEq* to draw some of the level curves for the following surfaces $z = F(t, x)$:

(a) $z = \dfrac{tx}{t-x}$

(b) $z = \sin(x^2 + t^2) - tx$

(c) $z = (t^2 + x^2)/t^3$

(d) $z = e^{tx+1} - \cos(t + x)$

Exercises 2.7 Series Solutions

2.7#1. For the equation $x' = xt + t^2 \sin t$,

(a) Use the method of undetermined coefficients to find the power series solution.

(b) Write a general term for the coefficients.

(c) Give the radius of convergence for the power series. Hint: Use the ratio test.

(d) Check your results with a computer drawing from *DiffEq*.

2.7#2. In the manner of the previous exercise, find power series solutions for the following differential equation (which are also solvable by other methods, so you can check your results). It is not always possible to get a nice formula for the coefficients, though, so it may not be easy to find the radius of convergence of a solution. When this happens, see if you can use computer drawings from *DiffEq* to help.

(a) $x' = 3t^2 x$, with $x(0) = 2$

(b) $x' = (x - 1)/t^3$ with $x(1) = 0$

(c)° $(1 + t)x' - kx = 0$, with $x(0) = 1$. You should recognize this series.

2.7#3. Consider the first order differential equation $x' = f(t, x)$.

(a) Show how to find the first four successive derivatives (and recursively how to find more) from the differential equation, using implicit differentiation and substitution. That is, find the functions F_n such that

$$x^{(n)}(t) = F_n(t, x, x').$$

One reason for doing this is to find inflection points as in Exercises 1.1#13–16. Another reason is shown in the next part.

(b) The derivatives from part (a) allow us to find the *Taylor series* (Theorem A3.1 of the Appendix) for the solutions $x(t)$:

$$x(t) = x(t_0) + x'(t_0)(t - t_0) + \left(\frac{1}{2}\right)x''(t_0)(t - t_0)^2$$

$$+ \ldots + \left(\frac{1}{n!}\right)x^{(n)}(t_0)(t - t_0)^n + \ldots$$

If you need to find the solution only for a particular initial condition, and if the derivatives are sufficiently easy to calculate from $x' = f(t, x)$, *then* you may find this a more convenient route to the series for that particular solution than the method of undetermined coefficients.

Use this Taylor series method to find the solution for $x' = x^2$ through $x(0) = 1$, and determine the radius of convergence of the resulting series.

2.7#4. Find the first five terms of the power series solution of $x' = x^2 + t^2$ with $x(0) = 1$. Do this problem both by the Taylor series method of the previous exercise and by the method of undetermined coefficients from this section. Which do you find easiest?

2.7#5. Using any method you like, find the first three terms of the power series solution of $x' = \sin tx$ with $x(0) = \pi/2$.

2.7#6. In the manner of Example 2.7.4, study the behavior of the differential equation $x' = -x/t^2$ at the origin of the t, x-plane. A power series solution is appropriate and okay in this case, because there *is* a special "nicest" element of the "pony tail." Use the computer program *DiffEq* to demonstrate.

2.7#7°. Power series solutions can often be helpful for higher order differential equations. The method of undetermined coefficients applies exactly as for a first order equation; you simply have to take more derivatives before you substitute in the differential equation, and the resulting relations between the coefficients involve more ordinary equations. Keep in mind that we expect n arbitrary constants for an nth order differential equation. Find power series solutions to the following differential equations, with the given conditions:

(a) Find a formula for the coefficients of a power series solution to $x'' + x = 0$.

   (i) Find a solution satisfying $x(0) = 0$, $x'(0) = 1$.

   (ii) Find a solution satisfying $x(0) = 1$, $x'(0) = 0$.

(Pretend that you never heard of sine and cosine.)

(b) $x'' + xt = 0$, with $x(0) = 0$, $x'(0) = 1$. This is an equation that has no solution in terms of elementary functions, yet you can (and must) show that this particular solution indeed converges for every t.

(c) $x'' + tx' - x = 0$, with $x(0) = 1$, $x'(0) = 0$. Find a recursion formula for the coefficients of the power series for x. Estimate the value of $x(0.5)$ to within 0.01 and show that your estimate is this good.

(d) $x'' = 2tx' + 4x$, with $x(0) = 0$, $x'(0) = 1$. Express your answer in terms of elementary functions.

(e) $x'' + (2/t)x' + x = 0$, with $x(0) = 1$, $x'(0) = 0$. Express your answer in terms of elementary functions.

(f) $x''' - tx'' + x = 0$, with $x(0) = 1$, $x'(0) = -1$, $x''(0) = 0$. This equation indeed has a power series solution. Find the terms through x^4.

2.7#8. Find power series solutions for the following equations:

(a) $x' = x^2$

(b) $x'' = xx'$

(c) $x' = x \sin t$

(d) $x' = x \sin t - (1+t)^2$

Exercises 2   Miscellaneous Problems

2misc.#1.

(i) Solve the following linear or separable differential equations.

    (a) $(1 - t^2)x' = x - 1$

    (b) $x' - tx = 3t + e^t$

    (c) $x' = 3t^2(x + 2)$

    (d) $2e^t x^2 x' = t + 2$

    (e) $x' = t^2 x + t$

    (f) $x' = (x - 2t)/(2x - t)$, with $x(1) = 2$

    (g) $x' = \dfrac{2x}{t} + t \tan\left(\dfrac{x}{t^2}\right)$ Hint: Substitute $y = x^2 g(x)$

    (h) $\sin t \left(\dfrac{dx}{dt}\right) - x \cos t + \sin^2 t$, with $x(\pi/2) = \pi/2$.

(ii) Graph the slope field and some solutions for each part of (i). You may find the computer program *DiffEq* a help. If any solutions stand out, identify them, both by formula and on your picture.

2misc.#2°. A certain class of nonlinear first order differential equations is traditionally called *"homogeneous"* in a very particular sense that is *not* related to our use of the word homogeneous with regard to linear equations, in Section 2.2. A nonlinear differential equation $x' = f(t,x)$ is called "homogeneous" if $f(t,x)$ can be written as a function of (x/t).

(i) Show that if indeed you can write $x' = f(x/t)$, then the substitution $v = x/t$ will always cause the variables v and t to separate, thus giving an equation that is easy to solve if the integrals exist.

(ii) Solve the following differential equations, which are "homogeneous" as defined in this exercise, by using the method of substitution presented in (i):

    (a) $(x - t)dt + (x + t)dx = 0$

    (b) $(t + x)dt + t\,dx = 0$

    (c) $(t + x)dt + (x - t)dx = 0$

    (d) $t\,dx - x\,dt = \sqrt{t^2 + x^2}dt$

    (e) $(8x + 10t)dt + (5x + 7t)dx = 0$

    (f) $(2\sqrt{st} - s)dt + t\,ds = 0$

3

Numerical Methods

We have been emphasizing that most differential equations do not have solutions that can be written in elementary terms. Despite this, the computer program *DiffEq* does draw something that purports to be an approximation to a solution, and you should wonder how.

The answer is by *numerical methods*, by which the computer approximates a solution step by step. Numerical solutions of differential equations are of such importance for applied mathematics that there are many books on the subject, but the basic ideas are simple. Most methods tell you to "follow your nose," but the fancier ones do some "sniffing ahead."

In this chapter we explain the schemes used in our computer drawings, but we give only an introduction to the subject of numerical approximation. We begin in Section 3.1 with Euler's method, followed in Section 3.2 with some other methods that are numerically superior to Euler's method but for which the theory is more cumbersome.

Next we try to throw additional light on these computational methods by an experimental approach for analyzing the error that occurs in numerical approximation. This error is due to two sources: truncation of the Taylor series for a solution (with the degree of the truncation caused by the method of numerical approximation), and the limitations of finite accuracy (due to computing on actual machines). In Section 3.3 we analyze the differences in errors that occur using the different methods, corresponding to the different truncations of the Taylor series. In Section 3.4 we discuss the finite accuracy effects of rounding down, up, or round.

In Section 3.5 we finish the experimental and computational discussion with other practical considerations. Later in Section 4.6 we will return to the theoretical side of numerical methods and show that, at least for the simpler methods, we can indeed justify bounds of the form illustrated in our experiments.

3.1 Euler's Method

Euler's method for approximating solutions of the differential equation $x' = f(t, x)$ can be summed up by the instruction *"follow your nose."*

Suppose you are at some point (t_0, x_0), representing *initial conditions*. At this point, the differential equation specifies some *slope* $f(t_0, x_0)$. As t increases by a small *step* h, you can move along the tangent line in the

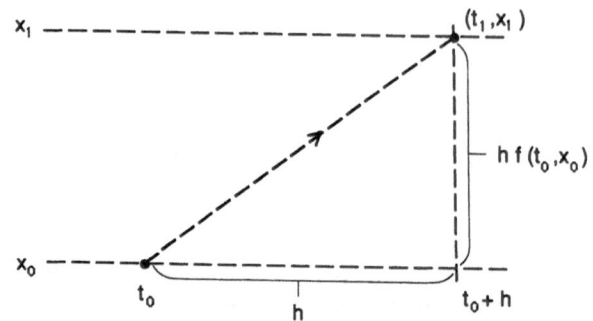

FIGURE 3.1.1. Euler's method. Single step, starting at (t_0, x_0).

direction of that slope to

$$(t_1, x_1) = (t_0 + h, x_0 + hf(t_0, x_0)),$$

as shown in Figure 3.1.1.

This is based on the Fundamental Theorem of Calculus, in the form

$$x(t_0 + h) = x(t_0) + \int_0^h x'(t_0 + s)ds.$$

If the step size h is small and *if* the slope is not changing too drastically near (t_0, x_0), the value x_1 will be close to $u(t_1)$, where $x = u(t)$ denotes the solution through (t_0, x_0).

The *Euler approximate solution* between the two points (t_0, x_0) and (t_1, x_1) is the straight line segment between those points.

The Euler approximate solution can be extended to additional points in a piecewise linear fashion. You can start from (t_1, x_1), using the slope given by $f(t_1, x_1)$, to get

$$(t_2, x_2) = (t_1 + h, x_1 + hf(t_1, x_1)).$$

In like manner, you can then use (t_2, x_2) to get (t_3, x_3), and so on. Figure 3.1.2 shows the result of using the Euler method over three successive steps.

It seems reasonable to expect that a smaller step, such as $h/2$, will give a closer approximation to the solution, that is, we might expect an improvement such as shown in Figure 3.1.3, p. 114, for the interval t_0 to $t_0 + 3h$.

Thus in an Euler approximation for a given stepsize h, we move through the following sequence of points:

$$(t_0, x_0)$$
$$(t_1, x_1) \quad \text{with } x_1 = x_0 + hf(t_0, x_0),$$
$$\vdots$$
$$(t_n, x_n) \quad \text{with } x_n = x_{n-1} + hf(t_{n-1}, x_{n-1}).$$

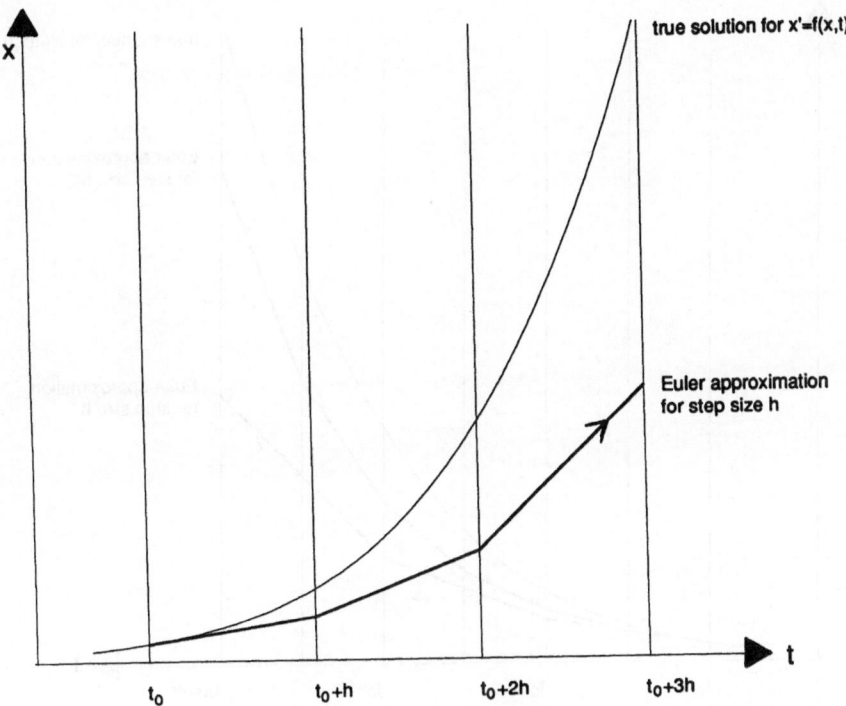

FIGURE 3.1.2. Euler approximate solution for three steps, stepsize h.

A more formal mathematical statement of Euler's method is the following:

Definition 3.1.1. Consider the differential equation $x' = f(t, x)$ with f a function defined in some rectangle $R = [a, b] \times [c, d]$. Choose a point $(t_0, x_0) \in R$ and a given stepsize h. Define a sequence of points (t_n, x_n) recursively by

$$\left. \begin{array}{c} t_n = t_{n-1} + h = t_0 + nh, \\[2mm] x_n = x_{n-1} + hf(t_{n-1}, x_{n-1}) \end{array} \right\} \qquad (1)$$

as long as $(t_n, x_n) \in R$. Then the *Euler approximate solution* $u_h(t)$ through (t_0, x_0) is the piecewise linear function joining all the (t_n, x_n), where each piece has formula

$$u_h(t) = x_n + (t - t_n)f(t_n, x_n)$$

for $t \in [t_n, t_{n+1}]$.

Definition 3.1.1 gives an approximate solution moving to the right if h is positive, and to the left if h is negative.

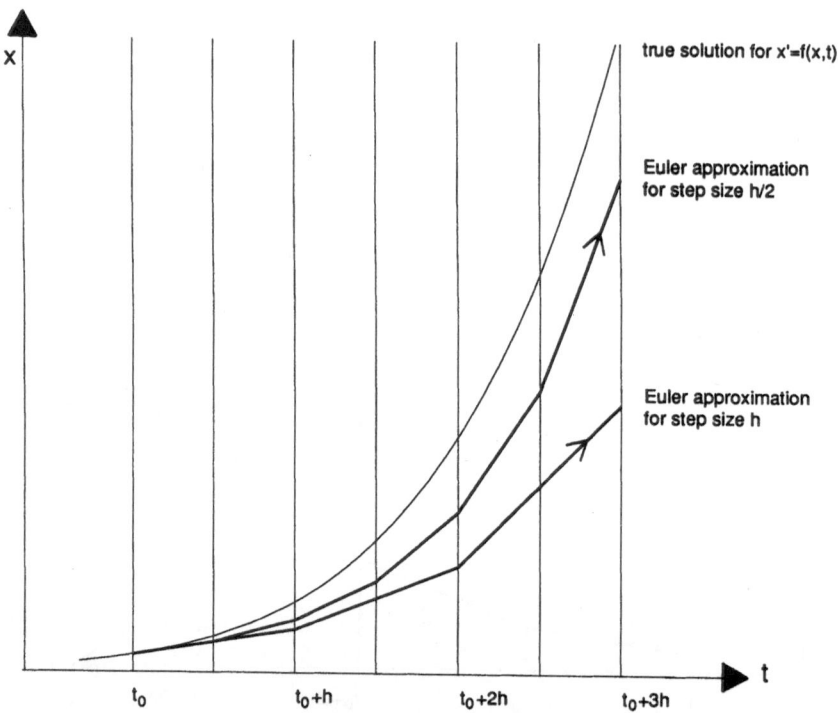

FIGURE 3.1.3. Smaller-step Euler approximate solution.

The idea underlying this method of numerical approximation is so intuitive that it is hard to find to whom to attribute it. The first formal description is generally attributed to Euler, and the first proof that as the step gets small the approximate solution does indeed converge to a solution is due to Cauchy. But the method was used without comment by Newton in the very first book using differential equations. Furthermore, this algorithm gives approximate solutions to $x' = rx$, which match calculations dating back to the Babylonians of 2000 B.C., as we shall discuss in Example 3.1.4.

First, however, let us pause to look at a sample calculation illustrating the use of the equations (1).

Example 3.1.2. For $x' = \sin(tx)$, start at $(t_0, x_0) = (0, 3)$ and construct an Euler approximate solution to the right with step size $h = 0.1$. An organized calculation for the x_n's proceeds as follows, moving from left to right along each row in turn:

Table 3.1.1

t_n	x_n	$f(t_n, x_n)$	$x_{n+1} = x_n + hf(t_n, x_n)$
$t_0 = 0.0$	$x_0 = 3.000$	$\sin[(0)(3)] = 0$	$x_1 = 3 + (.1)(0) = 3.000$
$t_1 = 0.1$	$x_1 = 3.000$	$\sin[(.1)(3)] = .296$	$x_2 = 3 + (.1)(.296) = 3.030$
$t_2 = 0.2$	$x_2 = 3.030$	$\sin[(.2)(3.03)] = .570$	$x_3 = 3.030 + (.1)(.570) = 3.087$
$t_3 = 0.3$	$x_3 = 3.087$	$\sin[(.3)(3.087)] = .799$	$x_4 = 3.087 + (.1)(.799) = 3.167$
$t_4 = 0.4$	$x_4 = 3.167$	continue in this manner	▲

Programming a computer for Euler's method allows extremely fast calculation, and expanded accuracy without effort (up to the limits of the
machine).

Example 3.1.3. For the equation of Example 3.1.2, we show a computer
tabulation listing the results of t_n and x_n (without showing the intermediate steps), with a smaller stepsize, $h = 0.05$. You can see how the numerical
results are refined by a smaller stepsize.

Table 3.1.2. $x' = \sin(tx)$

t_n	x_n
0.00	3.00000
0.05	3.00000
0.10	3.00747
0.15	3.02228
0.20	3.04418
0.25	3.07278
0.30	3.10752
0.35	3.14767
0.40	3.19227 ▲

The Euler method provides the simplest scheme to study the essentials
of numerical approximation to solutions of differential equations. The next
two examples are familiar cases where you will see that we already know
the Euler method from previous experience, and furthermore we can see
that when the stepsize $h \to 0$, these Euler approximate solutions converge
to solutions to the differential equation in question.

Example 3.1.4. Apply Euler's method to the differential equation that
represents bank interest continuously compounded,

$$x' = rx, \quad \text{where } x(t) = \text{amount of savings.}$$

(The interest rate r is annual if t is measured in years.) To apply Euler's method, let $h = 1/n$, where n is the number of compounding periods per year. Then

$$x_1 = x_0 + \frac{r}{n}x_0 = \left(1 + \frac{r}{n}\right)x_0,$$

$$x_2 = x_1 + \frac{r}{n}x_1 = \left(1 + \frac{r}{n}\right)^2 x_0,$$

$$\vdots$$

$$x_n = x_{n-1} + \frac{r}{n}x_{n-1} = \left(1 + \frac{r}{n}\right)^n x_0.$$

This x_n is the value of the savings account after n periods of compound interest; the Euler's method formula corresponds precisely to the tables used by banks to calculate interest compounded at regular intervals! In fact, the earliest mathematical writings, those of the Babylonians 4000 years ago, are tables of interest that match these calculations.

For continuous compounding, over one year, Euler's method indicates that savings will grow to

$$\lim_{n \to \infty} \left(1 + \frac{r}{n}\right)^n x_0.$$

For continuous compounding over one year we also know that the analytic solution to $x' = rx$ is $x_0 e^r$. So when we will prove in Section 4.5 that Euler's method converges to the solution to the differential equation, we will be proving a famous formula that you probably already know:

$$e^r = \lim_{n \to \infty} \left(1 + \frac{r}{n}\right)^n. \quad \blacktriangle$$

Example 3.1.5. Consider the differential equation $x' = g(t)$, which has no explicit dependence on x. From elementary calculus, with initial condition $u(t_0) = x_0$, we know the solution is the function defined by

$$u(t) = x_0 + \int_{t_0}^t g(s)ds. \tag{2}$$

Apply to $x' = g(t)$ Euler's method with stepsize $h \equiv (t - t_0)/n$, starting at $u(t_0) = x_0$. This gives, for any $t \geq t_0$,

$$u_h(t) = x_0 + h \sum_{i=0}^{n-1} g(t_i) \tag{3}$$

For $x_0 = 0$, equation (3) is precisely a *Riemann sum* for numerical integration of a function $g(t)$. This Riemann sum is illustrated for a positive function by the area of the shaded rectangles in Figure 3.1.4. Each rectangle has base h and lies above $[t_i, t_{i+1}]$, from the t-axis to $g(t_i)$. ▲

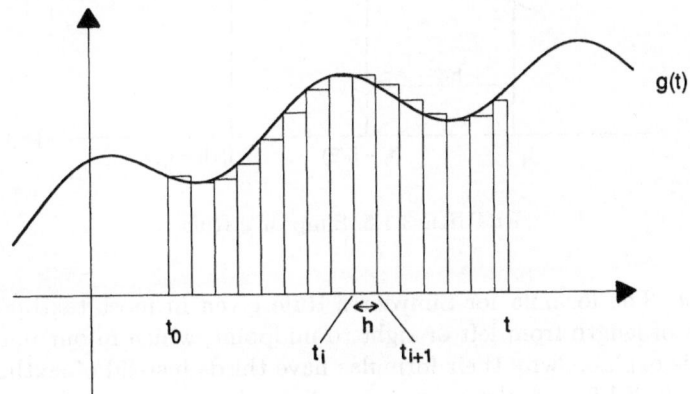

FIGURE 3.1.4. Riemann sum.

NUMERICAL METHODS AS APPROXIMATION OF AN INTEGRAL

As shown in Example 3.1.5 by equation (2), Euler's method amounts to *approximating the integral*, of $g(t)$. In the particular Rieman sum of that example, we are approximating the height of each vertical strip under $g(t)$ on an interval $[t_i, t_{i+1}]$ by the value

$$m_L = g(t_i) \qquad \text{value at left endpoint.}$$

This is often not a good approximation. Two reasonable improvements in most cases are obtained by using other values or averages of $g(t)$ in each subinterval:

$$m_M = g((t_i + t_{i+1})/2) \qquad \text{value at midpoint;}$$

$$m_T = \left(\tfrac{1}{2}\right)[g(t_i) + g(t_{i+1})] \quad \text{average of values at endpoints.}$$

In the computation of integrals, these improvements are called the *midpoint Riemann sum* and the *trapezoidal rule*, respectively.

A fancier method to approximate an integral or get a slope for the approximate solution to the differential equation $x' = g(t)$ is *Simpson's Rule*, using a weighted average of values of $g(t)$ at three points in each subinterval:

$$m_S = \left(\frac{1}{6}\right)[g(t_i) + 4g((t_i + t_{i+1})/2) + g(t_{i+1})],$$

as taught in first year calculus. There is a unique *parabola* passing through the three points (t_i, x_i), $(t_i + h/2, g(t_i + h/2))$, and $(t_i + h, g(t_i + h))$, as shown in Figure 3.1.5. Simpson's Rule gives exactly the area beneath that parabola.

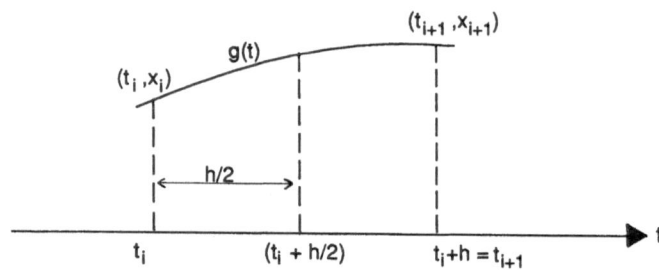

FIGURE 3.1.5. Simpson's Rule.

Remark. The formula for Simpson's Rule given in most textbooks uses intervals of length from left or right to midpoint, which in our notation is $h/2$. This explains why their formulas have thirds instead of sixths.

Example 3.1.5 and the subsequent discussion are more than a happy coincidence. In fact, it is precisely the *approximation of integrals* that has motivated and produced the numerical methods we are discussing in this book.

When slope $f(t, x)$ depends on x as well as on t, there are smarter ways than Euler's of approximating solutions to differential equations, corresponding to the improved schemes for approximating integrals: midpoint Euler, trapezoidal rule, and Simpson's rule. We shall study two of these in the next section, and the third in Exercise 3.1–3.2#7.

3.2   Better Numerical Methods

Two other numerical methods for solving a differential equation $x' = f(t, x)$ are based on the same idea as Euler's method, in that using intervals of step size h,

$$t_{i+1} = t_i + h \text{ and } x_{i+1} = x_i + hm, \text{ where } m = slope.$$

For Euler's method we simply use the slope, $f(t_i, x_i)$, available at the point where we begin to "follow our noses," the left endpoint of the interval. For fancier methods we "sniff ahead," and then can do a better job of "following."

1. *Midpoint Euler.* For the *midpoint Euler* method (also called *modified* Euler) we use the slope m_M at the *midpoint* of the segment we would have obtained with Euler's method, as shown in Figure 3.2.1.

This method takes into account how the slope is changing over the interval, and as we shall see, it converges to a solution considerably more quickly than the straight Euler method.

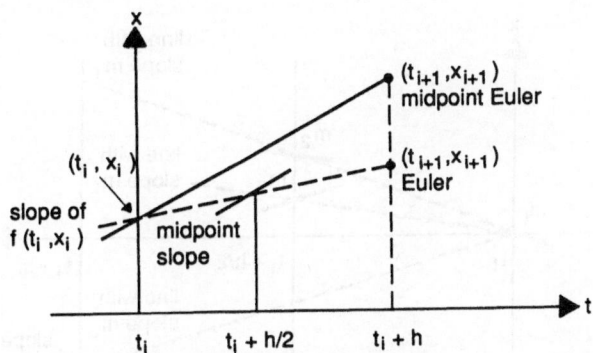

FIGURE 3.2.1. Midpoint slope $= m_M = f(t_i + \frac{h}{2}, x_i + \frac{h}{2}f(t_i, x_i))$.

If the midpoint Euler method is used in the case $x' = g(t)$, it reduces exactly to the midpoint Riemann sum mentioned at the end of the previous section.

2. *Runge–Kutta.* The *Runge–Kutta* numerical method converges considerably more rapidly than the Euler methods, and is what was used to make your *DiffEq* programs. The method was developed by two German mathematicians, C. Runge and W. Kutta, at the end of the nineteenth century. Without discussing the complexities of how these gentlemen arrived at their conclusions, we hereby describe the most commonly used fourth-order version, where the Runge–Kutta "slope" m_{RK} is a weighted average of four slopes:

$$m_1 \ = f(t_i, x_i) \qquad\qquad \text{slope at } \textit{beginning} \text{ of interval}$$

$$m_2 \ = f(t_i + \tfrac{h}{2}, x_i + \tfrac{h}{2}m_1) \qquad \text{slope at } \textit{midpoint} \text{ of a segment with slope } m_1$$

$$m_3 \ = f(t_i + \tfrac{h}{2}, x_i + \tfrac{h}{2}m_2) \qquad \text{slope at } \textit{midpoint} \text{ of a segment with slope } m_2$$

$$m_4 \ = f(t_i + h, x_i + hm_3) \qquad \text{slope at } \textit{end} \text{ of a segment with slope } m_3$$

$$m_{RK} = \tfrac{1}{6}(m_1 + 2m_2 + 2m_3 + m_4).$$

The Runge–Kutta method makes a linear combination of four slopes, illustrated in Figure 3.2.2, which you might think of as follows: m_1 is the Euler's method slope, m_2 is the midpoint Euler slope, m_3 "corrects the shot," and m_4 brings in the slope at the right-hand end of the interval. The weighted average m_{RK} is used to calculate $x_{i+1} = x_i + hm$.

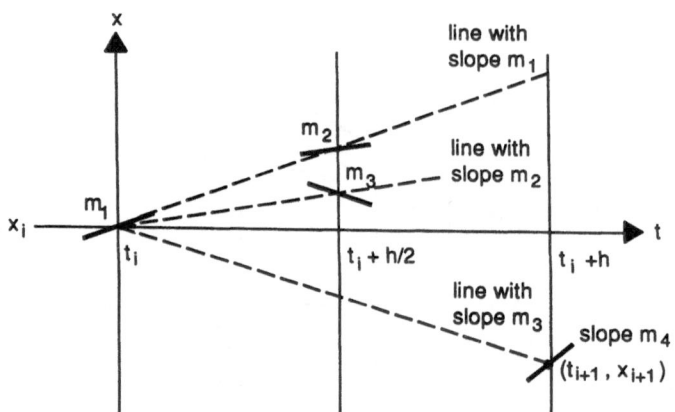

FIGURE 3.2.2. Runge–Kutta makes a linear combination of these four slopes using $m_{RK} = \left(\frac{1}{6}\right)(m_1 + 2m_2 + 2m_3 + m_4)$.

A special case of the Runge–Kutta method is the following: if $x' = g(t)$, then the slope depends only on t, not on x, so $m_2 = m_3$ (see Figure 3.2.2) and

$$m_{RK} = \frac{1}{6}(m_1 + 4m_2 + m_4) = m_S,$$

exactly Simpson's rule for numerical integration, as discussed in Section 3.1. Simpson's rule, in fact, was the original motivation for the fourth order Runge–Kutta scheme.

We compare these three numerical methods—Euler's, midpoint Euler, and Runge–Kutta—in Examples 3.2.1 and 3.2.2, using two different presentations. In the first we fix the stepsize and follow the approximate solution through a number of steps for each method.

Example 3.2.1. We return to $x' = \sin tx$, with $x(0) = 3$, and stepsize $h = 0.1$, and tabulate the computations in Table 3.2.1.

Table 3.2.1. $x' = \sin tx$.

t_n	Euler x_n	Midpoint Euler x_n	Runge–Kutta x_n
0.0	3.00000	3.00000	3.00000
0.1	3.00000	3.01494	3.01492
0.2	3.02955	3.05884	3.05874
0.3	3.08650	3.12859	3.12829
0.4	3.16642	3.21812	3.21744
0.5	3.26183	3.31761	3.31637
0.6	3.36164	3.41368	3.41185
0.7	3.45185	3.49163	3.48947
0.8	3.51819	3.53946	3.53746
0.9	3.55031	3.55144	3.55003
1.0	3.54495	3.52867	3.52803
1.1	3.50570	3.47694	3.47698
1.2	3.44016	3.40379	3.40433
1.3	3.35674	3.31643	3.31729
1.4	3.26276	3.22081	3.22187
1.5	3.16380	3.12145	3.12263
1.6	3.06386	3.02157	3.02285
1.7	2.96565	2.92340	2.92479
1.8	2.87102	2.82844	2.82998
1.9	2.78122	2.73772	2.73946
2.0	2.69713	2.65196	2.65396 ▲

In the second version we fix the final value of t_f and calculate $x_f = u_h(t_f)$, for different numbers of steps on the interval $[t_0, t_f]$.

Example 3.2.2. We return to $x' = \sin tx$, approximating $x(2)$, with $x(0) = 3$. On each line the stepsize is half that on the line above, so the number of steps is 2^N and $h = (t_f - t_0)/2^N$.

Table 3.2.2. $x' = \sin tx$

No. of steps	Euler	Midpoint Euler	Runge–Kutta
1	3.00000	3.28224	$3.00186 = u_h(t_f)$
2	3.14112	3.24403	2.66813
4	2.84253	2.61378	2.65370
8	2.75703	2.64049	2.65387
16	2.70750	2.65078	2.65396
32	2.68128	2.65321	2.65397
64	2.66776	2.65379	2.65397
128	2.66090	2.65393	2.65397
256	2.65745	2.65396	2.65397
512	2.65571	2.65397	2.65397
1024	2.65484	2.65397	2.65397
2048	2.65441	2.65397	2.65397
4096	2.65419	2.65397	2.65397
8192	2.65408	2.65397	2.65397
16384	2.65403	2.65397	2.65397 ▲

You can see in this example that as the number of steps increase, and the stepsize $h \to 0$, the values for x_f soon seem to approach a limit. If the computer worked with infinite precision, these values for x_f would converge, although sometimes not too quickly, to the actual solution to the differential equation, as we shall prove in Section 4.5.

You can also see how the midpoint Euler method approaches this limit considerably sooner than the Euler method, and how the Runge–Kutta method approaches the limit considerably sooner than midpoint Euler.

Summary. We summarize with three computer programs that calculate one step of the approximate solution to a differential equation, by the methods presented in Sections 3.1 and 3.2. The programs are written in the computer language Pascal, but it is *not* necessary that you know this language to read the sequences of equations that are the center of each program.

Table 3.2.3

Procedure StepEuler (var t, x, h:real);	Euler's method
Begin $\quad x := x + h * \text{slope}(t, x);$ $\quad t := t + h;$ end;	
Procedure StepMid(var t, x, h:real);	midpoint Euler
var $m1, t1, x1$:real; **begin** $\quad t1 := t + h/2; x1 := x + (h/2) * \text{slope}(t, x);$ $\quad m1 := \text{slope}(t1, x1);$ $\quad t := t + h;$ $\quad x := x + h * m1;$ end:	
Procedure StepRK(var t, x, h:real);	Runge–Kutta
var $t1, x1, x2, x3, m1, m2, m3, m4, m$:real; **begin** $\quad m1 := \text{slope}(t, x);$ $\quad t1 := t + h/2; x1 := x + m1 * h/2;$ $\quad m2 := \text{slope}(t1, x1);$ $\quad x2 := x + m2 * h/2;$ $\quad m3 := \text{slope}(t1, x2);$ $\quad t := t + h; x3 := x + h * m3;$ $\quad m4 := \text{slope}(t, x3);$ $\quad m := (m1 + 2 * m2 + 2 * m3 + m4)/6;$ $\quad x := x + h * m;$ end;	

There exist *many* other numerical methods for approximating solutions to differential equations. A few others are introduced in the exercises; references are listed at the end of this volume. In this text proper we concentrate only on the three methods already presented; our purpose is to show as clearly as possible what is going on and what needs to be considered, so that you can evaluate other methods for your particular circumstances.

However, we cannot close this section without discussing two other methods. The first is very natural, but seldom used in computers, because it requires the evaluation of high order derivatives; this method may become more practical as symbolic differentiators become more common. The second is of such importance in practice that it cannot reasonably be omitted.

The "Naive" Taylor Series Method

We saw in Chapter 2 that there exists a unique proper series solution $x = u(t)$ for $x' = f(t, x)$ with $u(t_0) = x_0$. In fact, $u(t)$ is the Taylor series (as presented in Theorem A3.1 in the Appendix), so we can write it in that fashion and show each coefficient in terms of $f(t, x)$:

$$u(t) = u(t_0) + u'(t_0)(t - t_0) + (1/2)u''(t_0)(t - t_0)^2 + \ldots$$
$$= x_0 + h \underbrace{f(t_0, x_0)}_{\text{slope}} + (h^2/2)[\partial f/\partial t + f(\partial f/\partial x)]_{t_0, x_0} + \ldots$$

$$\underbrace{}_{\text{Euler's method}}$$

Let $P_n(t_0, x_0; t)$ be terms of this series up to degree n, the nth degree Taylor polynomial of the solution $x = u(t)$. Then the "naive" method (which we will see in the proofs of Chapter 4 to be of order n) can be found as follows:

$$t_{i+1} = t_i + h$$
$$x_{i+1} = P_n(t_i, x_i, t_{i+1}).$$

Comment 1. The case $n = 1$ is exactly Euler's method. The cases $n = 2$ and $n = 4$ have the same "precision" as midpoint Euler and Runge–Kutta respectively, in a manner to be discussed below.

Comment 2. An attempt to solve Exercise 2.7#5, $x' = \sin tx$, will show the reader the main weakness of this numerical scheme: the computation of the polynomials $P_n(t_0, x_0; t)$ can be quite cumbersome, even for $n = 4$. Nevertheless, in those cases where the coefficients can be expressed as reasonably simple recursion formulas, a Taylor series method of fairly high degree, perhaps 20, may be a very good choice. See Exercise 3.3#6.

One way of understanding the Runge–Kutta method is that, *as a function of* h, the function $v(t) = v(t_0 + h) = x_0 + hm_{t_0, x_0, h}$ has the same 4th degree Taylor polynomial as the solution $u(t)$ to $x' = f(t, x)$ with $x(t_0) = x_0$. However, finding $v(t)$ only requires evaluating f at 4 points, and not computing the partial derivatives of f up to order 4. So you can think of Runge–Kutta as a substitute for the naive 4th order method, a substitute that is usually much easier to implement.

The computation required to show the equivalence of Runge–Kutta to a 4th degree Taylor polynomial is quite long, but the analogous statement that midpoint Euler is equivalent to the 2nd degree Taylor polynomial is quite feasible to prove (Exercise 3.1–3.2#9).

Implicit Methods

There is a whole class of numerical methods which are particularly well adapted to solving differential equations which arise from models like discretizations of the heat equation. We will show only one example of such a method. It appears at first completely unreasonable, but the analysis in Section 5.4 should convince the reader that it might well be useful anyway.

The implicit Euler method consists of choosing a step h, and setting

$$t_{i+1} = t_i + h \quad \text{and} \quad x_{i+1} = x_i + hf(t_{i+1}, x_{i+1}).$$

Note that the second expression above is not a *formula* for x_{i+1}, but an *equation* for x_{i+1}, i.e., it expresses x_{i+1} *implicitly*. To carry out the method, this equation must be solved at each step. There is in general no formula for such solutions, but a variety of schemes exist to approximate solutions of equations, most of which are some variant of *Newton's method*, which will be discussed in Section 5.3. As the reader will find, these methods are always a little dicey. So the scheme appears *a priori* of little interest: each step requires the numerical solution of an equation, with all the attendant possible problems. The reader is referred to Examples 5.4.2, part (d) to see why it is useful anyway: the other methods may be simpler but they have their problems too. The implicit method avoids a breakdown of the numerical method when the stepsize gets large.

3.3 Analysis of Error, According to Approximation Method

For a differential equation with exact solution $x = u(t)$ and a particular numerical approximation, $x_n = u_h(t_n)$, the actual *error* at the nth step,

$$E_n(h) = u(t_n) - u_h(t_n),$$

depends on both the number of steps n and on the stepsize h.

This actual error in a numerical approximation has two sources: one source is the method of approximation, which tells how the Taylor series for the actual solution $x = u(t)$ has been truncated; that is what we shall discuss in this section, and this is what contributes the greatest effect on error. The other source of error is the finite numerical accuracy of the calculation, which depends on the computing machine and its method of rounding decimals; this we shall discuss in the subsequent section.

Examples 3.3.1 and 3.3.2 compute for our three numerical methods— Euler, midpoint Euler, and Runge–Kutta approximate solutions $u_h(t_f) = x_f$, for fixed t_f, to a differential equation $x' = f(t, x)$ that can be solved analytically. In these examples, the number of steps in the interval $[t_0, t_f]$ varies as 2^N, from $N = 0$ to $N = 13$, so that $h = (t_f - t_0)/2^N$. In other words, this arrangement of setting t_f makes the number of steps N directly related to stepsize h, and error can be studied as an effect of h alone.

For each value of N, and for each method, we list both the computation of $u_h(t_f) = x_f$ and the actual error $E(h)$. Note that we have written $E(h)$

with only five significant digits, because E is the difference of two numbers that are quite close, so additional digits would be meaningless.

Example 3.3.1. For the equation $x' = x$, find $x(2)$, with $x(0) = 1$. By separation of variables, the exact solution is found to be $u(t) = x_0 e^t$, so $u(2) = e^2 \approx 7.38905609893065$.

Table 3.3.1. Actual Error $E(h) = u(t_f) - u_h(t_f)$ for $x' = x$

No. of steps	Euler	Midpoint Euler	Runge–Kutta
1	3.00000000000000	5.00000000000000	$7.000000000000 = u_h(t_f)$
	4.3891×10^0	2.3891×10^0	$3.8906 \times 10^{-1} = E(h)$
2	4.00000000000000	6.25000000000000	7.33506944444444
	3.3891×10^0	1.1391×10^0	5.3987×10^{-2}
4	5.06250000000000	6.97290039062500	7.38397032396005
	2.3266×10^0	4.1616×10^{-1}	5.0858×10^{-3}
8	5.96046447753906	7.26224718998853	7.38866527357286
	1.4286×10^0	1.2681×10^{-1}	3.9083×10^{-4}
16	6.58325017202742	7.35408290311116	7.38902900289220
	8.0581×10^{-1}	3.4973×10^{-2}	2.7096×10^{-5}
32	6.95866675721881	7.37988036635186	7.38905431509387
	4.3039×10^{-1}	9.1757×10^{-3}	1.7838×10^{-6}
64	7.16627615278822	7.38670685035460	7.38905598450266
	2.2278×10^{-1}	2.3492×10^{-3}	1.1443×10^{-7}
128	7.27566979312842	7.38846180262122	7.38905609168522
	1.1339×10^{-1}	5.9430×10^{-4}	7.2454×10^{-9}
256	7.33185059874104	7.38890664780243	7.38905609847485
	5.7206×10^{-2}	1.4945×10^{-4}	4.5580×10^{-10}
512	7.36032355326928	7.38901862627635	7.38905609890206
	2.8733×10^{-2}	3.7473×10^{-5}	2.8586×10^{-11}
1024	7.37465716034184	7.38904671701835	7.38905609892886
	1.4399×10^{-2}	9.3819×10^{-6}	1.7897×10^{-12}
2048	7.38184843588050	7.38905375173306	7.38905609893053
	7.2077×10^{-3}	2.3472×10^{-6}	1.1546×10^{-13}
4096	7.38545021553901	7.38905551191629	7.38905609893065
	3.6059×10^{-3}	5.8701×10^{-7}	-1.776×10^{-15}
8192	7.38725264383889	7.38905595215017	7.38905609893063
	1.8035×10^{-3}	1.4678×10^{-7}	1.6875×10^{-14} (∗)
16384	7.38815424298207	7.38905606223213	7.38905609893059
	9.0186×10^{-4}	3.6699×10^{-8}	6.4837×10^{-14} (∗) ▲

Example 3.3.2. For the equation $x' = x^2 \sin t$, finding $x(\pi)$, with $x(0) = 0.3$. By separation of variables, the actual solution is found to be $u(t) = 1/(\cos t + C)$, which gives $u(\pi) = 0.75$.

Table 3.3.2. Actual Error $E(h) = u(t_f) - u_h(t_f)$ for $x' = x^2 \sin t$

No. of steps	Euler	Midpoint Euler	Runge–Kutta
1	.300000000000000	.582743338823081	$.59825123558439 = u_h(t_f)$
	4.5000×10^{-1}	1.6726×10^{-1}	$1.5175 \times 10^{-1} = E(h)$
2	.441371669411541	.706815009470264	.735925509628509
	3.0863×10^{-1}	4.3185×10^{-2}	1.4074×10^{-2}
4	.556745307152106	.735645886225717	.749522811050199
	1.9325×10^{-1}	1.4354×10^{-2}	4.7719×10^{-4}
8	.634637578818475	.745355944609261	.749984292895033
	1.1536×10^{-1}	4.6441×10^{-3}	1.5707×10^{-5}
16	.684465875258715	.748779351548380	.749999711167907
	6.5534×10^{-2}	1.2206×10^{-3}	2.8883×10^{-7}
32	.714500155244513	.749695668237506	.750000010105716
	3.5500×10^{-2}	3.0433×10^{-4}	-1.011×10^{-8}
64	.731422748597318	.749924661781383	.750000001620961
	1.8577×10^{-2}	7.5338×10^{-5}	-1.621×10^{-9}
128	.740481836801215	.749981303141273	.750000000133992
	9.5182×10^{-3}	1.8697×10^{-5}	-1.340×10^{-10}
256	.745180317032169	.749995345938535	.750000000009427
	4.8197×10^{-3}	4.6541×10^{-6}	-9.427×10^{-12}
512	.747574578840291	.749998839192566	.750000000000615
	2.4254×10^{-3}	1.1608×10^{-6}	-6.153×10^{-13}
1024	.748783339005838	.749999710148590	.7500000000000036
	1.2167×10^{-3}	2.8985×10^{-7}	-3.608×10^{-14}
2048	.749390674843791	.749999927581687	.7499999999999986
	6.0933×10^{-4}	7.2418×10^{-8}	1.4211×10^{-14}
4096	.749695087867540	.749999981900960	.7499999999999919
	3.0491×10^{-4}	1.8099×10^{-8}	8.1157×10^{-14} (∗)
8192	.749847481433661	.749999995476055	.7500000000000086
	1.5252×10^{-4}	4.5239×10^{-9}	-8.560×10^{-14} (∗)
16384	.749923725077763	.749999998869095	.7500000000000008
	7.6275×10^{-5}	1.1309×10^{-9}	-7.661×10^{-15} (∗) ▲

Notice that, as expected, in general the error decreases as you run down each column. However, something funny (∗) can happen in the lower right-hand corner of both Examples 3.3.1 and 3.3.2 under the Runge–Kutta method, with the error and the last digits. Actually, the same funny business can eventually happen under *any* method that converges. The fact that the error begins to *increase* again, and even to wobble, is due to the finiteness of the computation, to be discussed at length in Section 3.4.

The overall behavior of $E(h)$ (ignoring the final wobbly ($*$) values) is different for each of the three methods. Considering Examples 3.3.1 and 3.3.2 as numerical experiments, it seems that Runge–Kutta is much better than midpoint Euler, which in turn seems much better than straight Euler; the better methods converge more quickly and give smaller errors for a given N.

It is *not* obvious from the tables what makes the difference. However, the fact, which we will support by numerical experiments (in this section) and Taylor series approximation (to come later, in Section 4.6), is that *the error $E(h)$ varies as a power of h*, where that power is 1, 2, and 4 respectively in these methods. That is, we claim

$$\left.\begin{array}{lll}\text{for Euler's method,} & E(h) \approx C_E h, \\ \text{for midpoint Euler,} & E(h) \approx C_M h^2, \\ \text{for Runge–Kutta,} & E(h) \approx C_{RK} h^4. \end{array}\right\} \qquad (4)$$

These formulas (4) produce the graphs for E versus h that are plotted in Figure 3.3.1, with the constants C_E, C_M, C_{RK} ignored in order to focus on the powers of h.

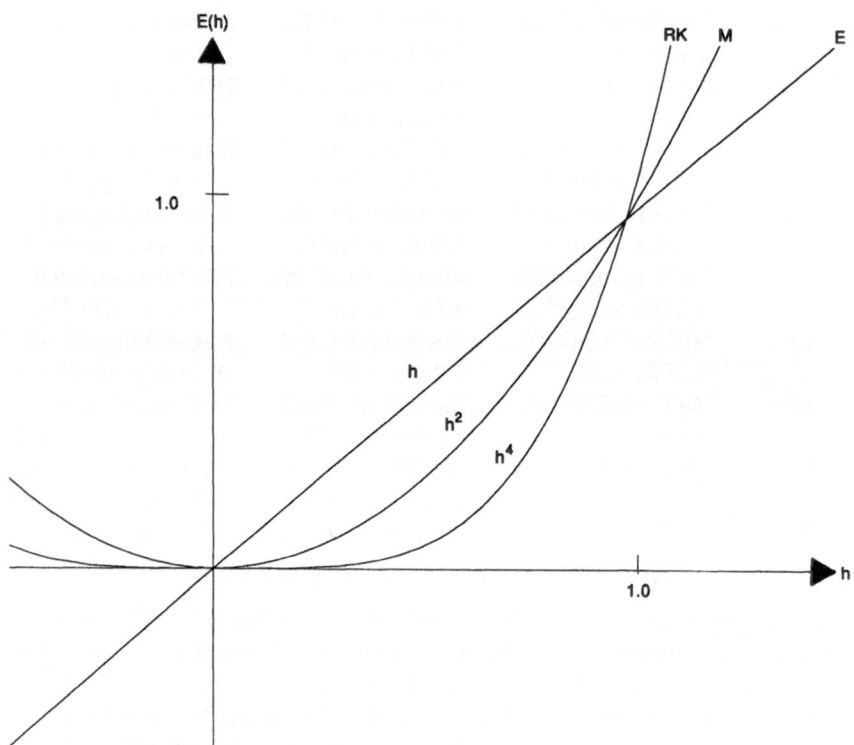

FIGURE 3.3.1. Claimed relationship between $E(h)$ and h.

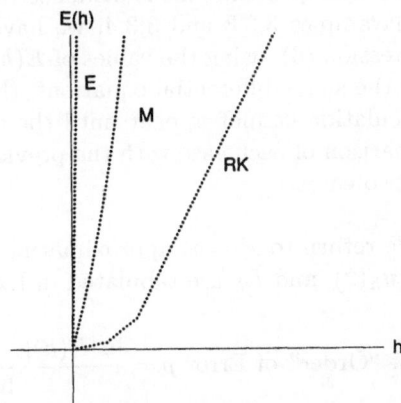

FIGURE 3.3.2. $E(h)$ versus h for $x' = x$.

In fact, Figure 3.3.2, which is a computer plot of $E(h)$ versus h for Example 3.3.1, gives a picture visually supporting the above claim. (To graph the actual values of the example, the scale must be skewed so that the 45° line for Euler's method in Figure 3.3.1 appears to be much steeper in Figure 3.3.2.)

Because of the formulas (4) for E in terms of powers of h, Euler's method is called a *first order method*, midpoint Euler is a *second order method*, and our version of Runge–Kutta is a *fourth order method*. There are in fact methods of all orders, and there are entire texts and courses on numerical methods. Our selection is just an introduction.

As the order of a numerical approximation increases, accuracy increases; for small h, h^4 is a much smaller error than h. But more computations are involved for each step of a higher order method, so each step of a higher order method is more costly to compute. We need to strike a balance between number of steps and cost per step; fourth order Runge–Kutta is often the practical choice. Later, in Section 4.6 of the next chapter, we shall give rigorous estimates that the errors are indeed *bounded* by such terms.

Meanwhile, if it is true that for order p

$$E(h) \approx Ch^p, \tag{5}$$

as in formulas (4), how can we exhibit this claim quantitatively in our numerical experiments? Using (5), on h and on $h/2$, we get

$$\ln|E(h)| - \ln|E(h/2)| \approx p \ln h - p \ln h + p \ln 2 = p \ln 2,$$

which says that the *order* of the error,

$$\frac{\ln|E(h)| - \ln|E(h/2)|}{\ln 2} \approx p, \tag{6}$$

approaches an integer as expected. This is evidence that $E(h) \approx Ch^p$.

In the following Examples 3.3.3 and 3.3.4, we have tabulated the "order" p from the expression (6), using the values of $E(h)$ as tabulated in the earlier examples for the same differential equations. (Note that this experimental "order" calculation cannot appear until the second case, because it represents a comparison of each case with the previous one and can only be computed after two cases.)

Example 3.3.3. We return to $x' = x$, approximating $x(2)$, with $x(0) = 1$. The values of $u(2)$, $u_h(2)$, and E_h are tabulated in Example 3.3.1.

Table 3.3.3. "Order" of Error $p = \dfrac{\ln|E(h)| - \ln|E(h/2)|}{\ln 2}$

Number of steps	Euler	Midpoint Euler	Runge–Kutta	
2	0.373	1.069	2.849	="order"
4	0.543	1.453	3.408	
8	0.704	1.714	3.702	
16	0.826	1.858	3.850	
32	0.905	1.930	3.925	
64	0.950	1.966	3.962	
128	0.974	1.983	3.981	
256	0.987	1.992	3.991	"order" ≈ 4
512	0.993	1.996	3.995	
1024	0.997	1.998	3.998	
2048	0.998	1.999	3.954	
4.96	0.999	1.999	6.022	
8192	1.000	2.000	-3.248	"order" $\approx ?$
16384	1.000	2.000	-1.942	▲

"order" ≈ 1 (Euler)

"order" ≈ 2 (Midpoint Euler)

In Example 3.3.3 we see the predicted tendency to

> 1 in the first column (Euler's method)
> 2 in the second column (midpoint Euler)
> 3 in the third column (Runge–Kutta).

Again something funny happens in the lower right due to finite accuracy. We see the same phenomena in Example 3.3.4.

Example 3.3.4. We return to $x' = x^2 \sin t$, approximating $x(\pi)$ with $x(0) = 0.3$. The values of $u(\pi)$, $u_h(\pi)$, and E_h are tabulated in Example 3.3.2.

Table 3.3.4. "Order" of Error $p = \dfrac{\ln|E(h)| - \ln|E(h/2)|}{\ln 2}$

Number of steps	Euler	Midpoint Euler	Runge–Kutta	
2	0.544	1.953	3.431	="order"
4	0.675	1.589	4.882	
8	0.744	1.628	4.925	
16	0.816	1.928	5.765	
32	0.884	2.004	4.837	
64	0.934	2.014	2.640	
128	0.965	2.011	3.597	
256	0.982	2.006	3.829	
512	0.991	2.003	3.937	"order" ≈ 4
1024	0.995	2.002	4.092	
2048	0.998	2.001	1.344	
4096	0.999	2.000	-2.514	"order" ≈?
8192	0.999	2.000	-0.077	
16384	1.000	2.000	3.482	▲

"order" ≈ 1

"order" ≈ 2

The form of the errors predicted in equations (4) and (5) is indeed true in general, as illustrated in Examples 3.3.3 and 3.3.4. However, *there are exceptions:*

In some cases the error E can be even smaller and can look like higher orders of h than we would expect. This phenomenon is nicely demonstrated in Example 3.3.5, where the solution is symmetric about the line $t = \pi$; such symmetry could reasonably lead to cancellation of dominant terms in the error (Exercise 3.3#7). So the errors for the midpoint Euler and Runge–Kutta methods vary as h^3 and h^5 respectively, instead of h^2 and h^4.

Example 3.3.5. We return to $x' = x^2 \sin t$, but change t_f from π to 2π. To find the actual solution $u(2\pi)$, recall that $u(t) = 1/(\cos t + C)$, which is periodic with period 2π. So without even calculating C, we know that $u(2\pi) = u(0) = 0.3$ for the actual solution when $t = 2\pi$. As in Examples 3.3.2 and 3.3.4, the computer can calculate $u_h(2\pi)$, $E(h)$, and "order" p; we tabulate here only the final result, p.

Table 3.3.5. "Order" of Error $p = \dfrac{\ln |E(h)| - \ln |E(h/2)|}{\ln 2}$

No. of steps	Euler	Midpoint Euler	Runge– Kutta	
2	-112.096	-53.649	-162.414	="order"
4	-51.397	2.223	1.915	
8	1.201	5.040	5.693	
16	0.831	3.992	5.088	
32	0.868	3.338	5.016	
64	0.917	3.097	5.004	"order"
				≈ 5
128	0.953	3.025	5.001	
256	0.975	3.006	5.000	
512	0.987	3.002	5.023	
1024	0.993	3.000	4.344	
		"order"		
2048	0.997	3.000 ≈ 3	0.279	
4096	0.998	3.001	-0.236	
	"order"			
8192	0.999 ≈ 1	2.935	-2.309	
16384	1.000	3.921	-0.336	▲

Compare Examples 3.3.4 and 3.3.5; both numerically solve the same differential equation. The first is over an interval on which the solution is *not* symmetric, so the expected orders 2 and 4 are observed for midpoint Euler and Runge–Kutta respectively. But the second is over an interval on which the solution *is* symmetric, so the observed order appears as 3 and 5 for midpoint Euler and Runge–Kutta respectively.

Another exception to the error predictions of equations (4) and (5) are cases where the partial derivatives of $f(t, x)$ are unbounded or do not exist. As we shall see in the next chapter, *bounds* on the error depend on the partial derivatives of order up to the order of the method. We shall present an example when we discuss estimating error in Section 4.6.

Meanwhile you are forewarned that such exceptions as these exist to the general rule $E(h) \approx Ch^p$.

3.4   Finite Accuracy

In practical computations, computers work with "real numbers" of finite accuracy. This is usually not a problem now that computers standardly calculate to 14 decimal places (IBM-PC with 8087 co-processor) or 18 decimal places (Macintosh). Nevertheless we shall discuss the effects of the finiteness of computation, and if you wish to explore these ideas, the program *NumMeths* is set up to artificially compute to fewer decimal places and allow you to more readily see the effects.

It is the phenomenon of *round-off* that affects finite accuracy, and the effects are different depending on whether the rounding is consistently "down" (towards $-\infty$), "up" (towards $+\infty$), or "round" (to the nearest grid point). We shall discuss these cases separately.

Rounding Down (or Up)

We will give evidence in Example 3.4.1 that if a computer systematically rounds *down* (or *up*), the error $E(h)$ will behave like

$$E(h) \approx C_1 h^p + \frac{C_2}{h}. \tag{7}$$

The first term, with order p according to the chosen method of numerical approximation, has been discussed in Section 3.3; we shall proceed to explain where a term like C_2/h might come from.

Typically, a number on a computer is something like 1.06209867E02, and the numbers between 1.06209867E02 and 1.06209868E02 simply do not exist. (Actually, the computer does this with *bits*, or numbers in base 2, but this will not affect the discussion.) The number of bits available depends on the computer; typically, it might be

24 bits, about 6 decimal digits	standard single precision
32 bits, about 8 decimal digits	Apple II Basic
52 bits, about 14 decimal digits	standard double precision; IBM-PC
64 bits, about 18 decimal digits	standard Macintosh numerics.

Clearly, it can make no sense to use a stepsize smaller than the smallest available increment, but we will show in this section that there are good reasons to stick to *much longer* stepsizes.

Consider the computer's "plane," which consists of an array of dots, spaced Δ apart. Even in floating point systems, this is true locally. For example, in standard double precision you can expect roughly that for numbers

$$\begin{array}{ll} \text{order } 1 & \Delta \approx 10^{-16}, \\ \text{order } .01 & \Delta \approx 10^{-18}, \\ \text{order } 1000 & \Delta \approx 10^{-13}. \end{array}$$

Suppose we apply Euler's method to $x' = f(t, x)$ starting at (t_0, x_0). We should land at the point (t_1, x_1) with

$$t_1 = t_0 + h, \qquad x_1 = x_0 + hf(t_0, x_0),$$

but the computer will have to choose a point $(\tilde{t}_1, \tilde{x}_1)$ which is a point of its grid, as shown in Figure 3.4.1. Optimistically, we might hope that the computer will choose the closest point (in the jargon, make an error of half a bit in the last place). This hope is justified on the IBM-PC with 8087 coprocessor, and you can choose your rounding on the Macintosh; other machines (including mainframes) might not make a smart choice.

Note in Figure 3.4.1, that the difference between the slope of the segment u_h of the "real" Euler's method and the segment $\tilde{u}_h$ actually computed by the computer is of order Δ/h, with a constant in front which might be about 0.5 in the best case.

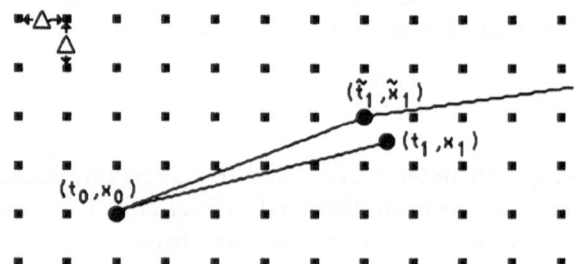

FIGURE 3.4.1. Grid of computer coordinates with spacing Δ.

Thus the computer systematically makes an error like Δ/h at each step, and moreover guesses low each time if it always *rounds down*. It seems reasonable that such errors will contribute a term like C_2/h to $E(h)$, and we will see evidence that this is so in Example 3.4.1 and Exercise 3.4#1. We will prove in Section 4.6 that a bound of this form is correct.

Remark. If instead of rounding down, the computer rounds to the nearest available number, there will probably be cancellation between the errors in the successive steps, and the contribution to $E(h)$ of round-off error is both smaller and harder to understand. We postpone that discussion to the second part of this section.

Returning to the case of rounding down (or up), we now have

$$E(h) = C_1 h^p + \frac{C_2}{h} \qquad\qquad (7,\text{ again})$$

which looks like Figure 3.4.2.

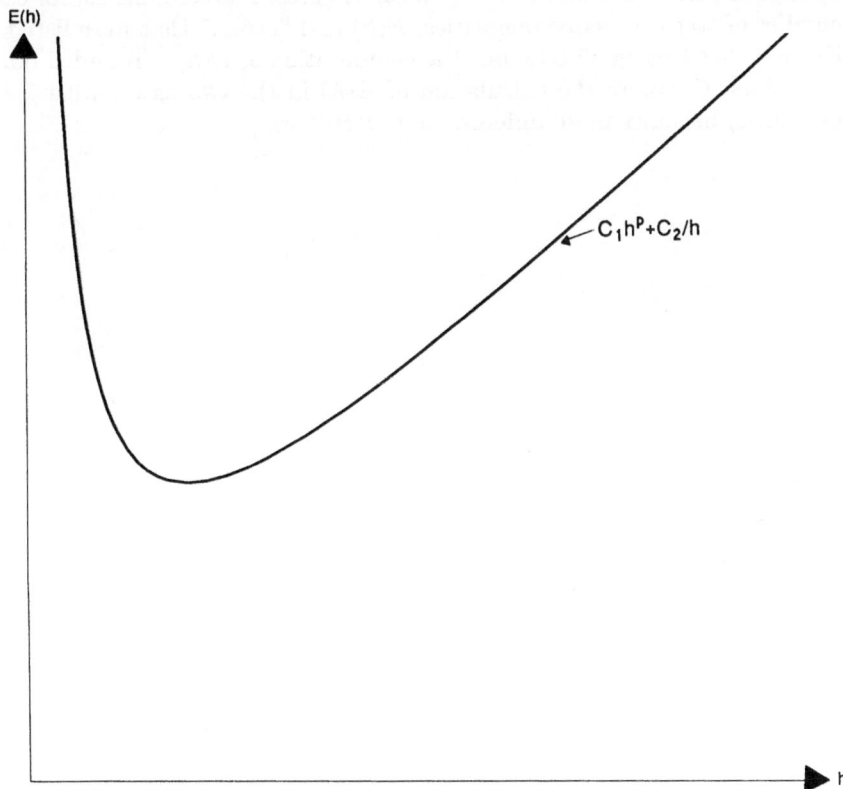

FIGURE 3.4.2.

We observe that if the error is going to behave as in equation (7), then only for a small range of h's will both terms be observable. For h large, the first term will swamp out the second; for h sufficiently small, the second will swamp out the first. Thus, if we print out as in Example 3.3.3 the "order" quantity

$$\frac{\ln|E(h)| - \ln|E(h/2)|}{\ln 2} \approx p,$$

we should see for large h's much the same as we saw in Section 3.3, namely a number close to the integer that gives the order of the method, but for small h's, we should see a number close to the integer -1, and this number should appear sooner if the number of bits you are using is small. Moreover, this dominance of -1 should occur much sooner for Runge–Kutta than for midpoint Euler, and much sooner for midpoint Euler than for Euler, because in each case more calculations are required for the second method. This is indeed what experiments demonstrate.

Example 3.4.1 has been computed using the program *NumMeths* where the user can decide with how many digits the computer should compute. For example 18 bits gives about five significant figures. The columns list for each number of steps the same quantities, $E(h)$ and "order," that were listed in Example 3.3.1 using 52 bits, but the computation of $E(h)$ is rounded down to 18 bits. Compare the calculation of $E(h)$ in the two cases; with fewer bits, $E(h)$ becomes more different as h decreases.

Example 3.4.1. We return to $x' = x$, approximating $x(2)$, with $x(0) = 1$. However, this time we round *down* using only 18 bits, which gives approximately five decimal digits.

Table 3.4.1. "Order" of Error $p = \dfrac{\ln |E(h)| - \ln |E(h/2)|}{\ln 2}$
for $x' = x$ (when rounded down)

No. of steps	Euler		Midpoint Euler		Runge–Kutta	
1	4.3891×10^0		2.3891×10^0		3.8906×10^{-1}	$=E(h)$
2	3.3891×10^0		1.1391×10^0		5.4004×10^{-2}	$=E(h)$
	0.373		1.069		2.849	="order"
4	2.3266×10^0		4.1616×10^{-1}		5.0992×10^{-3}	
	0.543		1.453		3.405	
8	1.4286×10^0		1.2685×10^{-1}		4.6052×10^{-3}	"order" ≈4?
	0.704		1.714		3.469	
16	8.0591×10^{-1}		3.5098×10^{-2}		1.8586×10^{-4}	
	0.826		1.854		1.309	
32	4.3062×10^{-1}		9.5242×10^{-3}	"order"	$\mathbf{2.9267 \times 10^{-4}}$	*
	0.904		1.882	≈2?	−0.655	
64	2.2335×10^{-1}		2.9935×10^{-3}		7.0466×10^{-4}	
	0.947		1.670		−1.268	
128	1.1457×10^{-1}	"order"	$\mathbf{1.7728 \times 10^{-3}}$	*	1.2235×10^{-3}	
	0.963	≈1	0.756		−0.796	
256	5.9695×10^{-2}		2.7799×10^{-3}		2.6883×10^{-3}	
	0.940		−0.649		−1.136	
512	3.3969×10^{-2}		5.1450×10^{-3}		5.0534×10^{-3}	
	0.813		−0.888		−0.911	
1024	$\mathbf{2.4371 \times 10^{-2}}$	*	1.0577×10^{-2}		1.0577×10^{-2}	
	0.479		−1.040		−1.066	
2048	2.8338×10^{-2}		2.1197×10^{-2}		2.1197×10^{-2}	"order"
	−0.218		−1.003		−1.003	≈−1
4096	4.5413×10^{-2}		4.1919×10^{-2}		4.1919×10^{-2}	
	−0.680		−0.984	"order"	−0.984	
8192	8.5437×10^{-2}		8.3575×10^{-2}	≈−1	8.3575×10^{-2}	
	−0.912	"order"	−0.995		−0.995	
16384	1.6709×10^{-1}	≈−1	1.6623×10^{-1}		1.6623×10^{-1}	
	−0.968		−0.992		−0.992	▲

*Notice that the *smallest* error (in boldface) for each method falls somewhere between the cases with "order" p and the cases with "order" ≈ -1.

If you wish to explore this phenomenon of rounding down, we leave it as Exercise 3.4#1 to construct a similar analysis for the equation $x' = x^2 \sin t$ of Examples 3.3.2 and 3.3.4.

ROUNDING ROUND

Most respectable modern computers do not round consistently down (or up); they round to the *nearest grid-point*. As a result, the bound C/h discussed in the previous subsection for the error contributed by finite calculation is exaggeratedly pessimistic. The round-off errors tend to cancel when rounding round, as we shall attempt to explain after a couple more examples.

Example 3.4.2 (and Exercise 3.4#1b if you wish to pursue this study) illustrate the smaller effects due to rounding *round* rather than consistently in one direction. The computer program *NumMeths* allows you to choose the number of bits, and whether you wish to round down, up, or round. As in the previous example, the two lines of information for each case give the values of the actual error $E(h)$ and the "order."

Example 3.4.2. We return to $x' = x$, approximating $x(2)$, with $x(0) = 1$. However this time we *round* the calculations to 18 bits, approximately five decimal digits.

Table 3.4.2. "Order" of Error $p = \dfrac{\ln|E(h)| - \ln|E(h/2)|}{\ln 2}$
for $x' = x$ (when rounded round)

No. of steps	Euler	Midpoint Euler	Runge-Kutta	
1	4.3891×10^{0}	2.3891×10^{0}	3.8906×10^{-1}	$= E_h$
2	3.3891×10^{0}	1.1391×10^{0}	5.3973×10^{-2}	$= E_h$
	0.373	1.069	2.850	$=$ "order"
4	2.3266×10^{0}	4.1616×10^{-1}	5.0992×10^{-3}	
	0.543	1.453	3.404	
8	1.4286×10^{0}	1.2682×10^{-1}	3.9948×10^{-4}	
	0.704	1.714	3.674	"order"
16	8.0582×10^{-1}	3.5006×10^{-2}	4.8531×10^{-5}	≈4
	0.826	1.857	3.041	
32	4.3042×10^{-1}	9.1428×10^{-3}	-1.250×10^{-5}	
	0.905	1.937	1.956	
64	2.2284×10^{-1}	2.3068×10^{-3}	3.3272×10^{-5}	
	0.950	1.987	-1.412	
128	1.1345×10^{-1}	6.1311×10^{-4} "order" ≈2	3.3272×10^{-5}	
	0.974	1.912	0.000	
256	5.7147×10^{-2}	1.5534×10^{-4}	7.9048×10^{-5}	
	0.989	1.981	-1.248	
512	2.8796×10^{-2}	-3.329×10^{-4}	-3.329×10^{-4}	"order"
	0.989 "order" ≈1	-1.100	-2.074	$\approx?$
1024	1.4254×10^{-2}	-1.804×10^{-4} "order" $\approx?$	-2.566×10^{-4}	
	1.014	0.884	0.375	
2048	7.5406×10^{-3}	2.0112×10^{-4}	2.0112×10^{-4}	
	0.919	-0.157	0.352	
4096	3.5733×10^{-3}	1.8013×10^{-5}	1.8013×10^{-5}	
	1.077	3.481	3.481	
8192	2.2153×10^{-3}	2.4689×10^{-4}	2.4689×10^{-4}	
	0.690 "order"	-3.777 random	-3.777	random
16384	2.2916×10^{-3} $\approx?$	1.3608×10^{-3}	1.3608×10^{-3}	
	-0.049	-2.462	-2.462	▲

Example 3.4.2 shows very nicely the most peculiar effects of rounding round. The exponents of h in the error do *not* appear to follow any patterns, quite unlike the systematic -1's that we found for rounding down. In fact, rounding round is best described by a *random process*, which we shall describe in two stages.

1. *Integrating in a noisy environment: random walk.* Consider the differential equation $x' = g(t)$, the solution of which,

$$u(t) = x_0 + \int_{t_0}^{t} g(s)ds,$$

is an ordinary integral. Suppose this integral is computed by any of the approximations discussed at the end of Section 3.1, with n steps. Then round-off error just adds something like $\varepsilon_i = \pm\Delta$ at each step to whatever would be computed if the arithmetic were exact, and we may think of the signs as random (in the absence of any good reason to think otherwise). Thus the cumulative round-off error

$$\sum_{i=1}^{n} \varepsilon_i$$

can be thought of *probabilistically*. For instance, you might think of a *random walk* where you toss a coin n times, and move each time Δ to the right if the toss comes out heads, and Δ to the left if it comes out tails.

What are the reasonable questions to ask about the cumulative error? It is perfectly reasonable to ask for its average (over all random walks $(\varepsilon_1, \ldots, \varepsilon_n)$ with n steps), but this is obviously zero since it is as often negative as positive. More interesting is the average of the absolute value, but this turns out to be very hard to compute. Almost as good is the *square root of the average of the squares* (which, because the *mean* is 0, is the *standard deviation* from statistical theory), and this turns out to be quite easy to study for a random walk.

Proposition 3.4.3. *The standard deviation of the random walk of n steps of length Δ, with equal probability of moving right or left, is $\Delta\sqrt{n}$.*

Proof. There are 2^n possible such random walks, so the expression to be evaluated is

$$\left[\frac{1}{2^n} \sum_{\text{random walks } (\varepsilon_1 \ldots \varepsilon_n)} \left(\sum_{i=1}^{n} \varepsilon_i\right)^2\right]^{1/2}. \tag{8}$$

Fortunately this computation is much easier than it seems at first glance. If the square is expanded, there are terms like ε_i^2 and terms like $\varepsilon_i \varepsilon_j$. The key point of the proof is that the terms like $\varepsilon_i \varepsilon_j$ cancel, when summed over all random walks: for half the random walks they give $+\Delta^2$ and for half $-\Delta^2$, as can easily be verified.

Thus our expression (8) becomes

$$\left[\frac{1}{2^n}\underbrace{\sum_{\text{random walks }(\varepsilon_1...\varepsilon_n)}\underbrace{\left(\sum_{i=1}^{n}\Delta^2\right)}_{\text{for each random walk}}}_{\text{there are }2^n\text{ possible random walks}}\right]^{1/2}=\Delta\sqrt{n}.$$

In probabilistic jargon, the cancellation of the cross terms follows from the fact that the ε_i are *independent, identically distributed, with mean 0*. In terms of coin tosses, this means that the ith and j^{th} toss are independent, each is the same experiment as the other, and that each of these experiments has the same probability of sending you to the left or to the right. This is all that is really used in the proof; the assumption that the ε_i are exactly $\pm\Delta$ is unnecessary, fortunately, since in our case it isn't true. □

The random walk result was fairly easy, but lies at the beginning of a whole chapter of probability theory, on which we will not touch. Instead, we will move on from $x' = g(t)$ to the more difficult stochastic problem of solving $x' = f(t, x)$, where f depends on x as well as t. We begin with the most important case, the simple linear equation $x' = \alpha(t)x$, and start with the simplest subcase, where $\alpha(t)$ is a constant, α.

2. *Solving $x' = \alpha x$ in a noisy environment.* We were discussing above the special differential equation $x' = g(t)$, with solutions given by indefinite integrals. Errors committed during the solution are not amplified, and so errors committed at the end can cancel those committed at the beginning. It is quite unclear whether anything similar should be true for a differential equation like

$$x' = \alpha x$$

where errors committed near the initial time t_0 are amplified by a factor of $e^{\alpha(t-t_0)}$, and will swamp out an error made at the end.

There is an intuitive way of understanding how the round-off error affects the numerical solution of $x' = \alpha x$ with $x(0) = x_0$, closely related to the discussion of the nonhomogeneous linear equation in Sections 2.2 and 2.3. Think that you have a bank account with interest α and initial deposit x_0 at time 0. At regular intervals, make either a deposit or a withdrawal of Δ, choosing whether to make a deposit or a withdrawal at random. How should you expect these random deposits and withdrawals to affect the value of the account? The equation becomes $x' = \alpha(x + \text{random deposits})$.

The variation of parameters formula of Section 2.3 is just the right tool to study this question. Looking back at the discussion in Section 2.3, we can see what to substitute in the parentheses of equation (8) for the standard deviation. Thus for $x' = \alpha x$, the *standard deviation* at time t to be evaluated is

$$\left[\frac{1}{2^n} \sum_{\text{random walks } (\varepsilon_1 \ldots \varepsilon_n)} \left(e^{\alpha t} x_0 - e^{\alpha t} \left(x_0 + \sum_{i=1}^{n} e^{-\alpha s_i} \varepsilon_i \right) \right)^2 \right]^{1/2}$$

$$= e^{\alpha t} \left[\frac{1}{2^n} \sum_{\text{random walks } (\varepsilon_1 \ldots \varepsilon_n)} \left(\sum_{i=1}^{n} e^{-\alpha s_i} \varepsilon_i \right)^2 \right]^{1/2}$$

with $s_i = (it)/n$ if there were n deposits or withdrawals. Again, upon expanding the square, we find terms like

$$\varepsilon_i^2 e^{-2\alpha s_i} \quad \text{and} \quad \varepsilon_i \varepsilon_j e^{-\alpha(s_i + s_j)}.$$

Just as before, the cross terms cancel in the sum. So the standard deviation is

$$\Delta e^{\alpha t} \left[\frac{1}{2^n} \sum_{\text{random walks } (\varepsilon_1 \ldots \varepsilon_n)} \sum_{i=1}^{n} e^{-2\alpha s_i} \right]^{1/2} = \Delta e^{\alpha t} \left(\sum_{i=1}^{n} e^{-2\alpha s_i} \right)^{1/2}.$$

The term

$$\sum e^{-2\alpha s_i} = \frac{n}{t} \sum e^{-2\alpha s_i} \frac{t}{n}$$

can be approximated for n large by observing that the sum on the right-hand side is a Riemann sum for

$$\int_0^t e^{-2\alpha s} ds = \frac{1}{2\alpha} (1 - e^{-2\alpha t}).$$

So the *standard deviation* is about

$$\Delta \sqrt{n} \left(\frac{e^{2\alpha t} - 1}{2\alpha t} \right)^{1/2}.$$

In particular, we find the term $\sqrt{n}$ again, and we see that the amplification of errors due to the exponential growth of the solutions just *changes the constants in front of $\sqrt{n}$ by a factor independent of the step, but which becomes very large when αt becomes large.*

This probability analysis says that if $u_h(t)$ is the approximate solution of $x' = f(t, x) = \alpha x$ computed with perfect arithmetic, then the values computed by rounding will be randomly distributed around this approximate solution, with standard deviation $C/\sqrt{h}$, where C is usually a small constant, of order Δ. This random distribution is illustrated in Figure 3.4.3.

The dots of Figure 3.4.3 represent the computed values of $u_h(t_f)$; most of these values lie closer to $u_h(t_f)$ than the standard deviation.

The discussion above was for the case of constant coefficients α, but the diligent reader will observe that it goes through without substantial change even if the constant α is replaced by a function $\alpha(t)$ (see Exercise 3.4#3).

In fact, the analysis above is more or less true in general, not just for linear equations, but exploration of that issue must be postponed until we discuss *linearization* of nonlinear functions, in Volume II, Chapter 8.

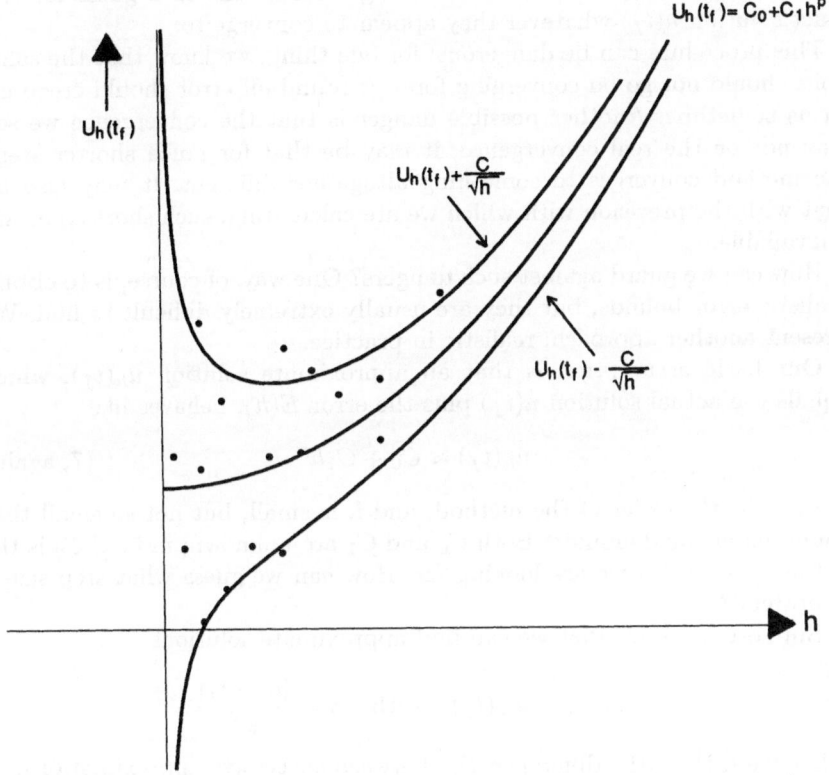

FIGURE 3.4.3. Random distribution of $u_h(t_f)$ versus h for $x' = \alpha x$.

3.5 What To Do in Practice

The examples of Sections 3.3 and 3.4 were made from differential equations whose solutions were analytically known, so that errors could be precisely analyzed. Of course, you will be interested in understanding errors primarily when the solution is not known ahead of time. The object of this section is to see how the discussion of the previous sections can help make good choices of method and of step-length.

Suppose we have a differential equation $x' = f(t, x)$ and that we are interested in finding a solution $u(t)$ for $t \in [t_0, t_f]$ with $u(t_0) = x_0$. Then it is useful to focus on the value of t in that interval for which you would expect the approximate error to be the worst, usually t_f, and use the program that calculates errors and "orders" for $u_h(t_f)$.

A first, naive (and not so bad) approach, quite adequate if nothing rides on the outcome, is to compute solutions for a succession of numbers-of-steps, for instance halving the step each time, and see whether the approximate solutions $u_h(t_f)$ appear to converge. Then take as a guess for the exact solution $u(t_f)$ whatever they appear to converge to.

This procedure can be dangerous: for one thing, we know that the solutions should not go on converging forever: round-off error should creep up on us sometime. Another possible danger is that the convergence we see may not be the real convergence; it may be that for much shorter steps the method converges to something altogether different. It may also be that with the precision with which we are calculating, such short steps are unavailable.

How can we guard against such dangers? One way, of course, is to obtain realistic error bounds, but they are usually extremely difficult to find. We present another approach, realistic in practice.

Our basic assumption is that an approximate solution $u_h(t_f)$, which equals the actual solution $u(t_f)$ plus the error $E(h)$, behaves like

$$u_h(t_f) \approx C_0 + C_1 h^p \qquad\qquad (7, \text{ again})$$

where p is the order of the method, and h is small, but not so small that round-off errors dominate. Both C_0 and C_1 are unknown; in fact, C_0 is the actual value $u(t_f)$ we are looking for. How can we guess what step size h is optimal?

Suppose as above that we can find approximate solutions

$$u_h(t_f) = u_N(t_f) \quad \text{with} \quad h = \frac{|t_0 - t_f|}{2^N}.$$

Then we look at the difference D_N between successive approximations,

$$D_N = u_N(t_f) - u_{N-1}(t_f) \approx C_1\big(2^{-pN} - 2^{-pN+p}\big)|t_0 - t_f|^p$$
$$= C_1 2^{-Np}(1 - 2^p)|t_0 - t_f|^p,$$

and the ratio

$$R_N = \frac{\ln|D_{N-1}| - \ln|D_N|}{\ln 2} = \frac{\ln|D_{N-1}/D_N|}{\ln 2} \approx p. \qquad (9)$$

This argument, similar to the one in Section 3.3, has the advantage that the numbers D_N can be computed without knowing the exact solution. As before, we expect the equivalence in the relation (9) to be destroyed by round-off error as soon as that becomes significant.

So one possible approach to the problem is to tabulate values of $u_N(t_f)$ with the corresponding R_N. Note that the R_N won't show up until the third case, since each R_N requires u_N, u_{N-1}, u_{N-2}.

Find, if it exists, the range of steps for which R_N is roughly the expected p. In this range, the solutions are converging as they should; the convergence *might* still be phony, but this is unlikely. *Then the optimal precision is probably achieved by the shortest step-length in that range, or perhaps the next shorter step-length below the range.*

The reason for saying maybe the *next* shorter is the following: recall from equation (7) of Section 3.4 that the error

$$E(h) \approx C_1 h^p + \frac{C_2}{h},$$

which gives graphs for $E(h)$ of the form shown in Figure 3.5.1. (Of course we do not know the values of the constants, but the behavior we wish to point out is not dependent on the values of the C_i's.)

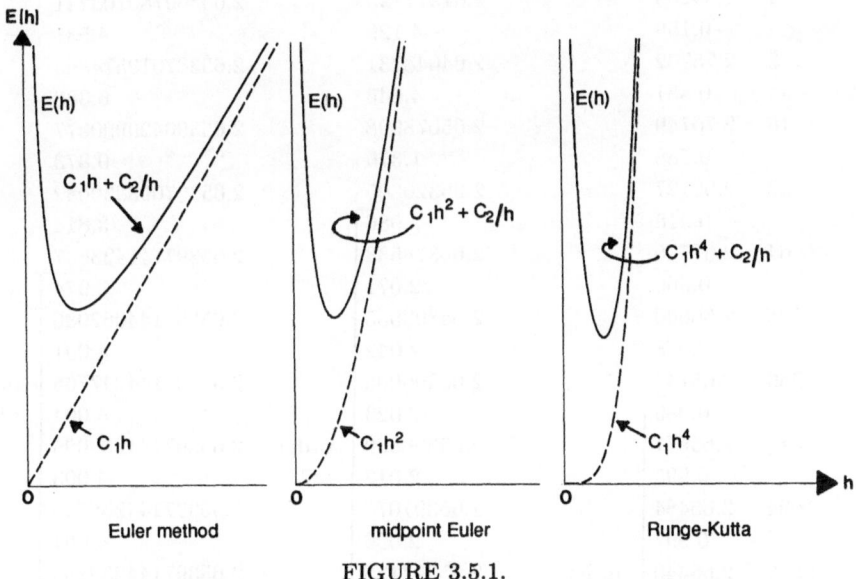

FIGURE 3.5.1.

As we move down numerical tables with increasing number N of steps and decreasing step size h, we move *left* along the curve for error. What we observe in Figure 3.5.1 is that the minimum for the error $E(h)$ occurs, moving left, soon *after* the error curve bends away from the dotted curve representing $C_1 h^p$.

The following Examples 3.5.1 and 3.5.2 show that for two differential equations *not* solvable in elementary terms, we observe the expected phenomena. The second line of the printout for each N *does*, in an appropriate range, tend to orders 1, 2, and 4 as we would expect.

We shall continue the discussion following a look at the examples, which were computed with the *Macmath* program *NumMeths*. The "curve-fitting" mentioned at the bottom of the printouts is explained in this later discussion.

Example 3.5.1. Consider again $x' = \sin tx$, the equation of Examples 1.5.2; 3.1.2,3; 3.2.1,2. We start at $t_0 = 0$, $x_0 = 3$, and end at $t_f = 2$ we approximate $u_h(t_f)$, using the full precision (64 bits) available on the Macintosh.

Table 3.5.1. $x' = \sin tx$

No. of Steps	Euler	Midpoint Euler	Runge-Kutta	
1	3.00000	3.28224001	3.00186186264321	$= u_N(t_f)$
2	3.14112	3.24403094	2.66813472585461	
4	2.84253	2.61377795	2.65369780703221	
	-0.158	-4.129	4.531	$= R_N$
8	2.75702	2.64048731	2.65387019514893	
	0.881	4.645	6.388	
16	2.70749	2.65078298	2.65396429900677	
	0.788	1.375	0.873	
32	2.68127	2.65320805	2.65397098850082	
	0.918	2.086	3.814	
64	2.63776	2.65378538	2.65397141423697	
	0.956	2.071	3.974	
128	2.66090	2.65392555	2.65397144082086	
	0.978	2.042	4.001	
256	2.65744	2.65396005	2.65397144247765	order
	0.989	2.023	4.004	≈ 4
512	2.65571	2.65396860	2.65397144258098	order
	0.995	2.012	4.003	≈ 2
1024	2.65484	2.65397073	2.65397144258743	
	0.997	2.006	4.001	
2048	2.65440	2.65397126	2.65397144258784	order
	0.999	2.003	3.997	≈ 1
4096	2.65418	2.65397139	2.65397144258787	*
	0.999	2.002	3.680	
8192	2.65408	2.65397143	2.65397144258788	
	1.000	2.001	1.757	

Curve fitting:

CF interpolation using method R and steps from 7 to 13 , is
$$2.65397144258792\text{E}+000 + -2.96461702367259\text{E}-002 \ h^4$$
CF interpolation using method M and steps from 6 to 13 , is
$$2.65397156047038\text{E}+000 + -1.90511410080302\text{E}-001 \ h^2$$
CF interpolation using method E and steps from 9 to 13 , is
$$2.65397177402202\text{E}+000 + 4.45219184726699\text{E}-001 \ h^1 \quad\quad \blacktriangle$$

Example 3.5.2. Consider again $x' = x^2 - t$, the famous equation of Examples 1.1.4, 1.3.3, 1.3.6, and 1.5.1. We start at $t_0 = 0$, $x_0 = 1$, and at $t_f = 1$ we approximate $u_h(t_f)$, using the full precision (64 bits) available on the Macintosh.

Table 3.5.2. $x' = x^2 - t$

No. of Steps	Euler		Midpoint Euler		Runge–Kutta	
1	2.0000		2.75000		5.2760823567	$= u_N(t_f)$
2	2.3750		3.73888		7.2621789573	
4	2.9654		5.19816		8.7088441194	
	−1.364		−1.308		−0.789	$= R_N$
8	3.8159		6.87271		9.2426947716	
	0.183		0.548		2.685	
16	4.9058		8.22320		9.3421536298	
	−0.358		0.310		2.424	
32	6.1089		8.95433		9.3529005110	
	−0.143		0.885		3.210	
64	7.2220		9.23519		9.3537418047	
	0.112		1.380		3.675	
128	8.0808		9.32167		9.3537987807	
	0.374		1.699		3.884	
256	8.6467		9.34546		9.3538024380	
	0.602		1.862		3.962	
512	8.9791		9.35168		9.3538026686	order
	0.768		1.937		3.987	≈4
1024	9.1606	order	9.35326	order	9.3538026831	
	0.873	≈1	1.970	≈2	3.995	
2048	9.2556		9.35366		9.3538026840	*
	−0.933		−1.986		3.998	

Curve fitting:

CF interpolation using method R and steps from 8 to 11 , is
        $9.35380268407649\text{E}+000 + -1.05668505587010\text{E}+003 \; h^4$

CF interpolation using method M and steps from 9 to 11 , is
        $9.35379926109397\text{E}+000 + -5.55196762347348\text{E}+002 \; h^2$

CF interpolation using method E and steps from 9 to 11 , is
        $9.34643244726150\text{E}+000 + -1.88376521685730\text{E}+002 \; h^1$

Note: The Euler interpolation is not very reliable since R_N is not yet very close to 1. Yet note also that despite this fact, the interpolation estimate is much closer to the solution than any of the Euler values calculated so far!   ▲

On the basis of the above discussion of the range of steps for which R_N is roughly the expected p, we expect that the $*$ approximations are, in each of the two preceding examples, the best obtainable by our three methods—Euler, midpoint Euler, and Runge–Kutta.

If the range does not exist, then with the precision with which you are computing, the method has not had a chance to converge, and you must use more steps, go to higher precision, or use a method of higher order. A shorter step-length is of course easiest, but it will not help if round-off error becomes predominant before convergence has really occurred.

If an appropriate range does exist, and if it includes several values of N, then it is reasonable to try *fitting a curve* of the form

$$C_0 + C_1 h^p$$

to the values of $u_N(t_f)$ in that range, for instance by the *method of least squares*. The value for $C_0 = u_h(t_f)$ is then the best guess available for the solution of the differential equation using that method, and the quality of the fit gives an estimate on the possible error.

In Examples 3.5.1 and 3.5.2 this curve-fitting has been done by the computer, with the results listed at the bottom of each printout.

None of this analysis using the difference of successive approximations is quite rigorous, but an equation has to be quite nasty for it not to work. Rigorous error bounds will be studied in Chapter 4; unfortunately, useful bounds are hard to get. Usually bounds, even if you can find them, are wildly pessimistic.

Summary. For a first order differential equation $x' = f(t, x)$ with initial condition (t_0, x_0) and a desired interval $[t_0, t_f]$, choose the value of t in that interval for which you would expect the approximate solution to be the worst, usually t_f.

A succession $u_N(t_f)$ of numerical approximations can be made using step size $h = \frac{|t_0 - t_f|}{2^N}$. For each N, calculate

$$D_N = u_N(t_f) - u_{N-1}(t_f)$$

and

$$R_N = \frac{\ln |D_{N-1}/D_N|}{\ln 2} \approx p.$$

Find, if it exists, the *range of steps* for which R_N is roughly the expected p, and expect the best value of $u_N(t_f)$ to be that with the shortest step length in that range, or perhaps the next shorter below that range.

If you desire more precision, use *curve fitting*, such as least squares, on the values $u_N(t_f)$ in that range of steps. The result should give an estimated $u(t_f)$ and an estimate on the error.

Then you can compute $u_h(t)$ over the entire interval $[t_0, t_f]$ using the stepsize you found best for $u_h(t_f)$.

Exercises 3.1–3.2 Basic Numerical Methods

3.1–3.2#1°. This exercise is not at all as trivial as it might seem at first glance, and it will teach you a great deal about the various numerical methods while keeping the computation to a minimum. (You will see that it's already a job just to keep all of the "formulas" straight, using proper t_i's and x_i's at each step, and you'll be able to figure out how you have to organize the necessary information.) We use a ridiculously large stepsize in order that you can really *see* the results.

Consider $x' = x$, starting at $t_0 = 0$, $x_0 = 1$.

(a) Using stepsize $h = 1$, calculate by hand an approximate solution for two steps, for *each* of the three numerical methods: Euler, midpoint Euler, and Runge–Kutta.

(b) Solve the equation analytically and calculate the exact solution for $t_1 = 1$ and $t_2 = 2$.

(c) Turn a sheet of graph paper sideways and mark off three graphs as shown, with large scale units (one inch units are good, with a horizontal domain from 0 to 2 and a vertical range from 0 to 1 to 8, for each graph).

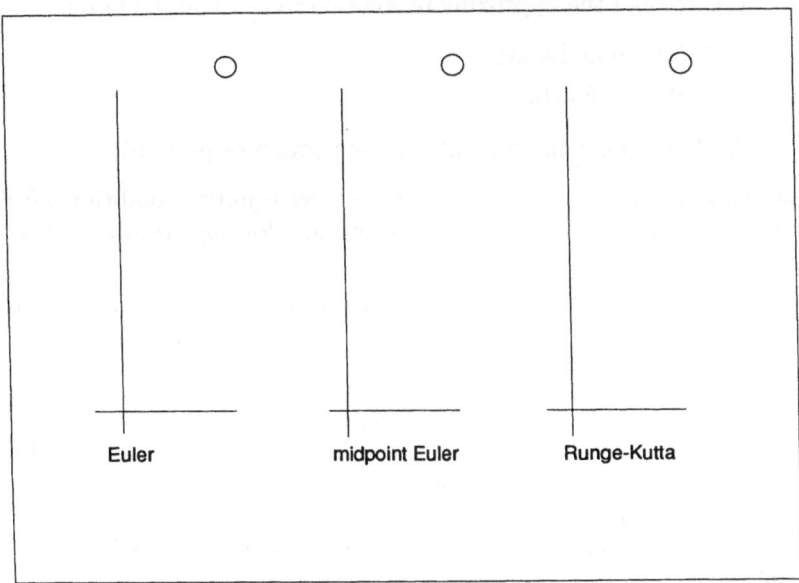

Using one graph for each of the three methods, construct the approximate solution graphically. That is, use the graph paper to make lines of exactly the proper slopes, as calculated in (a). For midpoint Euler and Runge–Kutta, make dotted lines for the trial runs showing the actual slope marks at beginning, midpoints, and endpoint, as in Figures 3.2.1 and 3.2.2. Confirm that the final slope for each step indeed *looks* like an average of the intermediate slopes from the trial runs. The graphical values for x_1 and x_2 should be close to the numerical approximation values calculated in part (a), and you should add for comparison the exact values calculated in part (b).

You may be surprised how well Runge–Kutta makes this approximation in two (very large) steps. This exercise was designed to emphasize graphically how the "better" methods indeed get closer to the true solution, by adjusting the slope rather than the stepsize.

3.1–3.2#2. Consider $x' = x^2 - t$, with $t_0 = 0$, $x_0 = 1$.

(a) With $x(0) = 1$ and $h = \frac{1}{2}$, calculate the forward (for positive h) and backward (for negative h) Euler approximate solution $x = u_h(t)$ in the interval $-1 \le t \le 1$.

(b) Make a drawing of the graph of $u_h(t)$ as found in part (a). Determine whether the approximate solution you have just formed is above or below the real solution.

(c) With $x(0) = 1$ and step $h = \frac{1}{2}$, calculate by hand (that is, without a program; you may certainly use a calculator to carry out the operations) the approximate solution $u_h\left(\frac{1}{2}\right)$ and $u_h(1)$ by

    (i) Midpoint Euler;

    (ii) Runge–Kutta.

Add and label these points on your graph of part (b).

3.1–3.2#3. Prove that for $x' = x^2 - t$, with initial condition $x(0) = 1$, that the forward Euler approximate solution, for any stepsize h, is a *lower fence*.

3.1–3.2#4. Actually work out, numerically, the first three steps of Euler's method and of Midpoint Euler for the differential equation $x' = x^2 - t^2$, $x(0) = 0$, with stepsize $h = 0.1$.

3.1–3.2#5. Euler's method can be run forwards or backwards in time, according to whether the stepsize h is positive or negative respectively. However, Euler's method run backwards is *not* the inverse of Euler's method running forward: running Euler's method backwards starting at (t_0, x_0) does not usually lead to a (t_{-1}, x_{-1}) such that if you start Euler's method at (t_{-1}, x_{-1}), you will get back to (t_0, x_0).

(a) Verify this fact for $x' = x^2 - t$, with $x(0) = 1$, $h = \frac{1}{2}$. Go forward one step, then backwards one step. You should not return to 1.

(b) Make a sketch to show why this is true.

3.1–3.2#6. A special case of the Runge–Kutta method is for $x' = g(t)$. In this case the slope of Runge–Kutta reduces from

$$m_{RK} = \frac{1}{6}(m_1 + 2m_2 + 2m_3 + m_4)$$

to

$$m_{RK} = \frac{1}{6}(m_1 + 4m_2 + m_4).$$

Show why this is true.

3.1–3.2#7°. Another numerical method, called the *improved Euler method*, is defined as follows for a given stepsize h:
For $n = 1, 2, \ldots$ set $t_{n+1} = t_n + h$

$$x_{n+1} = x_n + h\frac{f(t_n, x_n) + f(t_{n+1}, x_n + hf(t_n, x_n))}{2}.$$

(a) Explain on a drawing how x_{n+1} is defined from x_n.

(b) For the differential equations for the form $\frac{dx}{dt} = g(t)$ the improved Euler gives a numerical method for integration, another one with which you should already be familiar. Explain the method and give its name.

3.1–3.2#8°. For the differential equation $x' = x^2$, with initial condition $x(0) = 2$,

(a) Calculate (by hand, calculator, or computer) an Euler approximation with stepsize $h = 0.1$ to find $x(1)$. Does anything seem wrong numerically?

(b) Calculate a Runge–Kutta approximation with stepsize $h = 0.1$ to find $x(1)$. Does this agree at all with part (a)? *Now* do you think something might be wrong?

(c) To help find out what is going on, solve the equation analytically. State why parts (a) and (b) are both wrong, and why you might have missed the realization there was any problem.

(d) How could you avoid this problem? Do you think changing the stepsize would help? If so, try finding $x(1)$ with a smaller stepsize, and see if you can solve the problem. If not, can you explain why?

(e) This problem could be helped by a numerical method with a built-in (automatic) *variable stepsize;* that is, as the slope gets steeper, the stepsize is made smaller; e.g., make stepsize inversely proportional to slope. Explain how this can help.

(f) Yet another suggestion is to design the numerical method to take the step h along the approximate solution, instead of along t axis. Explain how this can help.

3.1–3.2#9. For $x' = f(t, x)$ with $x(t_0) = x_0$, consider the approximate solution

$$u_h(t) = u_h(t_0 + h) = x_0 + h m_{t_0, x_0, h},$$

with the *midpoint Euler* slope

$$m_{t_0, x_0, h} = f\left(t_0 + \frac{h}{2}, x_0 + \frac{h}{2} f(t_0, x_0)\right),$$

as a function of h. Show that $u_h(t)$ has the same 2nd degree Taylor polynomial as the exact solution $x = u(t)$. That is, expand $u_h(t)$ about (t_0, x_0), expand $u(t)$ about $u(t_0)$.

3.1–3.2#10°. Let the points (t_0, x_0), $(t_1, x_1), \ldots$ be constructed by the *implicit Euler method,* as introduced at the end of Section 3.2. That is, for $x' = f(t, x)$, use

$$t_{i+1} = t_i + h \quad \text{and} \quad x_{i+1} = x_i + h f(t_{i+1}, x_{i+1}).$$

This second expression is only implicit in x_{i+1}, and must be solved at each step to find x_{i+1} before proceeding to the next step. For the following four equations, write (t_i, x_i) for $0 \leq i \leq 3$, and $(t_0, x_0) = (0, 1)$. The equations are listed in order of increasing difficulty: the first can be done by hand, the second requires a calculator, and the third and fourth are best done using something like *Analyzer* to solve the cubic and transcendental equations which appear.

(a) $x' = 4 - x$

(b) $x' = 4x - x^2$

(c) $x' = 4x - x^3$

(d) $x' = \sin(tx)$.

3.1–3.2#11. We saw in Example 3.1.4 that if u_h is Euler's approximation to the solution of $x' = x$ with initial condition $x(0) = x_0$, then $u_h(t) = (1 + h)^{t/h}$, at least if t is an integral multiple of h.

(a) Find an analogous formula for the midpoint Euler method.

(b) Evaluate your formula at $t = 1$ with $h = 0.1$. How far is the answer from the number (e).

(c) Find an analogous formula for Runge–Kutta.

(d) Again evaluate your formula at $t = 1$ with $h = 0.1$, and compare the value with (e).

(e) What other name, in the case of this particular equation, would be appropriate for the Midpoint Euler and Runge–Kutta calculations?

Exercises 3.3 Error Due to Approximation Method

3.3#1 Consider the equation $x' = x$, solved for $0 \le t \le 2$ with $x(0) = 1$, as in Example 3.3.1.

(a) Using Table 3.3.1, show that the error for Euler's method does indeed behave like

$$E(h) \approx C_E h.$$

That is, show that the ratio $E(h)/h$ stabilizes as h gets small. Find an estimate of the constant C_E.

For this equation, the constant can be evaluated theoretically. As can be shown from Example 3.1.4, the Euler approximation $u_h(t)$ with step h gives

$$u_h(2) = (1 + h)^{2/h}.$$

(b) Find the Taylor polynomial to first degree of $(1 + h)^{2/h}$, by the following procedure:

(i) Writing $(1 + h)^{2/h}$ as $e^{(2/h)\ln(1+h)}$;

(ii) Expanding $\ln(1 + h)$ in its Taylor polynomial;

(iii) Use the Taylor polynomial of the exponential function, after carefully factoring out the constant term of the exponent.

(c) Find C_E exactly from part (b), as $\lim_{h\to 0}(E_h/h)$. Compare with the estimate in part (a).

3.3#2°. (harder) Consider the differential equation $x' = x$ as in the last exercise, and let $u_h(t)$ be the approximate solution using midpoint Euler with $u_h(0) = 1$.

(a) Again using Table 3.3.1, show that $E(h) \approx C_M h^2$ for some constant C_M, and estimate C_M.

(b) Use the explicit formulae from Exercise 3.1–3.2#11 to find an asymptotic development (see Appendix) of the error at $t = 2$; i.e., find an asymptotic development of the form

$$u_h(2) = e^2 + C_M h^2 + o(h^2),$$

and evaluate the constant C_M.

(c) Compare your theoretical value in part (b) with your estimate in part (a).

3.3#3. (lengthier) Again consider the differential equation $x' = x$, and do the same steps as the last two exercises for the Runge–Kutta method. Conceptually this is not harder than the problems above, but the computations are a bit awesome.

3.3#4. From Table 3.3.2 in Example 3.3.2, verify that the errors for Euler's method, Midpoint and Runge–Kutta do appear to behave as

$$\begin{aligned} E(h) &\approx C_E h & \text{for Euler's method,} \\ E(h) &\approx C_M h^2 & \text{for midpoint Euler and} \\ E(h) &\approx C_{RK} h^4 & \text{for Runge–Kutta} \end{aligned}$$

and find approximate values for C_E, C_M, C_{RK}.

3.3#5. For the following equations, run the program *NumMeths* over the given ranges, with a number of steps from 2^2 to 2^{10} (or 2^{12} if you have the patience, since the speed diminishes as number of steps increases). Observe which of these experiments give the predicted orders. For those cases where the "order" is unusual, try to find the reason why.

$$\begin{array}{llll} \text{(a)} & x' = -tx & 0 < t < 2 & x_0 = 2 \\ \text{(b)} & x' = x^2 - t & 0 < t < 2 & x_0 = 0.5 \\ \text{(c)} & x' = t^2 & 0 < t < 1 & x_0 = 1 \\ \text{(d)} & x' = t & 0 < t < 1 & x_0 = 1 \end{array}$$

3.3#6°. Consider $x' = 1/(x + t)$, where successive derivatives are not difficult to compute, and the exact analytic solution can be found as well.

(a) Calculate the 6th order Taylor series for the solution.

(b) Solve the equation analytically.

(c) Starting at $x(0) = 1$, compare the exact solution, the truncated sixth degree Taylor polynomial, and the fourth order Runge–Kutta approximation for two steps, with $h = 0.1$. Which approximation gives the better answer? Which is least costly to compute?

(d) Repeat part (c) for $x(0) = 0$. Explain any differences.

(e) Repeat part (c) for $x(0) = 2$. Explain any differences.

3.3#7. Consider Example 3.3.5: $x' = x^2 \sin t$, approximating $x(2\pi)$ with $x(0) = 0.3$. Show why symmetry about the line $t = \pi$ leads to cancellation of dominant terms in the error $E(h)$.

3.3#8°. Let $x' = f(t, x)$ be a differential equation, let $u(t)$ be the solution with initial condition $x(t_0) = x_0$, and let $u_h(t)$ be the Euler approximation to this solution. In this chapter we have given numerical evidence that the error

$$E(h) = u(t_1) - u_h(t_1)$$

should have an asymptotic development of the form $E(h) = Ch + o(h)$, and in Exercise 3.3#1 this is proved for the differential equation $x' = x$.

(a) Show that if the equation is simply $x' = f(t)$, then

$$E(h) = (f(t_1) - f(t_0))h/2 + o(h).$$

(b) Show that if the equation is linear; i.e., $x' = g(t)x$, then the error also has the form $E(h) = Ch + o(h)$, where this time the constant is given by the expression

$$E(h) = -\left(\frac{h}{2}\right)e^{\int_{t_0}^t g(s)ds}\int_{t_0}^t [g'(s) + (g(s))^2]ds + o(h).$$

(c) Show that this expression agrees with that found in Exercise 3.1–3.2#11.

(d) Set $t_0 = 0$, and find a number $t_1 > 0$ such that Euler's method as applied to $x' = -tx$ has order greater than 1. Use the computer to see what the order actually is.

(e) The formula in (b) tells us that the error in Euler's method will be of the form $Ch + o(h)$ for practically any function $g(t)$. In order to make this fail, we must find a sufficiently nasty function $g(t)$ such that $(g(t))^2$ does not have a finite integral. Show that $g(t) = 1/|t|^\alpha$ is such a function if $0 < \alpha < 1$.

(f) Show that nevertheless the differential equation $x' = x/|t|^\alpha$ has unique, continuous solutions through every point if $0 < \alpha < 1$.

(g) Use the program *Numerical Methods* to study this equation for $\alpha = 0.3$ and $\alpha = 0.8$, using Euler's method, for $-1 \le t \le 1$ and $x(-1) = 1$. Can you explain what you observe?

Exercises 3.4 Error Due to Finite Machines

3.4#1°. For $x' = x^2 \sin t$, the equation of Examples 3.3.2 and 3.3.4, construct an analysis like that of Examples 3.4.1 and 3.4.2. That is,

(a) make a calculation using only 18 bits, rounding down;

(b) make a calculation using only 18 bits, rounding round;

(c) for each of parts (a) and (b), state the range of stepsizes giving the proper "order"; also compare the magnitude of errors in parts (a) and (b).

3.4#2. Run the program *NumMeths*, for number of steps: 2^1 to 2^{10}, for the differential equation

$$x' = x, \quad 0 \le t \le 2, \quad x(0) = 0.5,$$

under *all* of the following conditions (a total of 18 runs):

> method: Euler, midpoint Euler, and Runge–Kutta
> number of bits: 10, 35, and 64.
> rounding: down, and round.

For each printout

(a) Describe the region in which the terms coming from integration error and those commonly from round-off error are dominant; and

(b) Say which step length gave the most accurate value and why.

3.4#3. Show that the analysis of Section 3.4.2, "solving $x' = \alpha x$ in a noisy environment," follows through in the case $x' = \alpha(t)x$, for α a function of t.

Exercises 3.5 What To Do in Practice

3.5#1. For the following differential equations, solve numerically by various methods. Use 20 bits so that you will see the effects of roundoff error. Find how closely your approximate solution near the bottom of a range of good numbers of steps (i.e., those with the proper order appearing) compares with an exact analytic solution (as found in Exercises 2.1#2b,d,f respectively).

(a) $x' = (1+x)/(1-t)$ $0 < t < 2$ $x(0) = 3$
(b) $(x - \alpha) + t^2 x' = 0$ $1 < t < 2$ $x(1) = 2e$
(c) $x' = (1+x^2)/(1+t^2)$ $0 < t < 1$ $x(0) = 3$.

3.5#2. For the following differential equations, which are not easy or possible to solve analytically, solve numerically by midpoint Euler and Runge–Kutta. Use 20 bits so that you will see the effects of roundoff error.

(a) $x' = \sin(x^2 - t)$ $0 < t < 2\pi$ $x(0) = 0$
(b) $x' = \sin(x^2 - t)$ $0 < t < 2\pi$ $x(0) = 0.7$
(c) $x' = x^2 - t$ $0 < t < 5$ $x(0) = 0$
(d) $x' = 1/(x^2 + t + 1)$ $0 < t < 2$ $x(0) = 0$.

4

Fundamental Inequality, Existence, and Uniqueness

To begin, we must emphasize *why* we need to get theoretical.

If we have a direction field for a differential equation, we need to know whether through a given point (i.e., with a given initial condition) there *is* a solution. Furthermore, if a solution does exist through such a point, when can we be sure it is the *only one*? These are the questions of *existence* and *uniqueness*, and they are not trivial.

Chapters on "existence and uniqueness" are usually viewed as absolutely essential and central by mathematicians, and absolutely useless or meaningless by everyone else. Students in particular have traditionally considered these chapters as a prime example of a mathematician talking to himself.

But we shall demonstrate the *need* for these concepts (in Sections 4.1 and 4.2) and develop useful *conditions* (in Sections 4.3 and 4.4) that will guarantee existence and uniqueness (in Section 4.5).

Then in Section 4.6 we use these results to *bound* the error $E(h)$ arising in numerical approximations of solutions, as discussed in Section 3.3. And finally in Section 4.7 we can prove general versions of the *fence, funnel,* and *antifunnel theorems* introduced in Sections 1.3 and 1.4.

4.1 Existence: Approximation and Error Estimate

Existence theorems fall in two classes, constructive and non-constructive. People wishing to use mathematics are probably right to disdain the non-constructive proofs; they correctly think that proving that something exists but giving no way of finding it will not help them much.

Constructive existence proofs are quite a different matter. They almost always consist of two parts: an *approximation procedure* and an *error estimate*. In other words,

> *constructive existence proofs give you a recipe for finding an approximation to what you want, and a formula for estimating how good an approximation it is.*

The existence is then proved by taking a sequence of better and better approximations, and showing that they converge using the error estimate.

Historically, the approximation procedure was mainly a theoretical tool. In most cases, carrying out an approximation involved an amount of computation that could only be handled by hand if the result were of sufficiently great interest to warrant the time and tedium, and in particular could not be carried out in a classroom. As a result, teachers were often careless about emphasizing the actual computability of their approximations, and the students then naturally felt that the entire subject was a prime example of mathematicians being impractical.

However in this day of easy computer access, a *constructive proof of existence and uniqueness* can be provided that will be useful as well as instructional.

In Chapter 3 we introduced the approximation procedures, which lead to existence. We finish in Section 4.4 with the error estimate, in the form of the *Fundamental Inequality,* which will immediately give uniqueness as well.

However, we should mention here that numerical accuracy of approximation is not the only problem to watch out for. Numerical methods are an *iterative* process, which will be discussed at length in Chapter 5. The pictures which result from numerical approximation do indeed demonstrate existence, but they can be misleading.

Peek ahead to Figure 5.4.6 for our favorite equation $x' = x^2 - t$. A slight increase in stepsize h for the approximate solutions that are drawn has led to *spurious* solutions. This drawing emphasizes the need for the fence, funnel, and antifunnel theorems to be given in Section 4.7. With these theorems we can *prove* which portions of the pictures are correct (the funnel and antifunnel on bottom and top respectively of the parabola $x^2 = t$) and recognize whether an apparent solution is in fact spurious (the seeming funnel above the lower curve) or incorrect (the jagged ones, or those that drop down below the parabola). Later, in Chapter 5, we'll discuss how to control the stepsize to avoid misleading pictures, but what you must remember (now and always) is that an apparent feature must be *proved* to be there before you can be sure.

4.2   Uniqueness: The Leaking Bucket Versus Radioactive Decay

You might take the attitude: "O.K., we can see on the computer screen that differential equations have solutions, determined by the initial conditions. So I believe in existence—let's get on to something practical."

It isn't quite that simple: for some equations, even some of practical importance, *uniqueness* does not hold. That is, there may be more than one solution for a given initial condition. Here we want to discuss this situation. What does non-uniqueness mean? How can you tell when this will be a problem?

A good example is the leaky bucket (Examples 4.2.1, 3 and 4.3.6), as contrasted with radioactive decay (Examples 4.2.2, 4 and 4.3.5).

Example 4.2.1. *Leaky bucket.* Consider a bucket with a hole in the bottom, as in Figure 4.2.1.

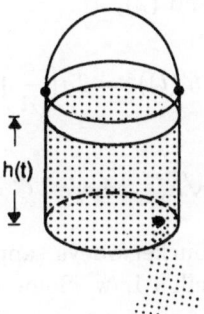

FIGURE 4.2.1. Leaky bucket.

If at a given time you see the bucket empty, can you figure out when (if ever) it was full? No, you (obviously) cannot!

We shall see (Example 4.2.3) that there simply is not a unique solution to the differential equation that states this problem. In order to find out *why* we have non-uniqueness and what is special about $f(t, x)$ in this example, we will develop its differential equation.

To study the rate at which the water level drops, we need a physical assumption about the velocity with which the water leaves the hole. Physically the following model may be an oversimplification for the situation in question, but it certainly is reasonable to assume that the velocity $v(t)$ will depend on the height $h(t)$ of the water remaining in the bucket at time t. After all, the water will flow faster when the bucket is full than when it is nearly empty (the greater depth of water exerts more pressure to push water out of the hole).

The volume of water leaving the hole equals the volume of water lost in the bucket, so

$$av(t) = Ah'(t), \tag{1}$$

where a = the cross-section of the hole and A = the cross-section of the bucket.

Furthermore, if one assumes no energy loss, then the potential energy lost at the top when a small amount of water has left the bucket must equal the kinetic energy of an equal amount of water leaving the bottom of the bucket through the hole. That is, if ρ is the density of the water, the mass of water in question is $(\Delta h)A\rho$, and

$$[(\Delta h)A\rho]gh = \left(\frac{1}{2}\right)[(\Delta h)A\rho]v^2,$$

so

$$v^2 = 2gh. \tag{2}$$

Combining equations (1) and (2),

$$(h'(t))^2 = 2g\left(\frac{a}{A}\right)^2 h(t)$$

or

$$h' = -C\sqrt{h}, \quad \text{where} \quad C = \sqrt{2g}\left(\frac{a}{A}\right). \tag{3}$$

The level of water in the bucket obeys (approximately) this equation (3), cited in physics as Torricelli's Law. Some specific examples are given in Exercises 4.1–4.2#2,3.

We shall study the differential equation (3) as $x' = -C\sqrt{x}$, for $0 \leq x \leq 1$, where $x = 0$ corresponds to an empty bucket and $x = 1$ to a full bucket.

By separation of variables,

$$\int \frac{dx}{\sqrt{x}} = -C \int dt,$$

and the solution with $u(0) = 1$ (full bucket) is

$$u(t) = \begin{cases} \frac{C^2}{4}(t - t_e)^2 & \text{for } 0 \leq t \leq t_e \\ 0 & \text{for } t > t_e, \end{cases} \tag{4}$$

where $t_e = 2/C$, the time required for the bucket to go from full to empty. You get to verify this solution in Exercises 4.1–4.2#1, where you will also verify that it is indeed differentiable—even at t_e—and that it does satisfy the differential equation. The graph of this full bucket solution occurs in Figure 4.2.2, the rightmost solution.

This solution for a *full* bucket initial condition $u(0) = 1$ uniquely determines the height $u(t)$ of the water at any time $t \geq 0$. The problem of non-uniqueness arises when we look *backwards* from an *empty* bucket initial condition.

For example, an *empty* bucket initial condition $u(t_e) = 0$ could have resulted from a full bucket at *any* $t \leq 0$. Figure 4.2.2 shows some solutions for this case. The highlighted curve corresponds to the full bucket solution (4) with $u(0) = 1$. The empty bucket solutions for $u(t_e) = 0$ are all horizontal translates (to the left) of that full bucket solution that reaches the t-axis at $t = t_e$. ▲

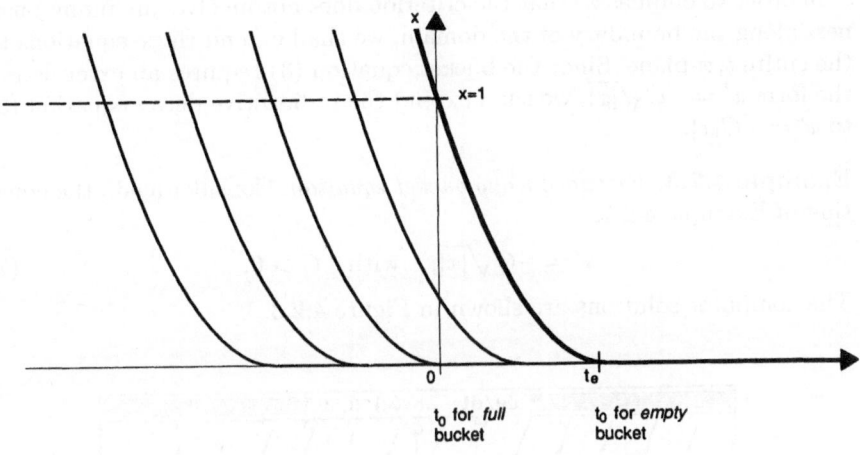

FIGURE 4.2.2. $x' = -C\sqrt{x}$.

Contrast the leaky bucket equation, $x' = -C\sqrt{x}$ of Example 4.2.1, with the deceptively similar equation $x' = -Cx$ of Example 4.2.2.

Example 4.2.2. *Radioactive decay.* Consider the disintegration of a radioactive substance over time, as discussed in Section 2.5. The differential equation is

$$x' = -Cx, \quad \text{with} \quad C > 0, \tag{5}$$

and initial condition $u(t_0) = x_0$; the solution is

$$u(t) = x_0 e^{-C(t-t_0)}. \tag{6}$$

Uniqueness holds for the solutions (6) to this differential equation (5), and working backwards from *any* initial condition is no problem, at least mathematically. For instance, for $t < 1,000,000$ years, then *carbon dating*, as discussed in Example 2.5.3, can determine, from the ratio of x/x_0 in a given sample of dead organic matter, exactly when that organism died. (From the physical point of view, for large t the right-hand side of (6) becomes zero within the limits of experimental error; this difficulty is explored in Exercise 4.1–4.2#4.) ▲

So, from current data the radioactive decay equation can tell you exactly what happened a given number of years ago, but the bucket equation cannot. We need a mathematical criterion to tell why these differential equations behave in such different ways. We shall prepare for this criterion by looking more closely at $f(t, x)$ for both differential equations in Examples 4.2.1 and 4.2.2.

In order to emphasize that the criterion does not involve any funny business along the boundary of the domain, we shall extend these equations to the entire t, x-plane. Since the bucket equation (3) requires an extension of the form $x' = -C\sqrt{|x|}$, we shall extend the radioactive decay equation (5) to $x' = -C|x|$.

Example 4.2.3. *Extended leaky bucket equation.* Consider again the equation of Example 4.2.1,

$$x' = -C\sqrt{|x|}, \quad \text{with} \quad C > 0. \tag{7}$$

The computer solutions are shown in Figure 4.2.3.

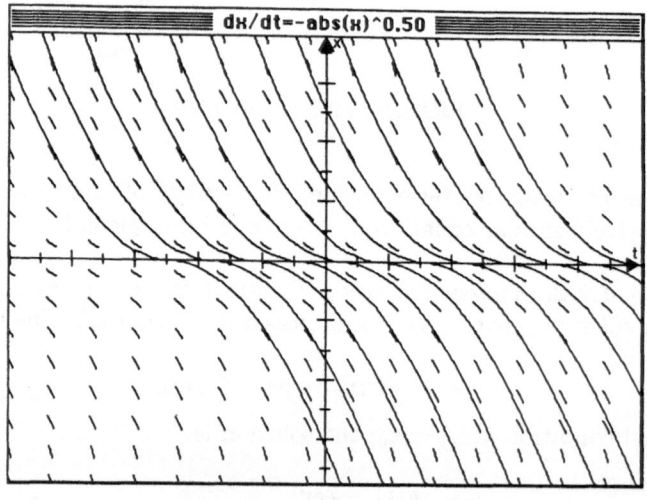

FIGURE 4.2.3. $x' = -C\sqrt{|x|}$.

Analytically, one solution to equation (7) is $u(t) = 0$, and from separation of variables (first for $x > 0$ and then for $x < 0$) we get the following general solution:

$$u(t) = \begin{cases} \dfrac{C^2}{4}(t - t_1)^2 & \text{for } t \le t_1 \\[2mm] 0 & \text{for } t_1 \le t \le t \\[2mm] -\dfrac{C^2}{4}(t - t_2)^2 & \text{for } t \ge t_2 \end{cases}$$

where t_1 and t_2 are the points where a solution meets and leaves (respectively, from left to right), the t-axis, as shown in Figure 4.2.4, and where possibly $t_1 = -\infty$ and/or $t_2 = \infty$.

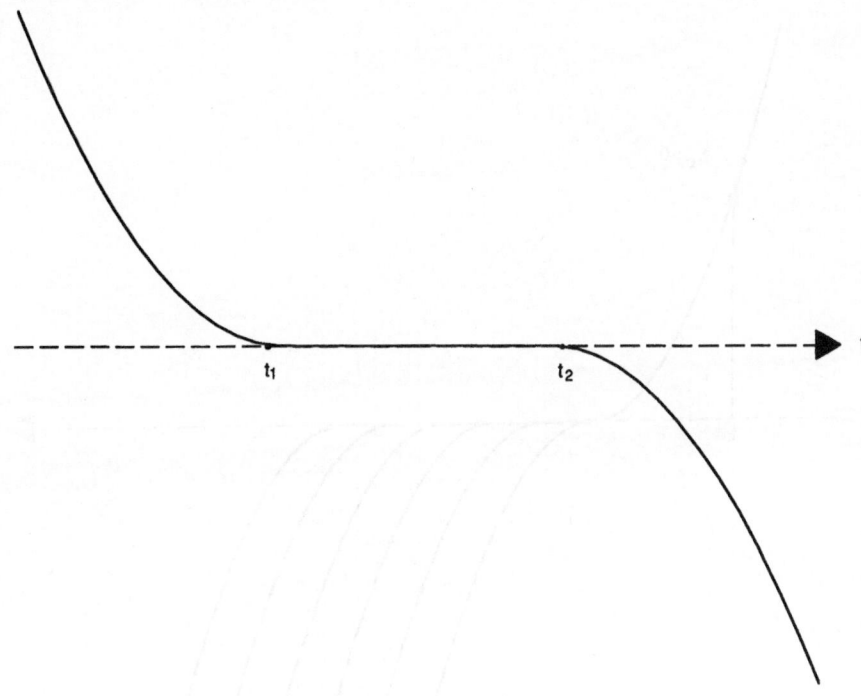

FIGURE 4.2.4. One solution for $x' = -C\sqrt{|x|}$.

If $x_0 \neq 0$, then for any initial condition (t_0, x_0),

$$t_1 = t_0 + (2/C)\sqrt{x_0} \qquad \text{(for } x_0 > 0\text{)},$$

$$t_2 = t_0 - (2/C)\sqrt{-x_0} \qquad \text{(for } x_0 < 0\text{)}.$$

Note that in Figure 4.2.3 and 4.2.4 the solutions for $x > 0$ and for $x < 0$ each show only half the parabola that you might have expected from the formula for the solutions—the half with non-positive slope, as required by the differential equation (7).

The big point illustrated in Figure 4.2.3 is that *all* the solutions to (7) meet or cross the t-axis, which is also a solution, illustrating graphically that there is *no uniqueness* in any region including the t-axis. For example, Figure 4.2.5 shows some of the infinite number of solutions for a given initial condition with x_0 above the axis—there is uniqueness only until the solution meets the axis.

Furthermore, the t-axis is a *weak* lower fence $\alpha(t)$, because it is true that

$$x' = f(t, 0) = 0 = \alpha'(t).$$

This illustrates the failure of the Fence Theorem 1.3.5 for weak fences, because solutions can slip right through, as in Figure 4.2.5.

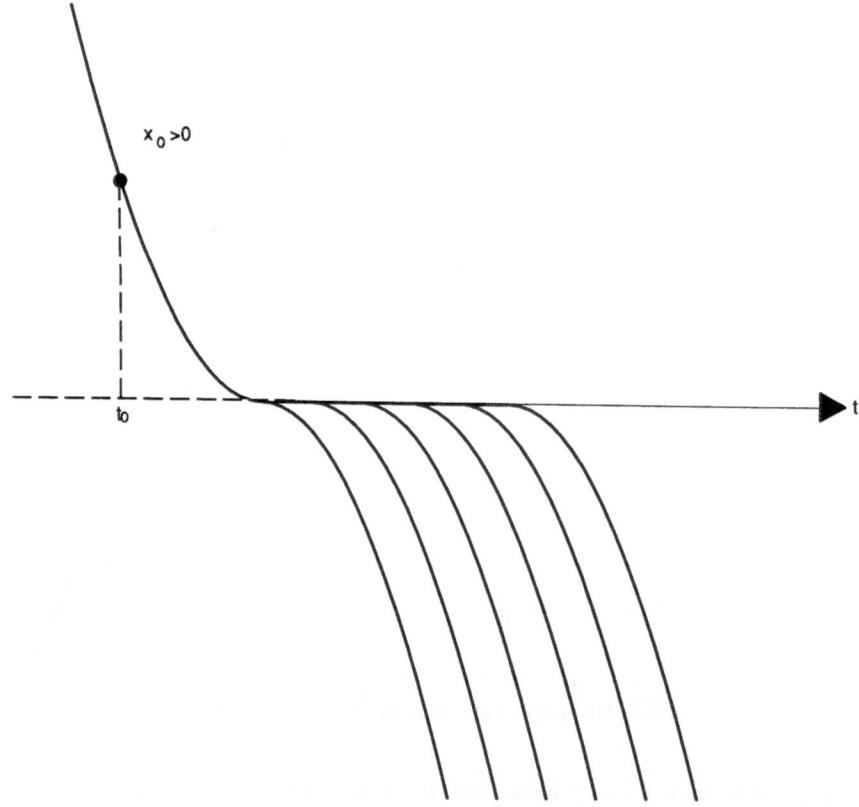

FIGURE 4.2.5. Many solutions for $x' = -C\sqrt{|x|}$, starting from same (t_0, x_0).

We shall return to this example in Sections 4.3 and 4.7, where a single reason for these difficulties will become apparent. ▲

Example 4.2.4. *Extended radioactive decay equation.* Consider again the equation of Example 4.2.2,

$$x' = -C|x|, \quad \text{with } C > 0. \tag{8}$$

Again, $u(t) = 0$ is a solution, and from separation of variables (first for $x > 0$ and then for $x < 0$) we get the following general solution:

$$u(t) = \begin{cases} x_0 e^{-C(t-t_0)} & \text{for } x_0 > 0 \\ 0 & \text{for } x_0 = 0 \\ x_0 e^{C(t-t_0)} & \text{for } x_0 < 0. \end{cases}$$

The solutions to the extended radioactive decay equation (8) are shown in Figure 4.2.6.

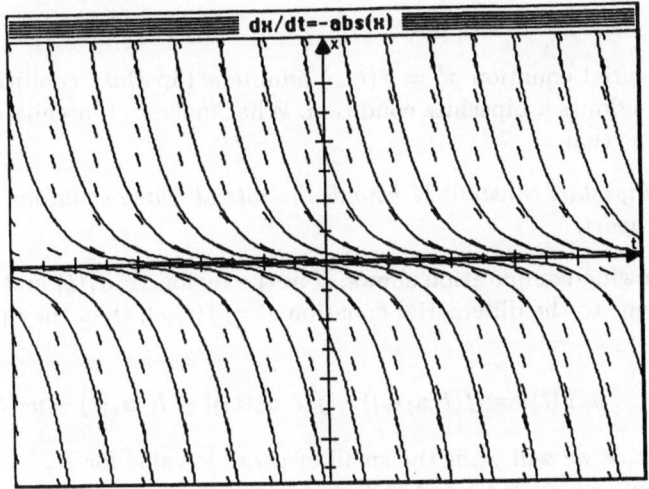

FIGURE 4.2.6. $x' = -C|x|$.

The difference between this example and the last is that this time we do have *uniqueness;* although $u(t) = 0$ is a solution to (8), the other solutions never meet or cross it. (You may not *see* that in the picture, but look at the formulas in the analytic solution!) ▲

We now have an example (4.2.1 and 3) with non-uniqueness, and another (4.2.2 and 4) with uniqueness. We shall proceed in the next section to examine the key to the difference.

4.3   The Lipschitz Condition

What makes the difference between the leaky bucket and radioactive decay examples is that Example 4.2.2 admits a *Lipschitz constant* while Example 4.2.1 does not. This quantity is named for Rudolf Otto Sigismund Lipschitz, a German mathematician who in the mid-19th century simplified and clarified Augustin-Louis Cauchy's original work, earlier in the same century, on existence and uniqueness of solutions to differential equations. Lipschitz' work provides a more general condition for existence and uniqueness than the requirement that $\partial f / \partial x$ be continuous, as quoted in most elementary texts.

Definition 4.3.1. A number K is a *Lipschitz constant* with respect to x for a function $f(t, x)$ defined on a region A of $\mathbb{R}^2$ (the t, x-plane) if

$$|f(t, x_1) - f(t, x_2)| \le K|x_1 - x_2|, \tag{9}$$

for all (t, x_1), (t, x_2) in A. We call this inequality a *Lipschitz condition* in x.

A differential equation $x' = f(t, x)$ admits a Lipschitz condition if the function f admits a Lipschitz condition. What makes a Lipschitz condition important is that

> *the Lipschitz constant K bounds the rate at which solutions can pull apart,*

as the following computation shows. If in the region A, $u_1(t)$ and $u_2(t)$ are two solutions to the differential equation $x' = f(t, x)$, then they pull apart at a rate

$$|(u_1 - u_2)'(t)| = |f(t, u_1(t)) - f(t, u_2(t))| \le K|u_1(t) - u_2(t)|.$$

So in practice we will want the smallest possible value for K.

As we will see, such a number K also controls for numerical solutions of a differential equation the rate at which errors compound.

The *existence* of a Lipschitz condition is often very easy to ascertain, if the function in question is *continuously differentiable*.

Theorem 4.3.2. *If on a rectangle $R = [a, b] \times [c, d]$ a function $f(t, x)$ is differentiable in x with continuous derivative $\partial f / \partial x$, then $f(t, x)$ satisfies a Lipschitz condition in x with best possible Lipschitz constant equal to the maximum value of $|\partial f / \partial x|$ achieved in R. That is,*

$$K \equiv \sup_{(t,x) \in R} \left| \frac{\partial f}{\partial x} \right|.$$

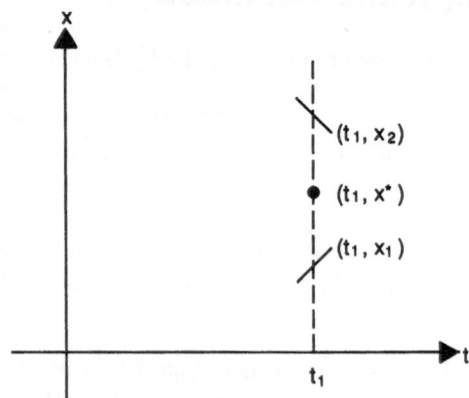

FIGURE 4.3.1.

Proof. If throughout R the function $f(t, x)$ is continuous and differentiable in x, the Mean Value Theorem says that

$$f(t, x_1) - f(t, x_2) = \left.\frac{\partial f}{\partial x}\right|_{(t, x^*)} (x_1 - x_2),$$

where $x_1 < x^* < x_2$, as shown in Figure 4.3.1.

Furthermore, since R is closed and bounded, if $\partial f / \partial x$ is continuous, the number $K = \sup |\partial f / \partial x|$ is finite. $\quad\square$

Example 4.3.3. The differential equation $x' = ax + b$ has Lipschitz constant $K = |a|$ everywhere. $\quad\blacktriangle$

Example 4.3.4. Consider $x' = \sin tx$ for $0 \le t \le 3$, $-5 \le x \le 5$. Since

$$\frac{\partial}{\partial x}(\sin tx) = t \cos tx, \quad \sup\left|\frac{\partial}{\partial x}(\sin tx)\right| \le |t| \le 3.$$

This value is realized at $t = 3$ and $\cos 3x = 1$, or $x = 0$, so 3 is the best Lipschitz constant on this rectangle. $\quad\blacktriangle$

The equation $x' = ax + b$ of Example 4.3.3 is *the* archetypal example. We will use the Fundamental Inequality in Section 4.4 to compare all other differential equations against this one because we know its explicit solution *and* its Lipschitz constant.

We shall see in the rest of this chapter that the Lipschitz condition is the key hypothesis for theorems giving uniqueness, fences and funnels, and the Fundamental Inequality. Since the Lipschitz condition is what is useful in proofs, we have isolated that rather than differentiability as the criterion for uniqueness.

Usually our functions will satisfy a Lipschitz condition because, by Theorem 4.3.2, they are continuously differentiable with respect to x. However $f(t, x)$ *may* satisfy a Lipschitz condition even if $f(t, x)$ is *not differentiable* in x, as the following examples show. Satisfying a Lipschitz condition simply means satisfying equation (9), and that equation by itself does not require a differentiable function.

Our familiar Examples 4.2.2,4 and 4.2.1,3 will give us a good look at the role of the Lipschitz condition.

Example 4.3.5. *Extended radioactive decay equation,* continued.
For $x' = -C|x|$,

$$\frac{\partial f}{\partial x} = \begin{cases} -C & \text{for } x > 0 \\ \text{not defined} & \text{for } x = 0 \\ C & \text{for } x < 0. \end{cases}$$

The function $f(t,x) = -C|x|$ is not differentiable in x at $x = 0$, but as $x \to 0$, $|\partial f/\partial x| = C$, as shown in Figure 4.3.2 (which is a graph of $f(t,x)$ versus x).

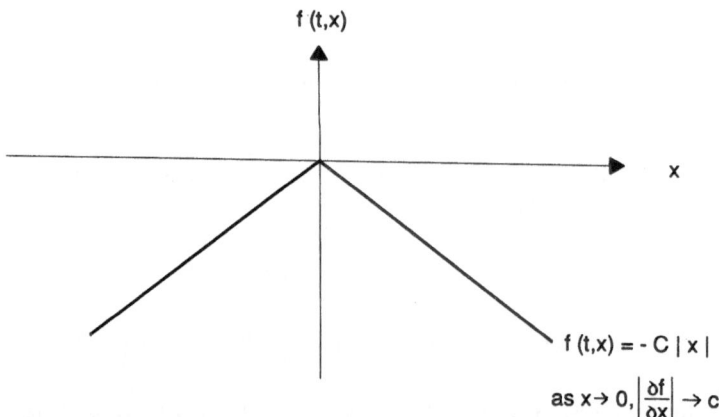

FIGURE 4.3.2. f versus x for $x' = -C|x| = f(t,x)$.

However, the following statement *is* true for $f(t,x) = -C|x|$:

$$|f(t,x_1) - f(t,x_2)| = C|(-|x_1| + |x_2|)| \le C|x_1 - x_2|.$$

Hence even though this $f(t,x)$ is not differentiable at $x = 0$, it nevertheless *does* satisfy a Lipschitz condition in the whole t, x-plane, with Lipschitz constant $K = C$.   ▲

Example 4.3.6. *Extended leaky bucket equation*, continued.
For $x' = -C\sqrt{|x|}$,

$$\frac{\partial f}{\partial x} = \begin{cases} 1/(2\sqrt{x}) & \text{for } x > 0 \\ \text{not defined} & \text{for } x = 0 \\ -1/(2\sqrt{|x|}) & \text{for } x < 0. \end{cases}$$

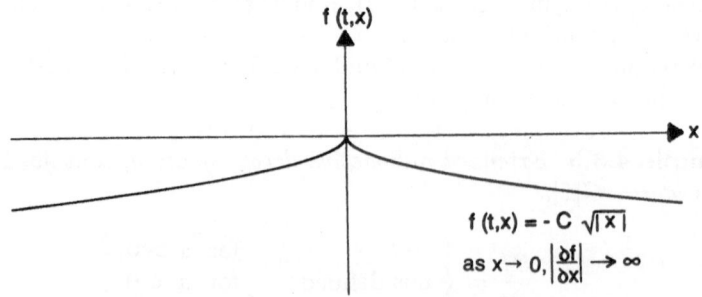

FIGURE 4.3.3. f versus x for $x' = -C\sqrt{x} = f(t,x)$.

The function $f(t, x) = -C\sqrt{|x|}$ is not differentiable in x at $x = 0$, but it also, as we shall see, does not satisfy a Lipschitz condition in any region which contains a point $(t, 0)$. This is because as $x \to 0$, $|\partial f/\partial x| \to \infty$, as shown in Figure 4.3.3 (another graph in $f(t, x)$ versus x).

Therefore there can be *no* finite Lipschitz constant K. ▲

Thus we now have the essential difference between the leaky bucket situation of Example 4.2.1 and the radioactive decay situation of Example 4.2.2. In the first case, there is *no* Lipschitz condition in any region A including the t-axis, and there is *no* uniqueness; in the second case, there *is* a Lipschitz condition throughout the entire plane, and there *is* uniqueness. We shall actually prove the theorem in Section 4.5.

Because of Theorem 4.3.2, in our Examples 4.3.6 and 4.3.5 we only needed to make actual calculations for K around $x = 0$, where $\partial f/\partial x$ is not differentiable. Throughout the rest of the t, x-plane in both cases, a *local* Lipschitz condition holds.

Definition 4.3.7. A function $f(t, x)$ defined on an *open* subset U of $\mathbb{R}^2$ is *locally Lipschitz* if about every point there exists a neighborhood on which f is Lipschitz.

For a function f, where the partial derivative $\partial f/\partial x$ is *continuous*, then f is locally Lipschitz, as a consequence of Theorem 4.3.2.

We shall show subsequently in Theorem 4.5.1 that the existence of a local Lipschitz condition is sufficient to assure uniqueness. Example 4.2.2 in the previous section, shows clearly that without a Lipschitz condition, you may *not* have uniqueness, as this example does not.

We shall proceed in the remainder of this chapter to show how the Lipschitz condition is used in proving the important theorems.

4.4   The Fundamental Inequality

The inequality that we shall state in this section and prove in the next contains most of the general theory of differential equations. We will show that two functions u_1 and u_2 which both approximately solve the equation, and which have approximately the same value at some t_0, are close.

Let $R = [a, b] \times [c, d]$ be a rectangle in the t, x-plane, as shown in Figure 4.4.1. Consider the differential equation $x' = f(t, x)$, where f is a continuous function satisfying a Lipschitz condition with respect to x in R. That is,

$$|f(t, x_1) - f(t, x_2)| \leq K|x_1 - x_2|$$

for all $t \in [a, b]$ and $x_1, x_2 \in [c, d]$.

Suppose $u_1(t)$ and $u_2(t)$ are two *piecewise differentiable functions,* the graphs of which lie in R and which are approximate solutions to the differential equation in the sense that for nonnegative numbers ε_1 and ε_2,

$$|u_1'(t) - f(t, u_1(t))| \le \varepsilon_1, \quad \text{and} \quad |u_2'(t) - f(t, u_2(t))| \le \varepsilon_2,$$

for all $t \in [a, b]$.

Suppose furthermore that $u_1(t)$ and $u_2(t)$ have approximately the same value at some $t_0 \in [a, b]$. That is, for some nonnegative number δ,

$$|u_1(t_0) - u_2(t_0)| \le \delta.$$

Figure 4.4.1 shows how all these conditions relate to the slope field for $x' = f(t, x)$.

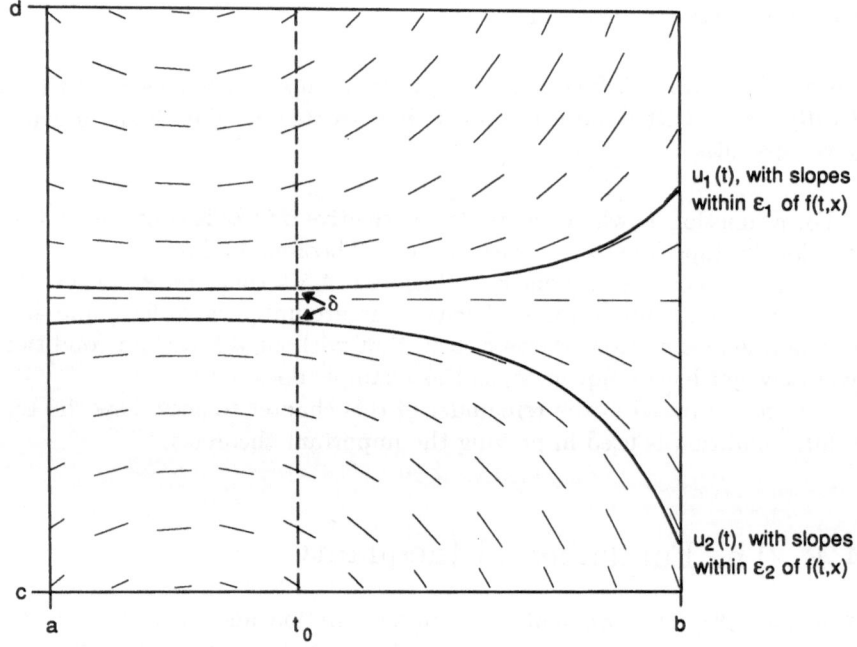

FIGURE 4.4.1. Slope field for $x' = f(t, x)$.

The diagram of Figure 4.4.1 is not the most general possible, which could include approximate solutions that cross and/or are piecewise differentiable, but it gives the clearest picture for our current purpose. In particular, this diagram sums up all the hypotheses for the following major result.

Theorem 4.4.1 (Fundamental Inequality). *If, on a rectangle $R = [a, b] \times [c, d]$, the differential equation $x' = f(t, x)$ satisfies a Lipschitz condition with respect to x, with Lipschitz constant $K \neq 0$, and if $u_1(t)$ and $u_2(t)$ are two continuous, piecewise differentiable, functions satisfying*

$$|u_1'(t) - f(t, u_1(t))| \leq \varepsilon_1$$
$$|u_2'(t) - f(t, u_2(t))| \leq \varepsilon_2$$

for all $t \in [a, b]$ at which $u_1(t)$ and $u_2(t)$ are differentiable; and if for some $t_0 \in [a, b]$

$$|u_1(t_0) - u_2(t_0)| \leq \delta;$$

then for all $t \in [a, b]$,

$$|u_1(t) - u_2(t)| \leq \delta e^{K|t-t_0|} + \left(\frac{\varepsilon}{K}\right)(e^{K|t-t_0|} - 1), \tag{10}$$

where $\varepsilon = \varepsilon_1 + \varepsilon_2$.

Before we proceed to the proof, let us study the result. The boxed formula (10) is the *Fundamental Inequality,* and it certainly looks formidable. Nevertheless, looking at it carefully will make it seem friendlier. First, notice that it does say the u_1 and u_2 remain close, or at least it gives a bound on $|u_1 - u_2|$, over the entire interval $[a, b]$, not just at t_0.

Second, notice that the bound is the sum of two terms, one having to do with δ and the other with ε. You can think of the first as the contribution of the difference or error *in the initial conditions,* and the second error *in solving the equation numerically.*

Both terms have an exponential with a K in the exponent, this means that *the bounds on all errors grow exponentially* in time, with the Lipschitz constant K controlling the growth rate. Note that this exponential growth with respect to t is in contrast to the error discussion in Chapter 3 for fixed t; there error functions are polynomials in h.

From some points of view, the Fundamental Inequality (10) is a wonderful tool. First, it gives *existence* and *uniqueness* of solutions to differential equations, as we shall prove in Section 4.5.

Second, even if a differential equation itself is only known approximately, the Fundamental Inequality tells us that an approximate solution to an approximate equation is an approximate solution to the real equation. This happens, for instance, in mathematical modeling when the coefficients that appear in a differential equation describing some situation are measured experimentally and as such are never known exactly. See Exercise 4.4#6.

And third, the Fundamental Inequality will hold even if a real differential equation is stochastic. This is the situation where the differential equation that is written down corresponds to an ideal system, but in reality the system is constantly disturbed by random (or at least unknown) noise. Essentially all real systems are of this type.

On the other hand, although the Fundamental Inequality gives all the wonderful results cited above, it is quite discouraging. Exponential growth is very rapid, which means we should not trust numerical approximations over long periods of time. The fact that the Fundamental Inequality correctly describes the behavior of differential equations in bad cases is a large part of why a theory of such equations is necessary: *numerical methods are inherently untrustworthy, and anything that you might guess using them, especially if it concerns long term behavior, must be checked by other methods.*

PROVING THE FUNDAMENTAL INEQUALITY

The Fundamental Inequality makes a statement about the error approximation $E(t) = |u_1(t) - u_2(t)|$ for all t, given its value at t_0. So we have

$$E(t_0) \leq \delta,$$

and we want to prove that for all t,

$$E(t) \leq \delta e^{K|t-t_0|} + \left(\frac{\varepsilon}{K}\right)\left(e^{K|t-t_0|} - 1\right). \qquad (10, \text{ again})$$

This is reminiscent of a fence situation, because we know that at t_0, $E(t) \leq$ right-hand side of (10), and we want to prove that $E(t)$ *stays* below the right-hand expression for all $t \geq t_0$, and then (which must be proved separately) for all $t \leq t_0$. So indeed we shall use a fence to prove it!

Proof. Any continuous piecewise differentiable function $u(t)$ can be approximated, together with its derivative, by a piecewise linear function $v(t)$, simply by replacing its graph with line segments between a sequence of points along the graph, and being sure to include among that sequence of points all those points of discontinuity of the derivative $u'(t)$, as in Figure 4.4.2.

FIGURE 4.4.2. Piecewise linear approximation $v(t)$.

So, for any pair of approximate solutions $u_1(t)$ and $u_2(t)$ and any positive number η, there exist piecewise linear functions $v_1(t)$ and $v_2(t)$ such that

$$|v_i(t) - u_i(t)| < \eta$$

and
$$|v_i'(t) - u_i'(t)| < \eta$$
wherever both $u_i(t)$ and $v_i(t)$ are differentiable. Let
$$\gamma(t) \equiv |v_1(t) - v_2(t)|;$$
the function $\gamma(t)$ is continuous and *piecewise linear*.

Then
$$
\begin{aligned}
\gamma'(t) &\le |v_1'(t) - v_2'(t)| \\
&< |u_1'(t) - u_2'(t)| + 2\eta \\
&\le |f(t, u_1(t)) - f(t, u_2(t))| + \varepsilon_1 + \varepsilon_2 + 2\eta \\
&\le K|u_1(t) - u_2(t)| + \varepsilon_1 + \varepsilon_2 + 2\eta \\
&\le K\{|v_1(t) - v_2(t)| + 2\eta\} + \varepsilon_1 + \varepsilon_2 + 2\eta \\
&\le K\gamma(t) + \varepsilon_\eta,
\end{aligned}
$$

where $\varepsilon_\eta \equiv \varepsilon_1 + \varepsilon_2 + 2\eta(1 + K)$.

You should justify these steps as Exercise 4.4#1; each step is important, but not quite obvious. The first inequality is true with left-hand derivatives and right-hand derivatives at points where they differ. The result of this sequence of steps,
$$\gamma'(t) < K\gamma(t) + \varepsilon_\eta,$$
says that the piecewise linear curve $\gamma(t)$ is a strong lower fence for *another* differential equation,

$$x' = Kx + \varepsilon_\eta, \text{ with solution } x = \omega(t). \tag{11}$$

We can apply the Fence Theorem 1.3.5 for $t \ge t_0$, and for $K \ne 0$, we have solved equation (11) as a linear equation in Example 2.3.1

$$w(t) = x_0 e^{K(t-t_0)} + \left(\frac{\varepsilon_\eta}{K}\right)\left(e^{K(t-t_0)} - 1\right).$$

If we reverse time and solve for the case where $t \le t_0$, we can replace this equation by

$$w(t) = x_0 e^{K|t-t_0|} + \left(\frac{\varepsilon_\eta}{K}\right)\left(e^{K|t-t_0|} - 1\right).$$

Because we have $\gamma(t_0) \le \delta_\eta = \delta + 2\eta$, we know that $\gamma(t)$ will remain below the solution $w(t)$ with $w(t_0) = x_0 = \delta_\eta$, so

$$|v_1(t) - v_2(t)| \le \delta_\eta e^{K|t-t_0|} + \left(\frac{\varepsilon_\eta}{K}\right)\left(e^{K|t-t_0|} - 1\right),$$

and

$$|u_1(t) - u_2(t)| \le \delta_\eta e^{K|t-t_0|} + \left(\frac{\varepsilon_\eta}{K}\right)\left(e^{K|t-t_0|} - 1\right) + 2\eta. \tag{12}$$

Equation (12) holds for any $\eta > 0$, so in the limit as $\eta \to 0$, which implies that $\varepsilon_\eta \to \varepsilon$ and $\delta_\eta \to \delta$, this is the desired conclusion of the Fundamental Inequality.

Solving $x' = Kx + \varepsilon$ for the case where $K = 0$ and reapplying the Fence Theorem 1.3.5, is left to the reader in Exercise 4.4#2.   □

4.5   Existence and Uniqueness

The Fundamental Inequality

$$|u_1(t) - u_2(t)| \leq \delta e^{K|t-t_0|} + \left(\frac{\varepsilon}{K}\right)\left(e^{K|t-t_0|} - 1\right), \qquad \text{(10, again)}$$

gives *uniqueness* of solutions, the fact that on a direction field wherever uniqueness holds two solutions cannot cross.

Theorem 4.5.1 (Uniqueness). *Consider the differential equation $x' = f(t, x)$, where f is a function satisfying a Lipschitz condition with respect to x on a rectangle $R = [a, b] \times [c, d]$ in the t, x-plane. Then for any given initial condition (t_0, x_0), if there exists a solution, there is exactly one solution $u(t)$ with $u(t_0) = x_0$.*

Proof. Apply the Fundamental Inequality of Theorem 4.4.1. If two different solutions passed through the same point, their δ would be zero. But their ε's would also be zero because they were both actual solutions. Therefore, the difference between such solutions would have to be zero.   □

More generally, if the function $f(t, x)$ in Theorem 4.5.1 is *locally Lipschitz*, then *locally* $x' = f(t, x)$ will have uniqueness. E.g., the leaky bucket equation of Example 4.2.1 *will* have uniqueness of solutions for any (t, x) with $x \neq 0$.

Please note, however, that lack of a Lipschitz condition does *not* necessarily mean that we have no uniqueness of solutions. Good examples are $x' = \sqrt{|x - t|}$ and $x' = \sqrt{|x|} + 1$ (Exercises 4.5#1,2).

The Fundamental Inequality also will give, in Theorems 4.5.5 and 4.5.6, *existence* of solutions to differential equations, as well as the fact that our approximation schemes converge to solutions.

You should observe that the statement of the Fundamental Inequality makes *no* reference to actual *solutions* of differential equations, that is, to continuous functions $u(t)$ that satisfy $x' = f(t, x)$. Rather it refers to functions $u_h(t)$ that *approximately* solve the differential equation in the sense that the *slope error*, $|u_h'(t) - f(t, u_h(t))|$, is small.

It is reasonable to hope that such functions $u_h(t)$ approximate solutions, and this is true under appropriate circumstances. But to make such a statement meaningful, we would need to know that solutions exist. Instead, we intend to use the Fundamental Inequality to *prove* existence of solutions, so we must use a different approach, by means of the following three theorems.

Theorem 4.5.2 (Bound on slope error, Euler's method). *Consider the differential equation $x' = f(t, x)$, where f is a continuous function on a rectangle $R = [a, b] \times [c, d]$ in the t, x-plane. Let u_h be the Euler approximate solution with step h. Then*

(i) *for every h there is an ε_h such that u_h satisfies*

$$|u'_h(t) - f(t, u_h(t))| \le \varepsilon_h$$

at any point where u_h is differentiable (and the inequality holds for left- and right-hand derivatives elsewhere);

(ii) *$\varepsilon_h \to 0$ as $h \to 0$;*

(iii) *if furthermore f is a function on R with continuous derivatives with respect to x and t, with the following bounds over R:*

$$\sup |f| \le M; \quad \sup \left| \frac{\partial f}{\partial t} \right| \le P; \quad \sup \left| \frac{\partial f}{\partial x} \right| \le K,$$

then there is a specific bound on ε_h:

$$|u'_h(t) - f(t, u_h(t))| \le h(P + KM).$$

Proof. Parts (i) and (ii) are not difficult to prove, using the concept of uniform continuity, but we choose not to do so here. Such a proof is non-constructive (that is, it does not lead to a formula for computing ε_h), and therefore is not in the spirit of this book.

We proceed with the proof of Part (iii). Over a single interval of the Euler approximation, that is, for $t_i \le t \le t_{i+1}$,

$$
\begin{aligned}
|u'_h(t) - f(t, u_h(t))| &= |f(t_i, x_i) - f(t, u_h(t))| \\
&\le |f(t_i, x_i) - f(t, x_i)| + |f(t, x_i) - f(t, u_h(t))| \\
&\le |t - t_i|P + |u_h(t) - x_i|K \\
&\le hP + hMK \\
&= h(P + MK) = \varepsilon_h.
\end{aligned}
$$

The bounds P, K, M are introduced by two applications of the Mean Value Theorem. Thus we have an *explicit expression for an upper bound of* ε_h, for any differentiable u_h. $\square$

Example 4.5.3. Consider $x' = \sin tx$, first discussed in Example 1.5.2. Later in Example 3.2.1 we found at $t_f = 2$ the Euler approximate solution through $t_0 = 0$, $x_0 = 3$. Here we shall calculate ε_h, the bound on slope error appropriate for that example.

First we need a rectangle R on which to work, which is often the hardest part in calculating ε_h; that is, given an interval for t, we need to find an interval for x that will contain the solutions in question.

For this differential equation we already know that $t \in [0, 2]$, and that $x_0 = 3$; we can also see that the maximum slope is $+1$ and the minimum slope is -1; therefore we know that $u_h(t_f)$ cannot move further than 3 ± 2, so $u_h(t)$ will stay within $R = [0, 2] \times [1, 5]$. See Figure 4.5.1. We can now calculate the various bounds over $R = [0, 2] \times [1, 5]$:

$$\sup \left| \frac{\partial f}{\partial x} \right| = \sup |t \cos(tx)| \leq 2 = K$$

$$\sup \left| \frac{\partial f}{\partial t} \right| = \sup |x \cos(tx)| \leq 5 = P$$

$$\sup |f| = \sup |\sin(tx)| \leq 1 = M.$$

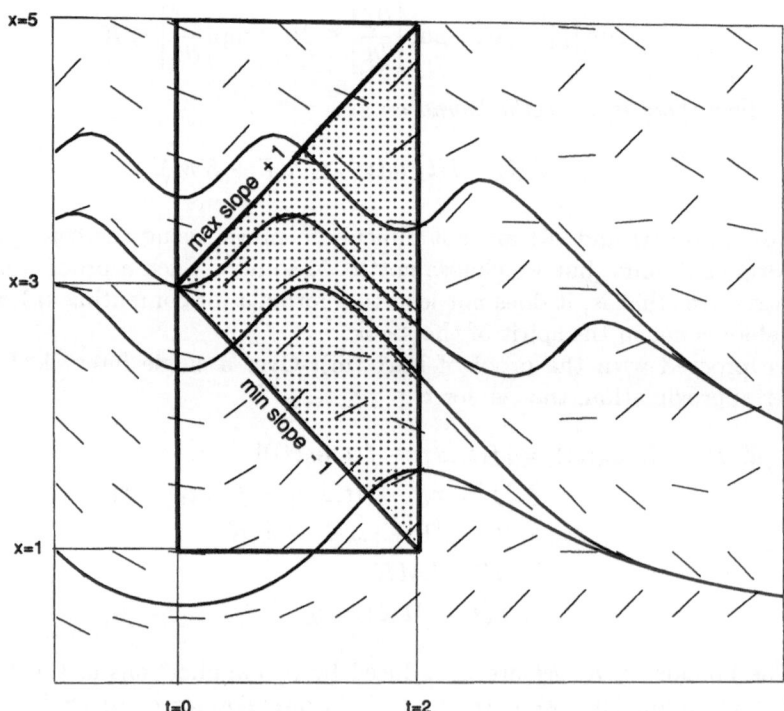

FIGURE 4.5.1. Rectangle for bounding slope error with Euler's method.

Therefore, by Theorem 4.5.2,

$$\varepsilon_h \leq h(P + MK) \leq 7h. \qquad \blacktriangle \qquad (13)$$

We can use the bound ε_h on *slope* error to bound the *actual* error $E(h)$. The Fundamental Inequality (11) says that through a given point (t_0, x_0), the actual error between the solution and an approximate solution is bounded as follows:

$$E(h) = |u(t) - u_h(t)| \leq \frac{\varepsilon_h}{K}\left(e^{K|t-t_0|} - 1\right). \qquad (14)$$

Equation (14) shows that the bounds on the actual error E depend on the slope error bound ε_h as a *factor* in an expression involving exponentials and a constant K. So the main thing to understand about numerical estimates is that:

> You cannot escape the exponential term in bounds on actual
> error $|u(t) - u_h(t)|$. All you can affect is ε_h, a bound on the
> slope error $|u'_h(t) - f(t, u_h(t))|$.

Furthermore, these formulas (13) and (14) for bounds on ε_h and $E(h)$ respectively are just that—only *bounds*, and overly pessimistic bounds at that; they are wildly overdone.

Example 4.5.4. Consider again $x' = \sin tx$, with $t_0 = 0$, $x_0 = 3$, $t_f = 2$ as in Example 4.5.3, where $K = 2$ and $\varepsilon_h = 7\,h$. Substituting these values into the estimate (14) gives

$$E(h) \leq \frac{\varepsilon_h}{K}\left(e^{K|t-t_0|} - 1\right) \leq \frac{7h}{2}\left(e^{2(2)} - 1\right) \approx 191\,h.$$

To compare this with an actual computation, we go back to Example 3.5.1 where, using the interpolated "solution" as $u(t_f)$, we find the actual constant of proportionality between $E(h)$ and h is more like 0.45 than 191.
▲

Theorem 4.5.5 (Convergence of Euler approximations). *Consider the differential equation $x' = f(t, x)$, where f is a continuous function satisfying a Lipschitz condition in x with constant k on a rectangle $R = [a, b] \times [c, d]$ in the t, x-plane. If u_h and u_k are two Euler approximate solutions, with steps h and k respectively, having graphs in R, and having the same initial conditions*

$$u_h(t_0) = u_k(t_0) = x_0,$$

then for all $t \in [a, b]$,

$$|u_h(t) - u_k(t)| \leq \frac{\varepsilon_h + \varepsilon_k}{K}\left(e^{K|t-t_0|} - 1\right),$$

where ε_h and ε_k both go to zero as h and k go to zero. Moreover, if $\partial f/\partial t$ and $\partial f/\partial x$ exist and are bounded as in Theorem 4.5.2, then,

$$|u_h(t) - u_k(t)| \le \frac{P + MK}{K}(h + k)\left(e^{K|t-t_0|} - 1\right).$$

Proof. The first part follows immediately from Theorem 4.4.1 (the Fundamental Inequality) and the first part of Theorem 4.5.2, since

$$\delta = |u_h(t_0) - u_k(t_0)| = 0,$$

because $u_h(t_0) = u_k(t_0) = x_0$.

The second part follows from the second part of Theorem 4.5.2, choosing

$$\varepsilon_h = (P + MK)h, \quad \varepsilon_k = (P + MK)k, \quad \varepsilon = \varepsilon_h + \varepsilon_k. \qquad \square$$

We now have all the ingredients for a proof of existence of solutions. Theorem 4.5.5 shows that the Euler approximations converge as the step tends to 0. Indeed, it says that once your step gets sufficiently small, the Euler approximations change very little. For instance, you can choose a sufficiently small step so that the first n digits always agree, where n is as large an integer as you like.

We have never met a student who doubted that if the Euler approximate solutions converge, they do in fact converge to a solution; in that sense the next theorem is really for the deeply skeptical, and we suggest skipping the proof in a first reading, for it is messy and not particularly illuminating. Still, if you think about it, you will see that the Euler approximate solutions with smaller and smaller step have angles at more and more points, so it is not quite obvious that the limit should be differentiable at all.

Theorem 4.5.6 (Existence of solutions). *Consider the differential equation $x' = f(t, x)$, where f is a continuously differentiable function on a rectangle $R = [a, b] \times [c, d]$ in the t, x-plane. If the Euler approximate solutions u_h, with step h and initial condition $u_h(t_0) = x_0$, have graphs which lie in R for sufficiently small h, then for all $t \in [a, b]$,*

(i) *The limit $u(t) = \lim_{h \to 0} u_h(t)$ exists;*

(ii) *The function $u(t)$ is differentiable, and is a solution of $x' = f(t, x)$.*

Proof. Part (i), the convergence, comes from Theorem 4.5.5.

For part (ii) (that convergence is to a *solution* to the differential equation, and that the result is differentiable) we need to prove a standard epsilon-delta statement, for which we shall use ε^* and δ^* to avoid confusion with our use of ε and δ in the Fundamental Inequality. This is provided by the following lemma:

Lemma. *For all t with $a < t < b$, for all $\varepsilon^* > 0$, there exists a $\delta^* > 0$ such that if $|\eta| < \delta^*$, then*

$$\left| \underbrace{\frac{u(t+\eta) - u(t)}{\eta}}_{\substack{\approx \text{slope of Euler} \\ \text{approximation}}} - \underbrace{f(t, u(t))}_{\substack{\text{slope of solution to} \\ \text{differential equation}}} \right| < \varepsilon^*.$$

Proof of Lemma. Choose $\varepsilon^* > 0$; we will show that $\delta^* = \frac{\varepsilon^*}{2(P+MK)}$ works. Since by Part (i)

$$u(t+\eta) - u(t) - \eta f(t, u(t)) = \lim_{h \to 0} [u_h(t+\eta) - u_h(t) - \eta f(t, u_h(t))],$$

we can work with u_h rather than with u. By the definition of u_h,

$$\left(\frac{du_h}{d\eta} \right)\bigg|_{(t+\eta)} = f(t_i, u_h(t_i)),$$

where, as illustrated in Figure 4.5.2, t_i is the t-coordinate of the left-hand end of the line segment that $(t+\eta, u_h(t+\eta))$ is on, at the points where u_h is differentiable. So

$$\left| \frac{d}{d\eta} [u_h(t+\eta) - u_h(t) - \eta f(t, u_h(t))] \right| = |f(t_i, u_h(t_i)) - f(t, u_h(t))|.$$

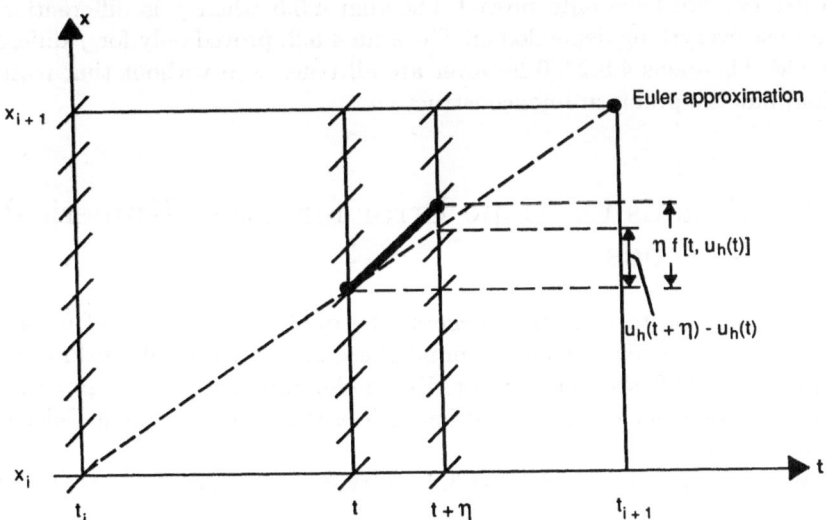

FIGURE 4.5.2. Between grid points of Euler approximation.

As in the proof of Theorem 4.5.2,

$$|f(t_i, u_h(t_i)) - f(t, u_h(t))| \leq |t - t_i|(P + MK).$$

Since if $|\eta| < \delta^*$ then $|t - t_i| < \delta^* + h$, if you take

$$\delta^* = \frac{\varepsilon^*}{2(P + MK)},$$

you find

$$\left| \frac{d}{d\eta} [u_h(t + \eta) - u_h(t) - \eta f(t, u_h(t))] \right| < \varepsilon^*.$$

at all points where u_h is differentiable, if $|\eta| < \delta^*$ and $h < \delta^*$.

Since the quantity in brackets is a continuous, piecewise differentiable function of η which vanishes when $\eta = 0$, this shows that

$$|u_h(t + \eta) - u_h(t) - \eta f(t, u_h(t))| \le \varepsilon^* |\eta|$$

if $\eta \le \delta^*$ and $h \le \delta^*$, by the Mean Value Theorem. Take the limit as h goes to zero, to get

$$|u(t + \eta) - u(t) - \eta f(t, u(t))| \le \varepsilon^* |\eta|.$$

Dividing this by η gives the desired result.   □

By proving the lemma, we have proved the theorem.   □

Remark. We have only proved Theorem 4.5.5 when f is differentiable, because everything depended on Theorem 4.5.2, proved only for f differentiable. Theorems 4.5.2,5,6 however are all true, even without that restriction, but the proofs are nonconstructive.

4.6 Bounds for Slope Error for Other Numerical Methods

Now it is time to discuss in general the qualitative aspects of errors in numerical methods, keeping in mind the quantitative results exhibited in Chapter 3. This section is devoted to understanding how the slope error bound ε_h depends on the step h for each of the Euler, midpoint Euler and Runge–Kutta methods.

We have presented in Section 3.3 pretty convincing evidence that *actual* error

$$\begin{aligned} E(h) &\approx C_E h &&\text{for Euler's method,} \\ E(h) &\approx C_M h^2 &&\text{for midpoint Euler,} \\ E(h) &\approx C_{RK} h^4 &&\text{for Runge–Kutta.} \end{aligned}$$

Now we shall discuss the fact that there exist constants B_E, B_M, and B_{RK} (different from the C_E, C_M, and C_{RK} above) such that the *slope* error is bounded by

$$\varepsilon_h = B_E h \qquad \text{for Euler's Method,}$$
$$\varepsilon_h = B_M h^2 \qquad \text{for midpoint Euler,}$$
$$\varepsilon_h = B_{RK} h^4 \qquad \text{for Runge–Kutta.}$$

The first was proved as Theorem 4.5.2; the second we shall now present (with a lengthier proof) as Theorem 4.6.1; for the third we shall omit an even more complicated proof in favor of the compelling experimental evidence cited above for $E(h) \approx C_{RK} h^4$, recalling that

$$E(h) = |u(t) - u_h(t)| \leq \frac{\varepsilon_h}{K} \left(e^{K|t-t_0|} - 1 \right). \tag{10, again}$$

For further reading, see References for numerical methods at the end of this volume.

Theorem 4.6.1 (Bound on slope error, midpoint Euler method).
Consider the differential equation $x' = f(t, x)$, where f is a continuously differentiable function on a rectangle $R = [a, b] \times [c, d]$ in the t, x-plane. Consider also the midpoint Euler approximate solution u_h, with step h. Then there is an ε_h such that u_h satisfies

$$|u_h'(t) - f(t, u_h(t))| \leq \varepsilon_h$$

at any point where u_h is differentiable (or has left- and right-hand derivatives elsewhere), and $\varepsilon_h \to 0$ as $h \to 0$.

Furthermore, if f is a function on R with continuous derivatives up to order two with respect to x and t, then there is a constant B_M such that

$$|u_h'(t) - f(t, u_h(t))| \leq B_M h^2.$$

This computation is not too difficult if you don't insist on *knowing* B_M, which is a fairly elaborate combination of sup's of the second order partial derivatives.

Proof. First we need to decide just what the mid-point approximation is. Of course we know what it is at the grid-points, as shown in Figure 4.6.1, but we can take any piecewise differentiable function joining them that we like. Segments of straight lines are the easiest choice, but you cannot get a slope error of order at most h^2 that way. Our choice will be the *quadratic function having slope $f(t_i, x_i)$ at (t_i, x_i) and passing through (t_{i+1}, x_{i+1})*, also shown in Figure 4.6.1. These properties do specify a unique function

$$v_h(t) = x_i + f(t_i, x_i)(t - t_i) + \alpha(t - t_i)^2 \quad \text{for} \quad t_i \leq t \leq t_{i+1}.$$

By the definition of the midpoint approximation scheme,

$$x_{i+1} = x_i + h f\left(t_i + \frac{h}{2}, x_i + \frac{h}{2} f(t_i, x_i) \right),$$

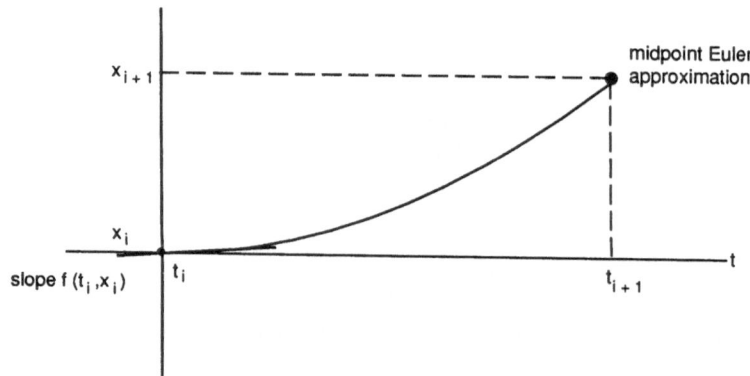

FIGURE 4.6.1. Between grid points of midpoint Euler approximation. Quadratic function $v_h(t)$.

so setting

$$v_h(t_i + h) = x_{i+1}$$

gives

$$\alpha = \left[f\left(t_i + \frac{h}{2}, x_i + \frac{h}{2} f(t_i, x_i) \right) - f(t_i, x_i) \right] / h. \tag{15}$$

Now that we know what $v_h(t)$ is, we need to evaluate

$$|v_h'(t) - f(t, v_h(t))|.$$

We only need to do this on one segment of the graph of v_h, and without loss of generality we may assume that $(t_i, x_i) = (0, 0)$.

First let us evaluate α to first order in h (you may refer to the Appendix on Asymptotic Development). Suppose that

$$f(t, x) = a + bt + cx + r_1(t, x),$$

where $a + bt + cx$ is the Taylor polynomial of f at $(0, 0)$ so that the remainder $r_1(t, x)$ satisfies

$$|r_1(t, x)| \le c_1(t^2 + x^2)$$

for a constant c_1 that can be evaluated in terms of the second partial derivatives of f. Substituting in equation (15) above, we find

$$\alpha = \frac{1}{2}(b + ac) + r_2(h),$$

where $|r_2(h)| \le c_2|h|$ with $c_2 = (c_1 + a)/4$, so that

$$|v_h'(t) - f(t, v_h(t))| = [a + (b + ac)t + 2tr_2(h)]$$
$$- \left[a + bt + c \left(at + \begin{array}{c} \text{second order} \\ \text{terms in } t \end{array} \right) \right]$$
$$= t \, (\text{first order terms in } h) + (\text{second order terms in } t).$$

Now on the segment of interest, $t \leq h$, so there exists a constant B_M such that
$$|v_h'(t) - f(t, v_h(t))| \leq B_M h^2.$$

The constant B_M can be evaluated in terms of the partial derivatives of f up to order two. In particular, B_M does not depend on the segment under consideration. $\square$

It is unclear from this derivation whether a wigglier curve joining grid points might give a higher order dependence on h; however, looking at the actual evidence in Section 3.3 shows that it won't.

Although we shall not take space to prove it, the Runge–Kutta method admits the slope error bound

$$|u_h'(t) - f(t, u_h(t))| \leq B_{RK} h^4.$$

In practice, only the *bounds* for B_E ($= P + MK$ by Theorem 4.5.2) can actually be evaluated (and even then only in special cases). Bounds for the constants B_M and B_{RK} are also sups of various partial derivatives of f, but partial derivatives usually become complicated so fast that the evaluation of their maxima is untractable. In any case, the bounds given by the theory are just that: *bounds,* and will usually be very pessimistic.

4.7   General Fence, Funnel, and Antifunnel Theorems

Once the Fundamental Inequality is proved, everything about fences, funnels, and antifunnels becomes more or less easy. We have actually used all the theorems in Chapter 1, but we can now prove them in greater generality. The basic result of the Fundamental Inequality is to extend Theorem 1.3.5, which was for strong fences, to weak fences by adding a Lipschitz condition.

Theorem 4.7.1 (Fence Theorem). *Consider the differential equation $x' = f(t, x)$ with f a function defined in some region A in $\mathbb{R}^2$ and satisfying a Lipschitz condition with respect to x in A. Then any fence (strong or weak) is nonporous in A.*

We shall prove this theorem for the case of a *lower* fence, $\alpha(t)$. The case of an upper fence would proceed similarly.

As we saw in Chapter 1, nonporosity is assured if the weak inequalities ($\leq$) are replaced by strong inequalities ($<$) as in Theorem 1.3.5. On the other hand, a fence could fail to be nonporous if we use weak inequalities and fail to impose a Lipschitz condition, as in the leaky bucket of Example 4.2.1.

The idea of the proof is to modify the situation slightly so that the weak inequalities become strong. The first temptation is to tilt α slightly. But a moment's reflection will show that this will probably not work, since in that case the tilted α will have a graph (top half of Figure 4.7.1) in parts of A in which we know absolutely nothing about the slopes f, since the restriction of the hypothesis is only for f along α. Instead we shall tilt the slopes f (bottom half of Figure 4.7.1).

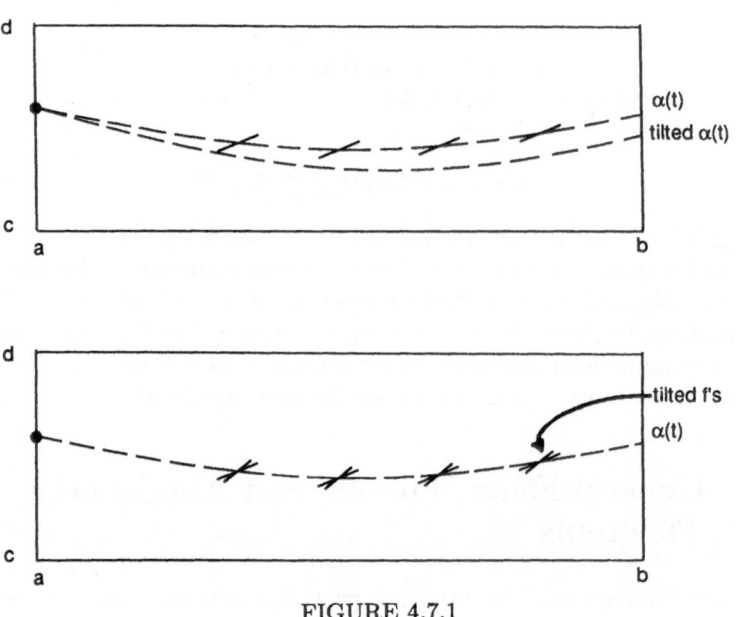

FIGURE 4.7.1

Proof. Let $f_\varepsilon(t,x) = f(t,x) + \varepsilon$. Then α is a strong lower fence for the differential equation $x' = f_\varepsilon(t,x)$ if $\varepsilon > 0$. Moreover, $f_\varepsilon(t,x)$ still satisfies a Lipschitz condition with respect to x (in fact with the same constant as for f).

Let $u_\varepsilon(t)$ be the solution of $x' = f_\varepsilon(t,x)$ with the same initial value $u_\varepsilon(a) = u(a)$. We showed in Theorem 4.5.5 that $u_\varepsilon(t)$ exists and is unique because the assumption of a Lipschitz condition for f implies the Fundamental Inequality. The solution $u_\varepsilon(t)$ to the differential equation $x' = f_\varepsilon(t,x)$ is an approximate solution to $x' = f(t,x)$, so

$$|u(t) - u_\varepsilon(t)| \leq \left(\frac{\varepsilon}{K}\right)\left(e^{K|t-a|} - 1\right),$$

and in particular for every fixed t we have that $u_\varepsilon(t) \to u(t)$ as $\varepsilon \to 0$.

Since $\alpha(t) < u_\varepsilon(t)$ for all $\varepsilon > 0$ and all $t > t_0$, we have $\alpha(t) \le u(t)$ for all t.   □

This proof is just the first instance of the benefits to be reaped by thinking of a solution to one differential equation as approximate solutions to another which we understand better.

Corollary 4.7.2 (Funnel Theorem). *Let $\alpha(t)$ and $\beta(t)$, $\alpha(t) \le \beta(t)$ be two fences defined for $t \in [a, b)$, where b might be infinite, defining a* **funnel** *for the differential equation $x' = f(t, x)$. Furthermore, let $f(t, x)$ satisfy a Lipschitz condition in the funnel.*

Then any solution $x = u(t)$ that starts in the funnel at $t = a$ remains in the funnel for all $t \in [a, b)$.

FIGURE 4.7.2. Funnel.

Proof. The result follows immediately from Definition 1.4.1 and Theorem 4.7.1.   □

Theorem 4.7.3 (Antifunnel Theorem; Existence). *Let $\alpha(t)$ and $\beta(t)$, $\beta(t) \le \alpha(t)$, be two fences defined for $t \in [a, b)$, where b might be infinite, that bound an* **antifunnel** *for the differential equation $x' = f(t, x)$. Furthermore, let $f(t, x)$ satisfy a Lipschitz condition in the antifunnel.*

Then there exists a solution $x = u(t)$ that remains in the antifunnel for all $t \in [a, b)$ where $u(t)$ is defined.

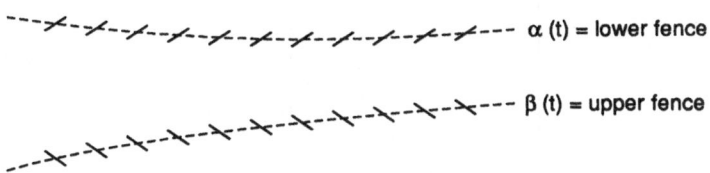

FIGURE 4.7.3. Antifunnel.

Proof. For any $s \in [a,b)$, consider the solutions $\nu_s(t)$ and $\eta_s(t)$ to $x' = f(t,x)$ satisfying $\nu_s(s) = \alpha(s)$ and $\eta_s(s) = \beta(s)$, as shown in Figure 4.7.4.

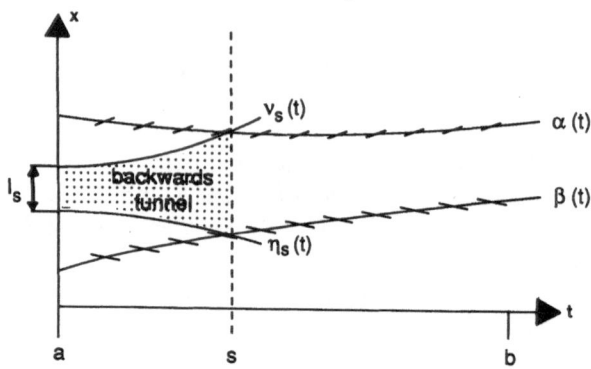

FIGURE 4.7.4. Backwards funnel inside antifunnel.

Using the funnel theorem backwards (i.e., reversing time), we see that $\nu_s(t)$ and $\eta_s(t)$ are defined for $t \in [a, s]$, and satisfy

$$\beta(t) \le \eta_s(t) < \nu_s(t) \le \alpha(t)$$

for $t \in [a, s]$. Let $I_s = [\eta_s(a), \nu_s(a)]$, as labelled in Figure 4.7.4. As $s \to b$, if there is a Lipschitz condition to guarantee uniqueness of the individual $\eta_s(t)$ and $\nu_s(t)$, the intervals I_s form a *nested family of closed intervals* (Figure 4.7.5). It is a basic and early theorem of advanced calculus that for a nested family of *closed* intervals, their common intersection must contain at least one point x_0. (See Exercise 4.7#4.)

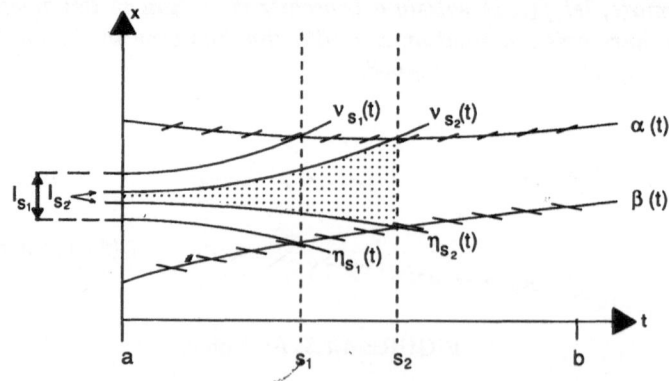

FIGURE 4.7.5. Nested family of closed intervals on vertical axis.

The solution $u(t)$ to $x' = f(t, x)$ with $u(a) = x_0$ is a solution which stays in the antifunnel. Indeed, for any $s \in [a, b)$, we have that $\eta_s(t)$ and $\nu_s(t)$ form a backwards funnel for $a \leq t \leq s$ and that $u(t)$ starts inside it, so $u(t)$ is defined for all $t \in [a, s]$ and satisfies

$$\beta(t) \leq \eta_s(t) \leq u(t) \leq \nu_s(t) \leq \alpha(t)$$

for $t \in [a, s]$. Since this is true for any $s \in [a, b)$, the result follows. $\square$

Theorem 4.7.3 is particularly useful when $b = \infty$, or if f is not defined when $t = b$.

The really interesting results about antifunnels are the ones which give properties which ensure that the solutions which stay in them are unique. We will give two such properties; the first is a special case of the second, but is so much easier to prove that it seems worthwhile to isolate it.

Theorem 4.7.4 (First uniqueness criterion for antifunnels). *Let $\alpha(t)$ and $\beta(t)$, $\beta(t) \leq \alpha(t)$, be two fences defined for $t \in [a, b)$ that bound an* **antifunnel** *for the differential equation $x' = f(t, x)$. Let $f(t, x)$ satisfy a Lipschitz condition in the antifunnel. Furthermore, let the antifunnel be* **narrowing,** *with*

$$\lim_{t \to b}(\alpha(t) - \beta(t)) = 0.$$

If $\partial f / \partial x \geq 0$ in the antifunnel, then there is a unique solution that stays in the antifunnel.

Proof. Theorem 4.7.4 is identical to Theorem 1.4.5, where the uniqueness was already proved; Theorem 4.7.3 provides the existence. $\square$

However, this first criterion for uniqueness does not give us all we can get. It is perfectly possible for an antifunnel to contain several solutions, for instance, an antifunnel may contain a funnel. (See Exercise 4.7#5.)

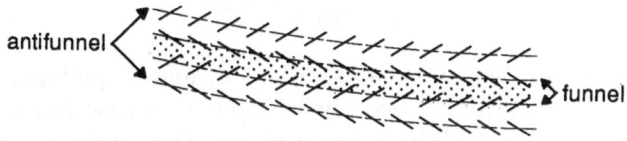

FIGURE 4.7.6. Funnel inside an antifunnel.

But in many important cases, antifunnels do contain unique solutions, without the condition $\partial f / \partial x \geq 0$ being satisfied, as we will presently see in Example 4.7.6 and also in Exercise 4.7#6. The following theorem gives a less restrictive criterion for uniqueness:

Theorem 4.7.5 (Second uniqueness criterion for antifunnels). *Let* $\alpha(t)$ *and* $\beta(t)$, $\beta(t) \leq \alpha(t)$, *be two fences defined for* $t \in [a, b]$ *that bound an* **antifunnel** *for the differential equation* $x' = f(t, x)$. *Let* $f(t, x)$ *satisfy a Lipschitz condition in the antifunnel. Furthermore, let the antifunnel be* **narrowing**, *with*

$$\lim_{t \to b}(\alpha(t) - \beta(t)) = 0.$$

If $(\partial f / \partial x)(t, x) \geq w(t)$ *in the antifunnel, where* $w(t)$ *is a function satisfying*

$$\int_a^b w(s)ds > -\infty,$$

then there is a unique solution which stays in the antifunnel.

Note that the first uniqueness criterion is a special case of the second, with $w(t) = 0$, since $\int_a^b 0\, ds = 0 > -\infty$.

Proof. Again, existence of solutions that stay in the antifunnel is provided by Theorem 4.7.3; let $u_1(t)$ and $u_2(t)$ be two such solutions, with $u_1(t) \geq u_2(t)$. Then as above,

$$(u_1 - u_2)'(t) = f(t, u_1(t)) - f(t, u_2(t)) = \int_{u_2(t)}^{u_1(t)} \frac{\partial f}{\partial x}(t, s(t))ds$$

$$\geq w(t)(u_1 - u_2)(t)$$

so that the difference between the solutions in the narrowing antifunnel is a function $\gamma(t) = (u_1 - u_2)(t)$ satisfying

$$\gamma'(t) \geq w(t)\gamma(t).$$

As such, $\gamma(t)$ is an upper fence for the differential equation $x' = w(t)x$, and, choosing the solution with $x = v(t)$ such that $v(a) = \gamma(a)$,

$$\gamma(t) \geq \gamma(a)e^{\int_a^t w(s)ds}.$$

Here we can see that if $w > 0$, solutions u_1 and u_2 pull apart; requiring the exponent $\int_a^t w(s)ds > -\infty$ gives a bound on how fast the solutions can pull together yet still have uniqueness in the antifunnel. Because the antifunnel is narrowing, we have $\gamma(t) \to 0$. Thus if $\int_a^b w(s)ds > -\infty$, we must have $\gamma(a) = 0$. This is the desired conclusion. $\square$

A further weakening of the uniqueness criterion of Theorems 4.7.4 and 4.7.5 is given in Exercise 4.7#3. (*Third uniqueness criterion for antifunnels.*)

We will finish this chapter with an example to show how much more powerful the second antifunnel criterion for uniqueness is than the first.

Example 4.7.6. Consider the differential equation of Example 1.5.3 (which we will later meet in Volume III when studying Bessel functions):

$$x' = 1 + (A/t^2) \cos^2 x,$$

with A a positive constant.

We have shown that for each C the two curves

$$\alpha(t) = t + C \quad \text{and} \quad \beta(t) = t + C - \frac{A}{t}$$

bound an antifunnel.

Of course,

$$\frac{\partial}{\partial x}\left[1 + (A/t^2) \cos^2 x\right] = -(A/t^2) \sin 2x$$

is both positive and negative in the antifunnel, but

$$-(A/t^2) \sin 2x \geq -A/t^2$$

and

$$\int_1^\infty (-A/t^2) dt = -A > -\infty,$$

so the solution in the antifunnel is unique.   ▲

Exercises 4.1–4.2 Uniqueness Examples

4.1–4.2#1.

(a) Derive by separation of variables the explicit solution to $x' = -c\sqrt{x}$ for the physical situation of the leaky bucket in Example 4.2.1. That is, show that indeed the solution is

$$x = \frac{c^2}{4}(t - t_0)^2 \quad \text{if} \quad t \leq t_0, \qquad 0 \quad \text{if} \quad t \geq t_0.$$

(b) Graph these solutions in the t, x-plane for different values of t_0.

(c) Verify that these solutions are differentiable everywhere, especially at the bottoms of the parabolas.

(d) Verify that the above solution (a) indeed satisfies the differential equation everywhere.

(e) Show how (b), despite (c) and (d), shows the nonuniqueness of solutions along $x = 0$.

4.1–4.2#2. A right circular cylinder of radius 10 ft and height 20 ft is filled with water. A small circular hole in the bottom is of 1-in diameter. How long will it take for the tank to empty?

4.1–4.2#3°. During what time T will the water flow out of an opening 0.5 cm^2 at the bottom of a conic funnel 10 cm high, with the vertex angle $\theta = 60°$?

Exercises 4.3 Lipschitz Condition

4.3#1. Find Lipschitz constants with respect to x for the indicated function in indicated regions:

$$\text{(a)} \quad f(t, x) = x^2 - t \qquad 0 \le t \le 2, \ -1 \le x \le 0$$

$$\text{(b)} \quad f(t, x) = \sin(tx) \qquad 0 \le t \le 3, \ 0 \le x \le 5$$

4.3#2°. For the differential equation $x' = \cos(x^2 + t) = f(t, x)$,

(a) Compute $\partial f / \partial x$.

(b) Where in the square $-10 \le t \le 10$, $-10 \le x \le 10$ is $\partial f / \partial x \ge 5$? Where in the square is $\partial f / \partial x \le -5$?

(c) Sketch the field of slopes and of solutions (perhaps with the computer), and mark the regions found above.

4.3#3. Find a Lipschitz constant for

(a) $x' = t^2 e^{-x}$ on $[-5, 5] \times [-5, 5]$

(b) $x' = |2x^3|$ on $[-2, 2] \times [-2, 2]$

(c) $c' = -x^{\arcsin x}$, for $0 \le t \le 2$, $-\frac{1}{2} \le x \le \frac{1}{2}$

(d) $x' = (2 + \cos t)x + 5\sqrt{|x|} + \frac{11}{2}x$, for $0 \le t \le \pi$, $1 \le x \le 2$

(e) $x' = e^t \sin t \cos x$, for $0 \le t \le \pi$, $1 \le x \le 2$

(f) $x' = e^t \sin(t + \frac{1}{3} \tan t) \cos x$, for $0 \le t \le \pi$, $0 \le x \le \pi$

Exercises 4.4 Fundamental Inequality

4.4#1. Consider in the proof of the Fundamental Inequality (10) the string of inequalities showing that $\gamma'(t) \le K\gamma(t) + \varepsilon_\eta$. Justify each step.

4.4#2. The Fundamental Inequality is derived under the assumption that the Lipschitz constant K is different from 0.

(a) Which functions $f(t,x)$ can have Lipschitz constant $K = 0$?

(b) Derive the Fundamental Inequality in the case where $K = 0$.

(c) Confirm, by computing $\lim_{K \to 0} \frac{1}{K}(e^{K|x-x_0|} - 1)$ that the Fundamental Inequality for $K = 0$ corresponds to the limit as $K \to 0$ of the Fundamental Inequality for $K \neq 0$.

4.4#3.

(a) Consider the very simple differential equation $x' = f(t)$. What is the best Lipschitz constant?

(b) Compute $\lim_{K \to 0} \frac{1}{K}(e^{K|x-x_0|} - 1)$.

(c) Use parts (a) and (b) to tell what the Fundamental Inequality says about the functions $u_i(t)$ that satisfy $|u_i'(t) - f(t)| < \varepsilon$.

4.4#4°. Suppose we have set up a differential equation, $x' = f(t,x)$ based on some physical system, but in so doing we have ignored some small forces. So we believe the actual physical system is governed by the differential equation $x' = f(t,x) + g(t,x)$, where all we know about g is that $|g(t,x)| < 0.1$.

Suppose f has Lipschitz constant 2, and $u(t)$ is a solution to $x' = f(t,x)$ with $u(0) = 0$ and $u(5) = 17$. If the physical system starts off with initial condition $x = 0 \pm 0.03$, what can we say about the x value of the physical system at $t = 5$? Hint: Use the Fundamental Inequality.

4.4#5. Show that the right-hand side of the Fundamental Inequality could not be made to grow any more slowly than it does (which is exponentially). In the jargon we say, "Show that the estimate is sharp."

4.4#6. Derive the changes to the Fundamental Inequality in a mathematical modeling situation where the coefficients in a differential equation have been derived experimentally rather than exactly.

That is, suppose a real system is described by a function $x(t)$ satisfying the differential equation $x' = f(t,x)$, where f (the forces of the system) should be measured. Measuring the forces in the system leads to the differential equation $x' = f_1(t,x)$. If η is the error of the measurements, then

$$|f_1(t,x) - f(t,x)| < \eta.$$

Suppose $u(t)$ is the "predicted" solution, i.e., the solution of $x' = f_1(t,x)$ with initial condition $u(0) = x_0$. Suppose also that $v(t)$ is the observed solution, when the system is prepared in state x_0 at time t_0. Find a bound for

$$|u(t) - v(t)|.$$

Exercises 4.5 Existence and Uniqueness Theorems

4.5#1°.

(a) Solve the differential equation $x' = \sqrt{|x|} + k$ for $k \neq 0$ explicitly.

(b) Show that $f(t, x) = \sqrt{|x|} + k$ does *not* satisfy a Lipschitz condition in x in a rectangle $R = [a, b] \times [-1, 1]$. What do you learn from this?

4.5#2. Consider the differential equation $x' = \sqrt{|x - t|} + c$ where c is a constant. Make the following coordinate change: $y = x - t$, $\tau = t$. Write down the corresponding differential equation for y: $dy/d\tau = f(\tau, y)$. For which values of c do we have uniqueness of solutions?

4.5#3. (a) Consider the differential equation $x' = \sqrt{|x|} + 1$ as in Exercise 4.5#1. Let $(t_0, x_0) = (0, -3)$. There is a unique solution through $(0, -3)$. For $x < 0$ the solution is

$$-2\sqrt{|x|} + 2 \ln(\sqrt{|x|} + 1) = t + (-2\sqrt{3} + 2 \ln(\sqrt{3} + 1)).$$

So the solution crosses the t-axis for $t = 2\sqrt{3} - 2 \ln(\sqrt{3} + 1)$. Let $t_f = 1$. Use the program *Numerical Methods* to calculate the errors and orders for Euler, Midpoint Euler, and Runge–Kutta with different numbers of steps. Graph the stepsize versus solution. Hint: Calculate the exact solution using the *Analyzer* program in "reverse."

(b) Choose another final time, $t_f > 2\sqrt{3} - 2 \log(\sqrt{3} + 1)$. Calculate the exact solution $u(t_f)$ with the same initial condition as before, $(t_0, x_0) = (0, -3)$. Use again the program *Numerical Methods* for Euler, Midpoint Euler, and Runge–Kutta with different numbers of steps to calculate the errors and orders. Why is the behavior different from part (a)?

4.5#4. Show (without needing to exactly evaluate the coefficients) that the Taylor series method described in Section 3.2 is in fact of order n.

4.5#5°. Consider Clairaut's differential equation, $x = tx' - (x')^2/2$, which is quite different from those we have been studying because the derivative is squared.

(a) Show that the straight lines $x = Ct - C^2/2$ are solutions.

(b) Show that the parabola $x = t^2/2$ is another solution.

(c) Show that the lines in part (a) are all tangent to the parabola in (b).

(d) Show why the ordinary existence and uniqueness theorem (for explicit differential equations) is not satisfied by this equation. That is, show where the hypotheses fail.

4.5#6.

(a) Describe the solutions of $x = tx' + (x')^2$. Hint: Look at the previous exercise.

(b) Can you describe the solutions of $x = tx' - f(x')$, where f is any function?

Exercises 4.6 Bound on Slope Error

4.6#1. For each of the following differential equations, find a step h such that if you solve the equations using Euler's method and the given initial condition, you can be sure that your solution at the given point is correct to three significant digits. (You have calculated Lipschitz constants in Exercises 4.3#2, but you should worry about whether the approximate solutions stay in the region in which those computations were valid.)

(a)° $x' = x^2 - t$ $x(0) = 0$; $x(2) = ?$

(b) $x' = \sin(tx)$ $x(0) = 5$; $x(3) = ?$

4.6#2. Let $u_h(t)$ be the Midpoint Euler approximation to the solution of $x' = -x$, with initial condition $u_h(0) = 1$.

(a) Show that $0 \leq u_h(t) \leq 1$ for $t \geq 0$ and $h < 1$.

(b) Using the same technique as in Theorem 4.6.1, find a function $C(t)$ such that
$$|u_h(t) - e^{-t}| \leq C(t)h^2.$$

(c) How short should you choose h to guarantee that $u_h(1)$ approximates $1/e$ to five significant digits?

4.6#3°. Let $u_h(t)$ be the Runge–Kutta approximation to the solution of $x' = -x$, with initial condition $u_h(0) = 1$.

(a) Show that $0 \leq u_h(t) \leq 1$ for $t \geq 0$ and $h < 1$.

(b) Find a function $C(t)$ such that $|u_h(t) - e^{-t}| \leq C(t)h^4$. This requires a bit of ingenuity. Let (t_i, x_i) and (t_{i+1}, x_{i+1}) be two successive grid points of $u_h(t)$. First find a formula for x_{i+1} in terms of x_i and h. This formula should suggest a curve joining the grid points; the result follows from evaluating the slope error of this curve, and applying the Fundamental Inequality.

(c) How short should you choose h to guarantee that $u_h(1)$ approximates $1/e$ to five significant digits? Check this using the program *Numerical Methods*.

4.6#4. In the text, we say that the slope errors which occur when approximating solutions of $x' = f(t, x)$ by the three methods we have discussed can be bounded by expressions of the form Ch^p, where p is the order of the method and C can be evaluated in terms of partial derivatives of f up to some order, which depends on the method. Use the program *Numerical Methods* on the following equations, for $-1 \leq t \leq 1$ and $x(-1) = 0$, to discover how many derivatives are needed for such a bound to exist.

(a) $x' = |x + t|$

(b) $x' = -(x + t)|x + t|$

(c) $x' = -(x + t)\sqrt{|x + t|}$

(d) $x' = -(x + t)^2 \sqrt{|x + t|}$

4.6#5. Using the program *Numerical Methods,* find the order of our three methods on the two differential equations $x' = .8|x - t|$ and $x' = .5|x - t|$ for $-1 < t < 1$ and $x(-1) = 0$. Can you explain when and why Runge–Kutta turns out to be less reliable here than, say, midpoint Euler? Hint: Make a sketch, using isoclines of slope 0, 1, and -1. Observe in which cases the approximate solution crosses $x = t$, and explain why that makes a difference.

4.6#6. For the differential equation $x' = \alpha|x - t|^{3/2}$, as in the previous exercise, we can expect some surprises if the approximate solution $u_h(t)$ crosses $t = 0$. Using the program *Numerical Methods,* try some different values of α that show different orders of some methods for $-1 < t < 1$ and $x(-1) = 0$. (Note: the computations will be faster if you enter $|x - t|^{3/2}$ as $|x - t|\sqrt{|x - t|}$, since the computer can then avoid computing logarithms.)

4.6#7. The following exercise shows that Euler's method does not necessarily converge to a solution if the differential equation does not satisfy a Lipschitz condition. Consider

$$x' = |x|^{-3/4}x + t\sin\left(\frac{\pi}{t}\right) \quad \text{with} \quad x(0) = 0.$$

Consider the Euler approximations $u_n(t)$ with $h = \left(n + \frac{1}{2}\right)^{-1}$, as $n \to \infty$.

(a) Use the computer program *DiffEq* for $n = 10, 11, 12,$ and 13, and observe that the approximations do *not* appear to converge. (Note: It is necessary for t_0 and x_0 to be *close,* but not equal, to zero. This can be arranged if you "uncenter" the domains; e.g., $-.5$ to $.5000000000001$.)

(b) Show that if n is *even,*

$$u_n(h) = 0;$$

$$u_n(2h) = h^2;$$

$$u_n(3h) > \frac{1}{2}h^{3/2} > \frac{1}{16}(3h)^{3/2}.$$

(c) Furthermore (for n even) show there exists a constant $c > 3h$ such that part (a) implies we also have

$$u_n(\pi h) > \frac{1}{16}(\pi h)^{3/2} \quad \text{for} \quad 3h < \pi h < c.$$

Hint: Reason by induction, showing that

$$f(\pi h, u_n(\pi h)) > (u_n(\pi h))^{1/4} - \pi h > \frac{1}{2}(\pi h)^{3/8} - \pi h > \frac{1}{10}(\pi h)^{3/8}$$

and note that for $t < c$ we have

$$\frac{1}{10}t^{3/8} > \frac{d}{dt}\left(\frac{1}{16}t^{3/2}\right).$$

(d) Reason similarly for n odd, showing that then

$$u_n(\pi h) < -\frac{1}{16}(\pi h)^{3/2} \quad \text{for} \quad 3h < \pi h < c.$$

Conclude that the approximate solutions $u_n(t)$ do not tend to any limit as $n \to \infty$.

(e) Verify that the equation does not satisfy a Lipschitz condition.

Exercises 4.7 General Fence, Funnel, and Antifunnel Theorems

4.7#1°. Consider $x' = \cos(t + x)$.

(a) Find funnels and antifunnels (they need not be narrowing). (Warning: the follow-up questions we ask assume that you found the same funnels and antifunnels that we did. If you find different ones, you may not be able to answer some of those questions.)

(b) Which of your funnels and antifunnels are strong and which are weak? For those that are weak, show which of the weak funnel and antifunnel theorems of Chapter 4 apply.

(c) For the antifunnels, show at least one solution which never leaves the antifunnel (big hint: the fences which form a funnel (resp. antifunnel) are considered part of the funnel (resp. antifunnel)).

(d) By drawing a computer solution with *DiffEq*, show that the solutions tend to be squeezing together, as if they were falling into a narrowing funnel. Even if your funnels are not narrowing, you may still be able to explain their behavior. Give it a try. Also, even if your antifunnels are not narrowing, you may be able to show that only one solution remains in an antifunnel for all time.

4.7#2. Consider $x' = \cos(xe^t)$. Find narrowing funnels and antifunnels.

4.7#3. Let $\alpha(t)$, $\beta(t)$ be defined for $t \geq t_0$, $\alpha(t) > \beta(t)$, and suppose that the region $t \geq t_0$, $\beta(t) \leq x \leq \alpha(t)$ is an antifunnel. Suppose there exists a function $w(t)$ such that $w(t) < (\partial f/\partial x)(t, x)$ for all x with $\beta(t) \leq x \leq \alpha(t)$, and that

$$\frac{\alpha(t) - \beta(t)}{e^{\int_{t_0}^t w(s)ds}} \to 0 \quad \text{as} \quad t \to \infty.$$

Then show there exists a unique solution in the antifunnel, thus replacing the restriction of Theorem 4.7.5 that an antifunnel narrow to get uniqueness.

Remark. You can consider the numerator $\alpha - \beta$ as the squeeze on the antifunnel, and the denominator $e^{\int_{t_0}^t w(s)ds}$ as the squeeze on the solutions.

Hint: Suppose $u_1(t) \geq u_2(t)$ are solutions in the antifunnel and define $\gamma(t) \equiv u_1(t) - u_2(t)$. Show that $\gamma(t)$ is an upper fence for the differential equation $x' = w(t)x$. Use the explicit solution of $x' = w(t)x$ with $x'(t_0) = \gamma(t_0)$ to show $\gamma(t_0) = 0$.

4.7#4. Prove, using decimal notation, that for a nested family of *closed* intervals, their common intersection must contain at least one point x_0. This confirms a fact used in the proof of Theorem 4.7.3.

4.7#5. Consider the equation $x' = \sin tx$, and refer back to the analysis of Example 1.5.2 where an antifunnel is formed by $\alpha_k(t)$ and $\beta_{-\ell}(t)$. Show that the first uniqueness criterion of Theorem 4.7.4 is insufficient to prevent *two* (or more) solutions from getting together fast enough to stay in the antifunnel.

5

Iteration

In this chapter we will study *iteration* in one dimension. (Iteration in two dimensions is far more complicated and comprises an important chapter in Volume II.) "Iterating" a function $f(x)$ consists of the following simple process:

Start with a *seed* x_0, and consider the sequence

$$x_1 = f(x_0), \quad x_2 = f(x_1), \quad x_3 = f(x_2), \dots . \tag{1}$$

The essence of iteration is to use the last output as the next input.

The sequence (1) is called the *orbit* of x_0, and the kinds of questions we will ask are:

What happens in the long run for a particular seed?

How does "what happens" depend on the seed?

How does "what happens" depend on the function $f(x)$?

How does "what happens" depend on parameters within $f(x)$?

Such questions are similar to the kinds of questions asked about differential equations, and indeed the two subjects are very closely related, as we hope to show in this chapter. In some sense, a differential equation describes evolution in *continuous time,* and an iteration describes evolution in *discrete time.* The continuous dependent variable t that has appeared in earlier chapters has here been "discretized" and appears as an index: x_i means the value of x after i units of "t."

Both differential equations and iteration are included in the area of mathematics called *dynamical systems.* In this book, when we say we will consider a function $f(x)$ as a dynamical system, we mean that we will iterate it.

The continuous-discrete dichotomy suggests one way in which iteration is related to differential equations. Since all the numerical methods we have seen, and in fact just about all the methods in existence, consist of "discretizing" time, we might expect that numerical methods are simply iterations. This is true, and is the real reason for this chapter: mathematicians are still trying to understand what numerical methods actually do.

Iteration arises in contexts other than differential equations. We have seen that computers find numerical methods for differential equations extremely congenial; a truer statement is that computers find iteration very congenial. As a result, a great many other fundamental algorithms of mathematics are simply iterations. This includes the most popular of them all, *Newton's method* for solving equations (nondifferential), to be described in Section 5.3. Other iterative algorithms occur in linear algebra, such as Jacobi's method and the QR method, discussed in Volume II, Appendices L8 and L7 respectively.

Even though iteration is not traditionally a part of the differential equations curriculum, in this day of interaction between computers and mathematics, it would be a mistake to ignore this important topic which is closely related to differential equations.

As you read on in this chapter, you will realize that iteration is more complicated than anything we have studied so far, and problems involving order versus chaos rapidly come to the fore. It is rather surprising that iteration, which looks easier at first view than differential equations, is really not so. The source of increased difficulty with iteration is that there are no simple analogs of the fence and funnel theorems, which require conditions only on the fences.

In Section 5.1 we shall consider how to represent graphically and analyze a function $f(x)$ under iteration. In the remaining sections we shall examine some specific iterative systems: In Section 5.2 we show how the famous logistic model behaves very differently in the cases where population growth is amenable to a difference equation rather than a differential equation. Section 5.3 delves into the complications of Newton's method for calculation of roots, the iterative scheme that is used within all our MacMath computer programs.

Section 5.4 examines numerical methods for differential equations. In Section 5.5 we look closely at periodic differential equations. This is where we introduce the work of Henri Poincaré, who in the late 1890's was the first mathematician to have established and exploited the connection between iteration and differential equations. His theory of Poincaré sections revolutionized differential equations, and this whole book consists largely of an attempt to bring Poincaré's ideas to the undergraduate audience. Here we will give the first introduction to Poincaré sections, but the subject only becomes really serious in higher dimensions; that will have to wait for Volume II.

Finally, in Section 5.6 we offer as diversion a peek at the delights and rewards of iterating in the complex numbers. This is one direction of current research action in dynamical systems.

5.1 Iteration: Representation and Analysis

THE ITERATES OF A FUNCTION

We need a notation for iteration of a function $f(x)$. We will denote the nth
iterate of f by

$$f^{\circ n} = \underbrace{f \circ f \circ \ldots \circ f}_{n \text{ functions } f},$$

so that

$$f^{\circ n}(x) = f(\ldots (f(x)) \ldots),$$

is the n-fold composition of $f(x)$ with itself.

Examples 5.1.1.

$$
\begin{array}{ll}
\text{If } f(x) = \alpha x, & \text{If } f(x) = x^2 + c, \\
f^{\circ 1}(x) = \alpha x & f^{\circ 1}(x) = x^2 + c \\
f^{\circ 2}(x) = \alpha(\alpha x) & f^{\circ 2}(x) = (x^2 + c)^2 + c \\
f^{\circ 3}(x) = \alpha(\alpha(\alpha x)) & f^{\circ 3}(x) = ((x^2 + c)^2 + c)^2 + c, \\
\quad\vdots & \quad\vdots
\end{array}
$$
▲

Remark. Many authors simply write f^n for the nth iterate, which invites
confusion with powers; consequently we will always use the composition
symbol. As you can see in the first example, we have $f^{\circ n}(x) = \alpha^n x$, whereas
$f^n(x) = \alpha^n x^n$; in the second example there is not even a formula for $f^{\circ n}(x)$,
but $f^n(x) = (x^2 + c)^n$.

Remark. Writing the nth iterate in closed form, that is, as an explicit
function of n, is analogous to solving a differential equation in terms of
elementary functions. (For example, for $x' = \alpha x$ we can write the solution
$x(t) = e^{\alpha t}x(0)$.) Such closed form is definitely *not* always possible, despite
the fact that our examples might seduce you into thinking it is.

The idea of iteration is very simple, but the results are not, largely because the iterates $f^{\circ n}$ can be very complicated. For instance, we shall look
long and hard at the quadratic polynomial $f(x) = x^2 + c$, the second example of 5.1.1. You might reasonably think that everything there is to say
about quadratic polynomials was said, long ago, and is easy. But note that
for a quadratic, the nth iterate $f^{\circ n}$ is a polynomial of degree 2^n, so that
studying quadratic polynomials as dynamical systems involves the study of
infinitely many polynomials, of arbitrarily high degree. This is the source
of the great complication of iteration.

Example 5.1.2. The computer program *Cascade* will automatically graph
for you any iterate of $x^2 + c$. Figure 5.1.1 on the next page represents the
16th iterate of $x^2 + c$, for $c = -1.39$. The function $f^{\circ 16}$ is of degree 2^{16}, so
this picture only begins to show how complicated it is. ▲

16th Iterate of x² - 1.39

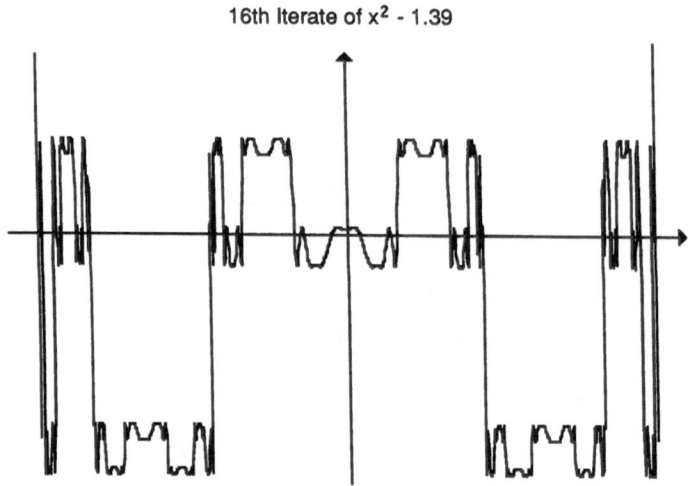

FIGURE 5.1.1. $f^{\circ 16}(x)$ for $f(x) = x^2 - 1.39$.

TIME SERIES

For the remainder of this chapter we will be iterating mappings $f: \mathbb{R} \to \mathbb{R}$ (except for Section 5.6, where we will show some examples of iterations of mappings $f: \mathbb{C} \to \mathbb{C}$).

There are many ways of representing an iterative process. Probably the simplest to understand is a *time series*. This simply means plotting the points $(n, f^{\circ n}(x_0))$, perhaps joining them. We already have a program that will do this for us, and it is good old *DiffEq* applied as follows:

Suppose we want to iterate a function f.

---

1. Enter the differential equation $x' = f(x) - x$.

2. Apply Euler's method, using step $h = 1$.

These two steps yield $x_{n+1} = x_n + (f(x_n) - x_n) = f(x_n)$.

Remark. Euler's method can be run forwards or backwards in time. However, Euler's method run backwards is not the inverse of Euler's method running forward: running Euler's method backwards starting at x_0 does not usually lead to an x_{-1} such that $f(x_{-1}) = x_0$. (See Exercise 3.1–3.2#5.) Consequently, the part of the orbit run backwards from your initial condition is not relevant to the function you are trying to iterate.

Let us give a few examples of such iterations.

Examples 5.1.3. Time series for $x^2 + c$, $c = -1$, -1.3, -1.8.
If $c = -1$, then the orbit of -1 is, from iterating $x^2 - 1$,

$$x_0 = -1, \quad x_1 = (-1)^2 - 1 = 0, \quad x_2 = 0^2 - 1 = -1, \ldots .$$

These results can be summarized by the formula

$$x_n = \frac{(-1)^{n+1} - 1}{2}.$$

We see this repetitive pattern, a cycle of period two, in Figure 5.1.2.

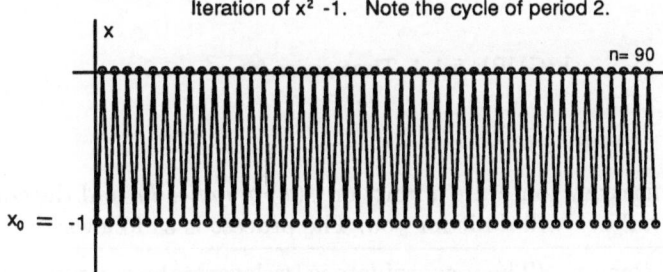

FIGURE 5.1.2. Time series for $x^2 - 1$.

If $c = -1.3$, iteration produces a cycle of period four, as in Figure 5.1.3.

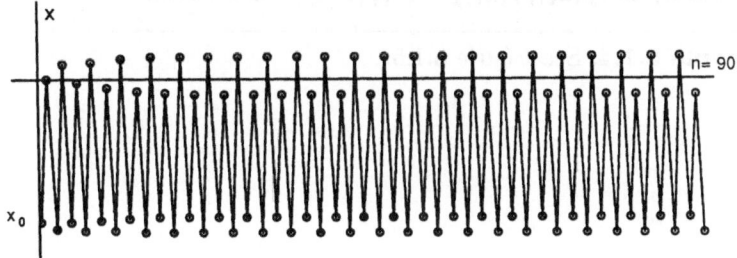

FIGURE 5.1.3. Time series for $x^2 - 1.3$.

If $c = -1.8$, the orbit appears to be *chaotic,* meaning simply "without apparent order," as shown by the time series in Figure 5.1.4.   ▲

Iteration of x^2 - 1.801.   Note that the orbit looks chaotic.

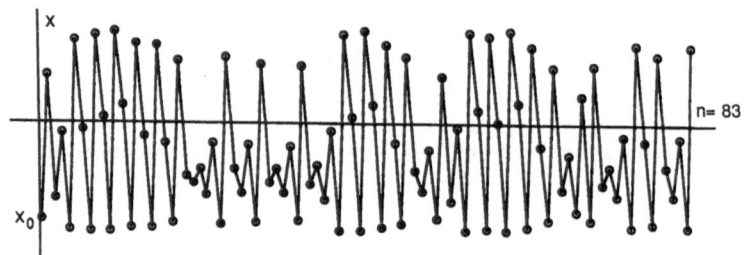

FIGURE 5.1.4. Time series for $x^2 - 1.8$.

GRAPHIC ITERATION

It is also quite easy to iterate a function $f(x)$ *graphically,* and the computer program *Analyzer* will do it for you. The process is as follows:

Begin with some x_0. (The y-coordinate is irrelevant; the computer program uses $(x_0, 0)$.) Then

(i) Go *vertically to the graph* of f: $(x_0, f(x_0))$

(ii) Go *horizontally to the diagonal:* $(f(x_0), f(x_0)) = (x_1, x_1)$

To iterate you repeat the construction, going vertically to $(x_i, f(x_i))$, then horizontally to $(f(x_i), f(x_i)) = (x_{i+1}, x_{i+1})$, and so on.

Example 5.1.4. See Figure 5.1.5.

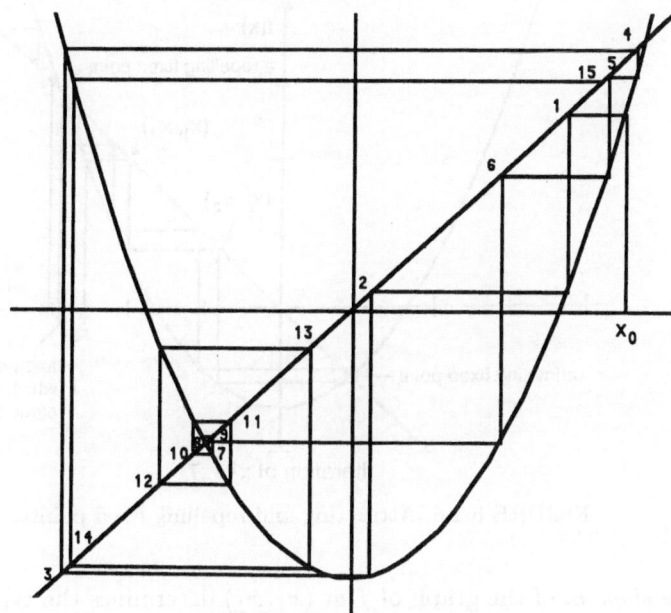

The first 15 iterates of 1.86 under $x^2 - 2$

FIGURE 5.1.5. Graphic iteration of $x^2 - 2$ for $x_0 = 1.86$.

FIXED POINTS

If you play with *Analyzer* a bit, you will see that the *intersections* (x_f, x_f) of the graph of $f(x)$ with the diagonal are important. Such points x_f are called *fixed points* of f, because $f(x_f) = x_f$.

You should think of fixed points as representing *equilibrium* behavior, which is usually classified as *stable* or *unstable*. Fixed points can be classified as "attracting" (corresponding to stable equilibrium), "repelling" (corresponding to unstable equilibrium), or "indifferent" (leading to various other behaviors).

A fixed point x_f is *attracting* if there is some neighborhood of x_f which is attracted to x_f; i.e., if x_0 is in that neighborhood, then the orbit of x_0 is a sequence converging to x_0. See Figure 5.1.6.

The fixed point x_f is called *repelling* if there is some neighborhood of x_f such that the orbit of any point x_0 in that neighborhood eventually leaves that neighborhood (though it might still later return).

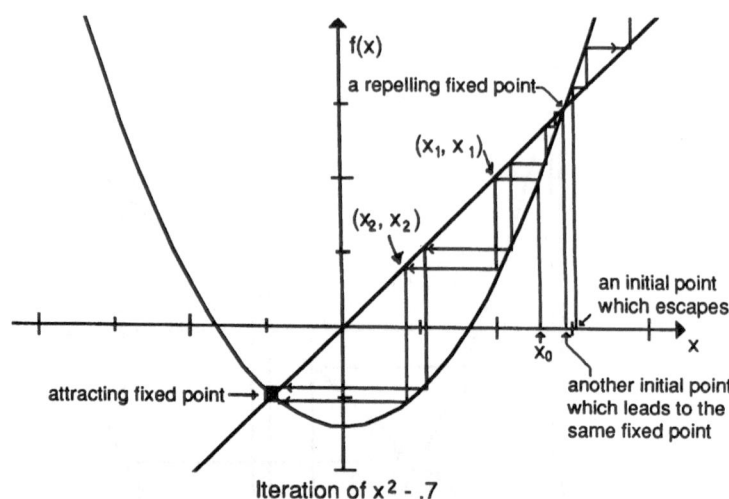

Iteration of x² - .7

FIGURE 5.1.6. Attracting and repelling fixed points.

The *slope* m of the graph of f at (x_f, x_f) determines the type of a fixed point, as follows: A fixed point x_f is

attracting	if	$|m| < 1$;
repelling	if	$|m| > 1$;
indifferent	if	$|m| = 1$;
superattracting	if	$m = 0$.

This derivative criterion for classifying fixed points can be seen from a Taylor expansion of $f(x)$, centered at x_f: Setting $x = x_f + h$ gives as the Taylor polynomial of f,

$$f(x) = f(x_f + h) = x_f + mh + \text{higher order terms in } h,$$

so the deviation from x_f is simply multiplied (to first order) by m, and the number m is called the *multiplier* of f at the fixed point x_f. Theoretical details are spelled out in Exercises 5.1#3-5, but you can directly observe the following.

As soon as you are sufficiently close to x_f for the linear approximation to be valid, the point x_f will repel you if $|m| > 1$, and suck you in if $|m| < 1$. By far the strongest attracting situation occurs when $m = 0$, when the first order term in h disappears entirely, which is the reason for the term *superattracting*.

If $m = \pm 1$, a fixed point is called *indifferent* because many things can happen, depending on higher terms in the Taylor polynomial of f. We will explore some possibilities in Exercises 5.1#2.

The graphic behavior of an iteration attracted to or repelled by a fixed point depends on the sign of the multiplier, as illustrated in the following:

Example 5.1.5. The polynomial $x^2 - 1.2$ has two fixed points, at

$$x_f \approx 1.7041 \ldots \quad \text{and at} \quad x_f \approx -.7041 \ldots .$$

Both fixed points are repelling, since the derivatives $2x$ at these points are approximately $3.4083\ldots$ and $-1.4083\ldots$ respectively. Figure 5.1.7 shows them both, with some orbits escaping from them. Note particularly the "outward spiralling" behavior of the graph of the orbits near a fixed point with negative multiplier < -1.

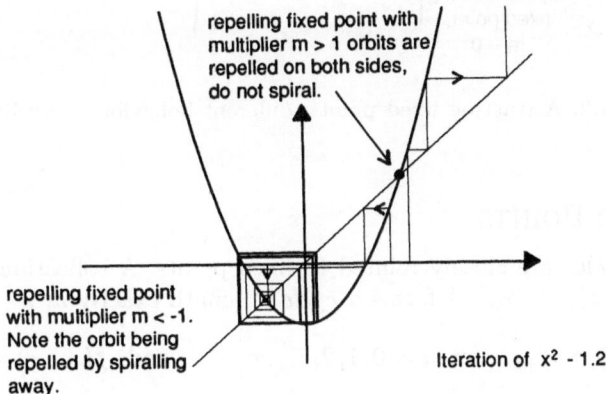

repelling fixed point with multiplier m > 1 orbits are repelled on both sides, do not spiral.

repelling fixed point with multiplier m < -1. Note the orbit being repelled by spiralling away.

Iteration of x^2 - 1.2

FIGURE 5.1.7. Repelling fixed points; different behaviors according to sign of multiplier.

Exercise 5.1#8 asks you to explore a question you should ask about Figure 5.1.7. What is going on between the two fixed points if both are repelling? Although the explanation belongs in the next subsection, guessing and experimenting with the computer provides valuable insight.   ▲

Example 5.1.6. A function like $-0.15x^5 + 0.4x + 0.7x^3 + 0.05$ exhibits various iterative behaviors, depending on the values of the multiplier at each fixed point, as shown in Figure 5.1.8 on the next page.   ▲

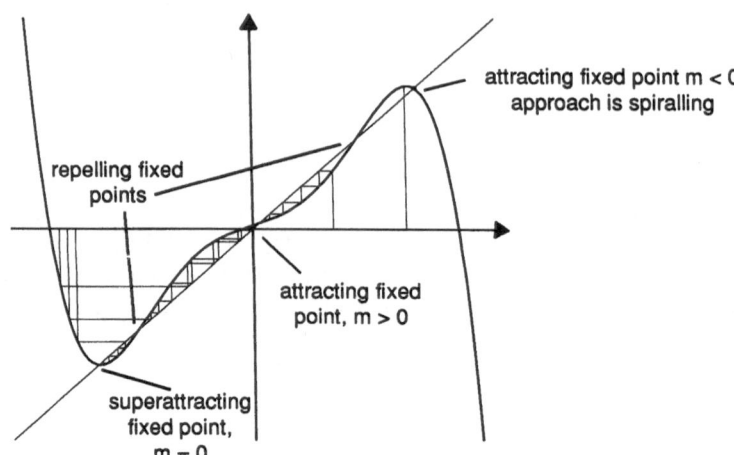

FIGURE 5.1.8. Attracting fixed points; different behaviors according to multiplier.

PERIODIC POINTS

Periodic cycles are closely related to fixed points. A collection of distinct points $\{x_0, x_1, \ldots, x_{n-1}\}$ forms a *cycle* of length exactly n, if

$$f(x_i) = x_{i+1}, \quad \text{for} \quad i = 0, 1, 2, \ldots, n-2, \quad \text{and} \quad f(x_{n-1}) = x_0.$$

This is equivalent to saying that each point of the cycle is a fixed point of the nth iterate $f^{\circ n}$, i.e., that $f^{\circ n}(x_i) = x_i$, and is not a fixed point of any lower iterate of f. The elements of the periodic cycle are called *periodic points* of *period* (or *order*) n.

Fixed points are a special case of periodic points, of period 1.

A cycle is *attracting* or *repelling* if each of its points is an attracting or repelling fixed point of $f^{\circ n}$. In particular, if the seed of an orbit is sufficiently close to a point of an attractive cycle, then the orbit will tend to a periodic sequence, with successive terms closer and closer to a point of the cycle.

Examples 5.1.7. Iteration of $x^2 - 1.75$ gives the two-cycle $.5 \to -1.5 \to .5$, as shown in the top half of Figure 5.1.9.

Iteration of $x^2 - 1.75488\ldots$ gives the three-cycle $0 \to -1.75488\ldots \to 1.3247\ldots \to 0$, as shown in the bottom half of Figure 5.1.9.

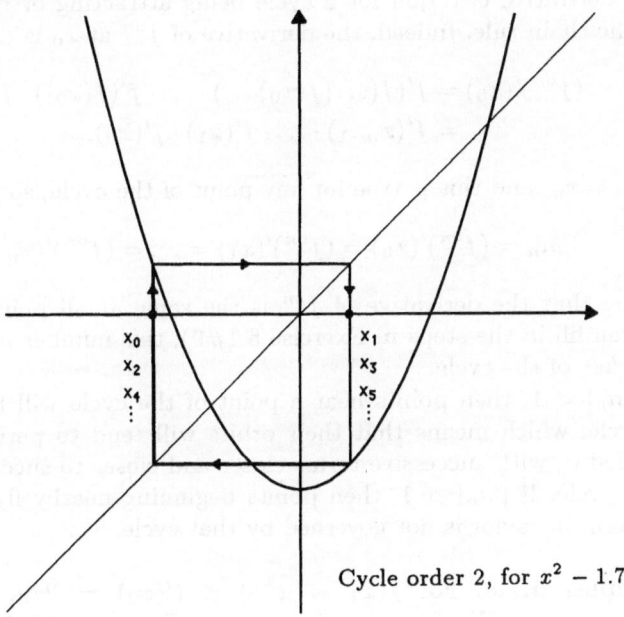

Cycle order 2, for $x^2 - 1.75$

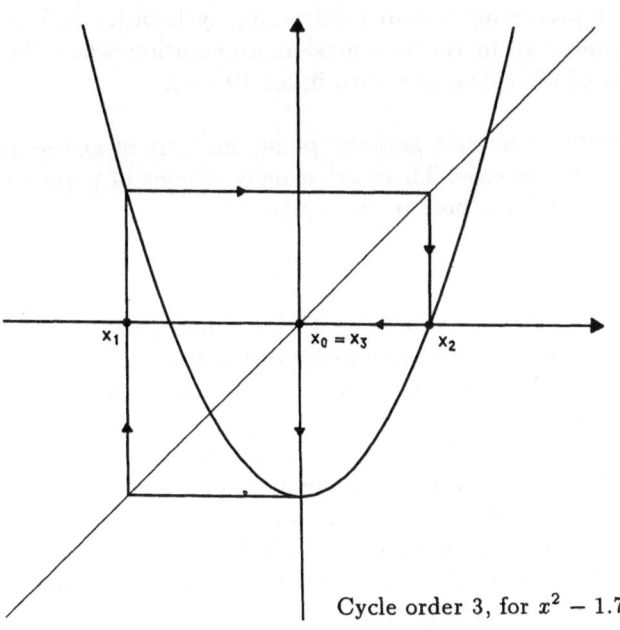

Cycle order 3, for $x^2 - 1.75488\ldots$

FIGURE 5.1.9. Periodic cycles

The derivative criterion for a cycle being attracting or repelling comes from the chain rule. Indeed, the derivative of $f^{\circ n}$ at x_0 is

$$(f^{\circ n})'(x_0) = f'(f(\ldots(f(x_0)\ldots)\cdot\ldots\cdot f'(f(x_0))\cdot f'(x_0)$$
$$= f'(x_{n-1})\cdot\ldots\cdot f'(x_1)\cdot f'(x_0).$$

But $x_0 = x_n$, and this is true for any point of the cycle, so we get

$$m_n = (f^{\circ n})'(x_0) = (f^{\circ n})'(x_1) = \ldots = (f^{\circ n})'(x_{n-1}),$$

showing that the derivative of $f^{\circ n}$ is the same at all points of the cycle (you can fill in the steps in Exercise 5.1#9); this number m_n is called the *multiplier* of the cycle.

If $|m_n| < 1$, then points near a point of the cycle will be attracted to the cycle, which means that their orbits will tend to periodic sequences of period n, with successive terms closer and closer to successive elements of the cycle. If $|m_n| > 1$, then points beginning nearby fly off, and their long-term behavior is not governed by that cycle.

Examples 5.1.8. For $f(x) = x^2 + c$, $f'(x_0) = 2x_0$, and $(f^{\circ n})' = 2^n x_0 x_1 \cdots x_{n-1}$. We return to the cycles of Examples 5.1.7. For $x^2 - 1.75$, $m_2 = (f^{\circ 2})' = 4x_0 x_1 = 4(.5)(-1.5) = -3$, predicting a repelling cycle order 2. For $x^2 - 1.7548\ldots$, $m_3 = (f^{\circ 3})' = 8x_0 x_1 = 8(0)(-1.75488\ldots)(1.3247 \ldots) = 0$, predicting a (super)attracting cycle order 3. The behavior of the cycle shows up in the graphical representation when the seed is not an element of the cycle, as shown in 5.1.10. ▲

It might seem that periodic points are sort of exotic and rare, but the opposite is the case. There are usually zillions of periodic cycles; periodic points x_p are the roots of the equation

$$f^{\circ n}(x) = x.$$

If f is a polynomial of degree d, this is an equation of degree d^n, with in general d^n roots. Of course many of the roots x_p may be complex, which will not correspond to a periodic cycle on the graph of $f(x)$, but many may also be real.

The first object of any analysis of an iterative system is to try to locate, among the zillions of periodic points or cycles, the attractive ones, because they are the only ones we can see. For specific equations, the program *Analyzer* can be an immense help by locating them graphically. Algebraic calculations proceed as in the next example, where we deal with a whole parameter family of equations.

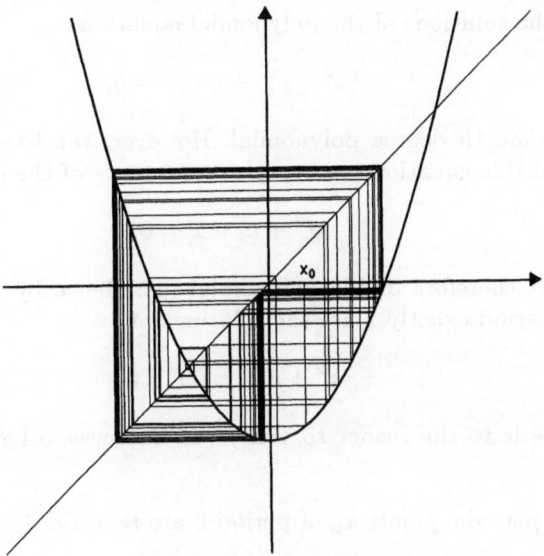

Repelling cycle order 2,
for $x^2 - 1.75$, with $x_0 = .4$

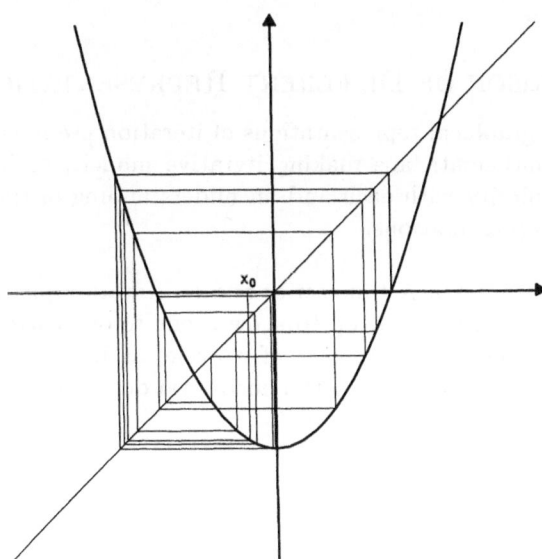

Attracting cycle order 3,
for $x^2 - 1.75488\ldots$, with $x_0 = -.3$

FIGURE 5.1.10. Repelling and attracting cycles.

Example 5.1.9. Let us look at the periodic points of period 2 of $x^2 + c$. They are the solutions of the polynomial equation

$$(x^2 + c)^2 + c - x = 0, \tag{2}$$

which is a fourth degree polynomial. However, the fixed points are also solutions of this equation, and they are the roots of the polynomial

$$(x^2 + c) - x = 0,$$

which must therefore divide polynomial (2). Hence by long division, the points of period exactly 2 are the solutions of

$$x^2 + x + c + 1 = 0.$$

We leave it to the reader to verify (as Exercises 5.1#11) the following facts:

(a) The periodic points x_p of period 2 are real if $c < -3/4$.

(b) The multiplier m_2 of the period 2 cycle is $4(c + 1)$.

(c) The cycle of period 2 is attracting if and only if $-5/4 < c < -3/4$.
▲

COMPARISON OF DIFFERENT REPRESENTATIONS

All of the graphical representations of iteration are in common use among applied mathematicians making iterative models, so it is useful to gain some familiarity with each and an understanding of the relation between different representations.

Examples 5.1.10. A time series and an iteration picture, if made to the same scale, can be lined up to show a one-to-one correspondence. This is particularly clear for the case of a cycle, as in Figure 5.1.11 where the correspondence is shown by the horizontal dotted lines.   ▲

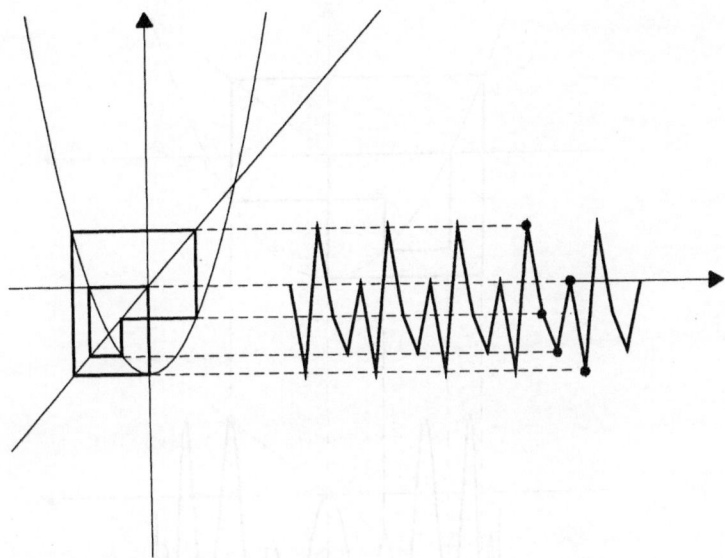

FIGURE 5.1.11. Iterating $x^2 - 1.62541$; cycle order five. Iteration matched with time series.

Example 5.1.11. For an equation that iterates to give a cycle of order n, the iteration picture showing that cycle can be graphically lined up with a picture of the nth iterate, again to show a one-to-one correspondence, between the points on the cycle in the first case and the points where the nth iterate in the second case is *tangent* to the diagonal, as in Figure 5.1.12 on the next page.

The two places where the nth iterate *crosses* the diagonal are *fixed points* of the nth iterate, i.e. periodic points of period n. Both are repelling. ▲

There is nothing magic about showing the first correspondence side-by-side and the second above-and-below. We just point out that you can work either horizontally *or* vertically, because of the role of the diagonal in graphical iteration.

Exercise 5.1#13 ask you to explore these comparisons for other functions.

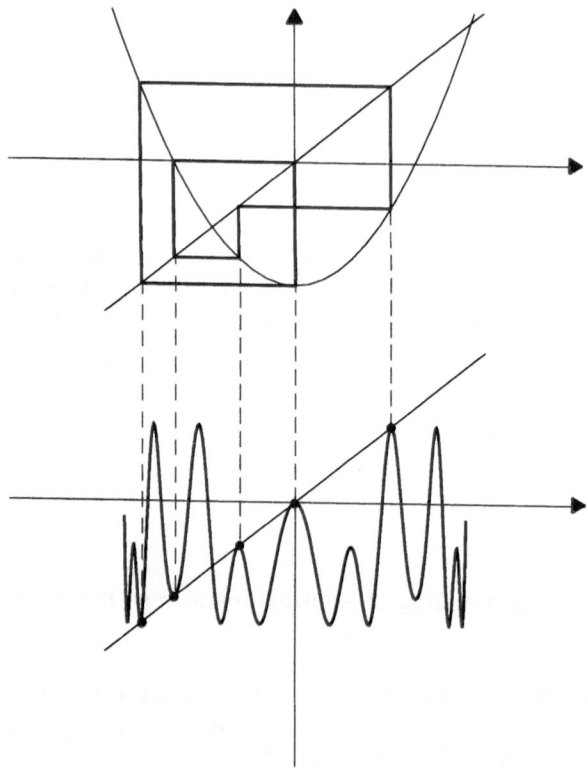

FIGURE 5.1.12. Iterating $x^2 - 1.62541$; cycle order five. Iteration matched with fixed points of 5th iterate.

CHANGES OF VARIABLES IN ITERATIVE SYSTEMS

We have deliberately underplayed changes of variables in the earlier chapters on differential equations. It is much easier, at least at first, to compute blindly rather than to try to understand geometrically what the computations mean, and we have tried to steer away from formal manipulation of symbols. Changes of variable in iterative systems are easier to understand. Each field of mathematics has its own scheme for "change of variables"; the one relevant here is *conjugation,* which is different, for instance, from the change of variables scheme encountered in integration. Conjugation proceeds as follows:

If $f: X \to X$ is a mapping we wish to iterate, and $\varphi: Y \to X$ is a 1–1 correspondence between elements of Y and elements of X, then there ought to be a mapping $g: Y \to Y$ which "corresponds" to f under φ.

It is not really surprising, but it is important, to realize that

$$g = \varphi^{-1} \circ f \circ \varphi; \tag{3}$$

we say that φ *conjugates* f *to* g.

Example 5.1.12 (Nonmathematical). Let X be the set of English nouns, Y be the set of French nouns, and φ the French-English dictionary. Let f be the "English plural" function, which associates to any English noun its plural. Then the corresponding function g in French is the "French plural" function. To go through the formula (3), start for instance with

"oeil", which the French-English dictionary φ will translate to "eye"; the English plural function f will transform this to "eyes". We now need to return to French using the English-French dictionary φ^{-1}; when applied to "eyes" this gives "yeux", the French plural of "oeil", and the desired function g. ▲

Example 5.1.13 (Mathematical). Let $f(x) = 2x - x^2$. We can rewrite this as

$$1 - f(x) = (1 - x)^2.$$

If we now set $\varphi(x) = 1 - x$, we see that $g(y) = y^2$ "corresponds" to f under φ. This can be computed out as follows to show that $g = \varphi^{-1} \circ f \circ \varphi$:

$$y \underset{\varphi}{\rightarrow} 1 - y \underset{f}{\rightarrow} 2(1-y) - (1-y)^2 = 1 - y^2 \underset{\varphi^{-1}}{\rightarrow} 1 - (1 - y^2) = y^2. \quad ▲$$

Equation (3) is exactly the same formula that occurs when doing changes of bases for linear transformations (Volume II, Appendix L2). Example 5.1.13 brings out the real purpose of changes of variables: things are often simpler in some other variable than the one which seems natural. In Volume II, we will see this in a different context in Chapter 6; when studying the central force problem we will pass to polar coordinates, where things will simplify. In Chapter 7, the passage to coordinates with respect to a basis of eigenvectors will be the main theme.

One way of seeing that formula (3) is an extremely useful relation to require between f and g is to see that if φ conjugates f to g, then φ also conjugates $f^{\circ n}$ to $g^{\circ n}$. A *conjugation diagram* like Figure 5.1.13 shows how the process can go on and on under iteration.

Another important aspect of conjugacy is to see that it preserves qualitative behavior of orbits under iteration. If y is a fixed or periodic point of g, then $x = \varphi(y)$ is a fixed (or periodic) point of f, so that periodic points of f and g "correspond" under φ, and moreover attracting (resp. repelling) periodic points correspond to attracting (resp. repelling) points.

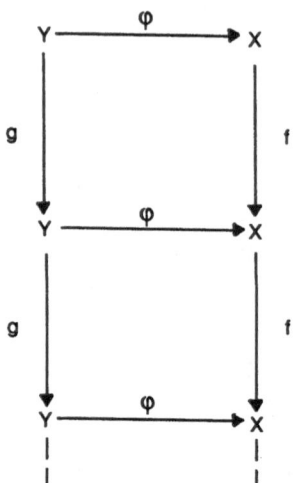

FIGURE 5.1.13. Conjugation of f to g. $g = \varphi^{-1} \circ f \circ \varphi$.

Furthermore, if φ and φ^{-1} are differentiable, the multipliers at corresponding points are equal. This is proved in Exercise 5.1#16.

See Exercises 5.1#15–19 for further exploration of conjugation.

5.2   The Logistic Model and Quadratic Polynomials

Second degree polynomials are about the simplest mappings next to linear ones we can think of. In this section we will study the iteration of quadratic polynomials as functions of a *real* variable; in Section 5.6 we will give some hints of what happens for quadratic polynomials of a complex variable.

Mathematical applications in physics, chemistry and engineering only occasionally give rise to iterative systems as models of real systems; these topics are usually modelled by differential equations. We will motivate the iteration of quadratic polynomials from an ecological model: in ecology, iterative systems are fairly common, and the logistic model to be described below is the granddaddy of them all.

As we will see in Sections 5.4 and 5.5, the logistic model is also useful in understanding some differential equations.

POPULATION GROWTH AND THE LOGISTIC MODEL

Suppose we have some species, whose population goes from generation to generation (you might think of gypsy moths, or any of many other species of insects which live one year, existing only as eggs during the winter). We will denote the population of the nth generation by $p(n)$; to say that the population of this species can be described by an iterative system in one variable is to say that there is a function f such that

$$p(n+1) = f(p(n)).$$

This can also be expressed by saying that the population in any generation depends only on the population in the previous generation.

A first possible such model is

$$p(n+1) = (1+\alpha)p(n) \tag{4}$$

where α is a number (presumably positive if the species is to survive) describing the fertility rate of the species.

This simplest model (4) is linear, like that of Example 5.1.1, and it is easy to solve explicitly:

$$p(n) = (1+\alpha)^n p(0).$$

(Solving an iterative system means giving a formula for x_n in terms of n and x_0, as above for $p(n) = x_n$.) In particular, if $\alpha > 0$ the population grows exponentially. This model describes growth only under ideal circumstances, for instance in a laboratory trying to raise a maximal number of a particular species, isolated from any disturbing influences. It can never describe for long a natural system, which is always subject to such factors as crowding, competition, and disease.

A more realistic model than (4) takes crowding into account. Assume that there is some "stable" population p_s which the environment can support, and that the fertility rate depends on how far the population is from p_s. One possible way to model this is

$$p(n+1) = \left(1 + \alpha\frac{(p_s - p(n))}{p_s}\right)p(n). \tag{5}$$

It is more convenient to use the stable population as the unit of population, i.e., to set $q(n) = p(n)/p_s$. Then equation (5) becomes

$$q(n+1) = (1+\alpha)q(n) - \alpha q(n)^2. \tag{6}$$

This equation (6) is called the *logistic model*. You should observe how similar this discrete logistic model is to the differential equations population model

$$x' = \alpha(x - x^2) \tag{7}$$

studied in Chapter 2.5. In fact, Euler's method (Chapter 3.1) with step $h = 1$ applied to the differential equation (7) leads exactly to the iteration of (6) (Exercise 5.2#2).

Back in Section 2.5 we found that the differential equation predicted something quite reasonable, namely that whatever the initial population (assumed positive), the population tended to the stable population. It may well come as a surprise that the iterative model has enormously more complicated behavior, especially if the fertility is large.

More precisely, if the fertility is "small," so that $0 < \alpha \leq 2$, and if the initial population is "small," meaning $0 < q(0) < 1+1/\alpha$, then the sequence $q(n)$ will be attracted to the stable population $q_s = 1$, as was the case for the differential equation (Exercise 5.2#3).

In fact, in this case of "small" fertility, $q_s = 1$ is a fixed point of (6), for any α, with multiplier $m = (1 - \alpha)$. In particular, $q_s = 1$ is an attracting fixed point if and only if $0 < \alpha \leq 2$ (Exercise 5.2#4).

But as soon as $\alpha > 2$ the fixed point $q_s = 1$ becomes repelling, and can attract nothing. We will show in detail what happens for $\alpha = 1$ and $\alpha = 3$; but the analysis we will make cannot be carried out for other values of α, and playing with a computer for a while is the best way to gain insight into this remarkable iterative system. It is even more remarkable if you notice that all this complicated behavior results just from iterating a second degree polynomial, which is about the simplest thing you can think of beyond a polynomial of degree one.

Example 5.2.1. *Analysis of logistic model* (6) *if* $\alpha = 1$. A change of variables will simplify the problem. Let us center q at $q_s = 1$, i.e., set $x(n) \equiv 1 - q(n)$. Then the mapping

$$q(n + 1) = (1 + \alpha)q(n) - \alpha q(n)^2 \qquad \text{(6, again)}$$

with $\alpha = 1$ becomes

$$x(n + 1) = x(n)^2. \qquad (8)$$

Equation (8) is very easy to analyze:

  if $|x| < 1$ then $x(n)$ tends to 0 (very fast);
  if $|x| > 1$ then $x(n)$ tends to infinity (very fast);
  if $x = 1$ it stays there;
  if $x = -1$ then it goes to 1 and stays there after that.   ▲

Example 5.2.2. *Analysis of logistic model* (6) *if* $\alpha = 3$. Again a change of variables helps, using the particular change (to be explained two pages hence) which gives the simplest result: an iteration of form $x^2 + c$. Setting $x(n) = 2 - 3q(n)$, we find the logistic equation (6) becomes

$$x(n + 1) = x(n)^2 - 2. \qquad (9)$$

Even in this simpler form (9), it still isn't easy to see what is going on. Furthermore, there could be a little problem here, which is explored in

Exercises 5.2#5. However, if $|x| \leq 2$, we can set $x = 2\cos(\theta)$ and transform equation (9) into

$$2\cos(\theta(n+1)) = 4\cos(\theta(n))^2 - 2,$$

which (remembering the formula for $\cos 2\theta$) becomes

$$\theta(n+1) = 2\theta(n). \tag{10}$$

This last clever change of variables, $x = 2\cos\theta$, permits us to view graphically the transformation law on x as in Figure 5.2.1.

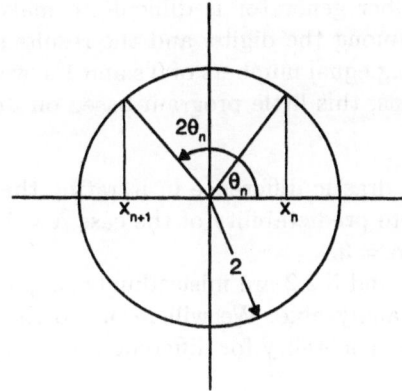

FIGURE 5.2.1. Circular representation of iteration after change of variables.

The interpretation of Figure 5.2.1 is as follows, using a circle of radius 2: Given x (between -2 and 2), find a point on the circle with x as its first coordinate; then double the polar angle of the point (i.e., use equation (10)). The first coordinate of the resulting point is $x^2 - 2$ (working backwards from equation (10) to equation (9)). ▲

It should be clear from this description that the iterative system of Example 5.2.2 with $\alpha = 3$ behaves in a completely different way from that of Example 5.2.1 with $\alpha = 1$. In particular, in the most recent case there is no x value which attracts others (especially not $x = -1$, which corresponds to $q = 1$). The values of x just seem to jump around rather randomly, at least from the point of view of someone who sees only the x-values rather than the entire "circular" pattern of iteration. This observation leads to the following diversion:

Example 5.2.3. *Pseudo-random number generator.* Actually, the function $x^2 - 2$, which was derived in Example 5.2.2, can be used to make quite a good *random number generator*. Start with some random x_0 in $[-2, 2]$, your *seed,* and write a sequence of 0's and 1's by checking whether the orbit of

x_0 is positive or negative. Here is a PASCAL implementation, which will print out an apparently random sequence of "true" and "false."

```
PROGRAM PSEUDORANDOM;
VAR x: real;
BEGIN
    Writeln('enter a seed x with −2 < x < 2'); Readln(x);
    REPEAT writeln (x > 0); x := x*x − 2 UNTIL keypressed;
END.
```

Remark. The computer program evaluates $x > 0$ (just as it would evaluate $x + 2$), only it evaluates it as a "*boolean*," i.e., as true or false.

A random number generator is difficult to make. There should be no obvious pattern among the digits, and the results should meet statistical tests, such as giving equal numbers of 0's and 1's overall. By empirical (and theoretical) reasons, this little program based on iteration seems to work.
▲

Note again the drastic difference in iterating the logistic model (6) between the complete predictability of the case $\alpha = 1$ and the random character of the case $\alpha = 3$.

Examples 5.2.1 and 5.2.2 are misleadingly simple, or perhaps we should say misleadingly analyzable. We will resort to the computer to see what happens in greater generality for different values of α.

A Change of Variables.

In both Examples 5.2.1 and 5.2.2, we made a *change of variables*. Actually, this change can be made in general. If we set

$$x = \frac{1 + \alpha}{2} - \alpha q,$$

the iterative system (6) when changed to the x variable becomes

$$x_{n+1} = x_n^2 + \frac{1 + \alpha}{2} - \frac{(1 + \alpha)^2}{4} = x_n^2 + c.$$

Thus we can always transform the logistic model to the simple quadratic $x^2 + c$.

Exercise 5.2#1b asks you to show that setting

$$x = \frac{1 + \alpha}{2} - \alpha q,$$

is equivalent to a conjugation (end of Section 5.1) with

$$\varphi(x) = \frac{1 + \alpha}{2\alpha} - \frac{x}{\alpha}.$$

We will usually study the iteration of $x^2 + c$, because the formulae are simpler; and then we can change variables back as necessary.

THE CASCADE PICTURE.

For a feeling of what happens in general to the logistic model (7) of iteration, for different values of c, the program *Cascade* is quite enlightening. Choose some interval of values of c. The program will then fit 200 values of c in that interval, along the vertical axis. Starting at the top of the c scale, for each value of c the program will iterate $x^2 + c$ a total of $k + \ell$ times, starting at 0. It will discard the first k (unmarked) iterates, and plot horizontally in the x-direction the next ℓ (marked) iterates, where k and ℓ are numbers you can choose. The idea in discarding the first k iterates is to "get rid of the noise"; if the sequence is going to settle down to some particular behavior, we hope it will have done so after k iterates. In practice, $k = 50$ is adequate for large-scale pictures, but blow-ups require larger values of k and ℓ. Larger k makes for cleaner pictures at any stage.

The picture we get for $-2 < c < 1/4$, with $k = 50$ and $\ell = 50$ is given as Figure 5.2.2. It certainly looks very complicated.

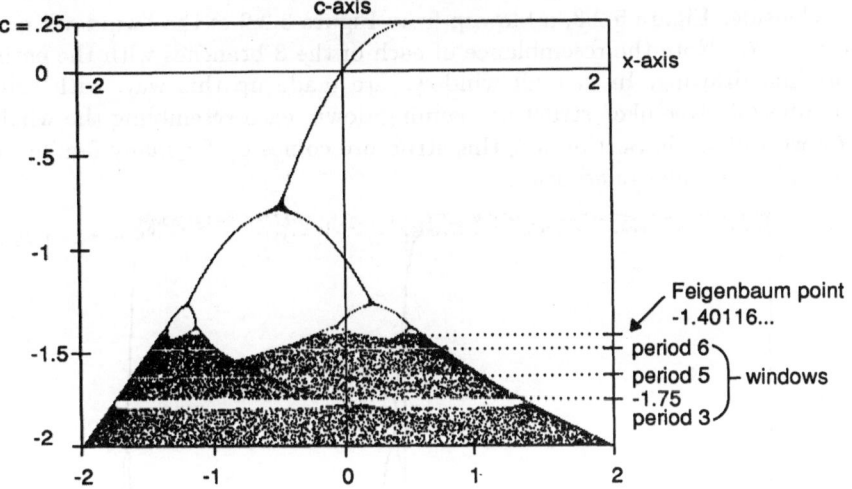

FIGURE 5.2.2. Cascade of bifurcations.

For $-3/4 < c < 1/4$, there is one point on each horizontal line. There should be 50 ($= \ell$); what happened? All fifty got marked on the same point, which is an attracting fixed point. Similarly, for $-5/4 < c < -3/4$, all fifty points are marked at just two points, which are the points of an attracting cycle of period 2. Between these two behaviors, at $c = -3/4$, we say that the cycle *bifurcates,* which means splitting in two.

At $c = -5/4$, this cycle bifurcates again, to give you an attractive cycle of period 4, which later bifurcates to give a cycle of period 8, and so on.

All these points of bifurcation accumulate to some value c approximately equal to $-1.40116\ldots$. This value of c is called the Feigenbaum point, after physicist Mitchell Feigenbaum who in the late 1970's produced reams of experimental evidence that such cascades are universal.

After the Feigenbaum point, the picture looks like a mess, with however occasional *windows of order*, the open white horizontal spaces. The most noticeable such window occurs near $c = -7/4$. There we see an attractive cycle of period 3, which itself bifurcates to one of period 6, then 12, and so on.

Actually, there are infinitely many such windows; if you start blowing up in any region, you will find some, with periodic points of higher and higher period, and each one ending in a cascade of bifurcations, with the period going up by multiplication by powers of 2 (Exercises 5.2#6 and 7).

Playing with the program for a while will convince you that there are windows all over the place; in mathematical language, that the windows are dense. The computer certainly suggests that this is true, but it has not been proved, despite an enormous amount of effort.

Consider Figure 5.2.3, a blowup from Figure 5.2.2 of the "window" near $c = -1.75$. Note the resemblance of each of the 3 branches with the entire original drawing. In fact all windows are made up this way, with some number of "tree-like" structures coming down, each resembling the whole. As we will see in Section 5.3, this structure comes up for many families of mappings besides quadratics.

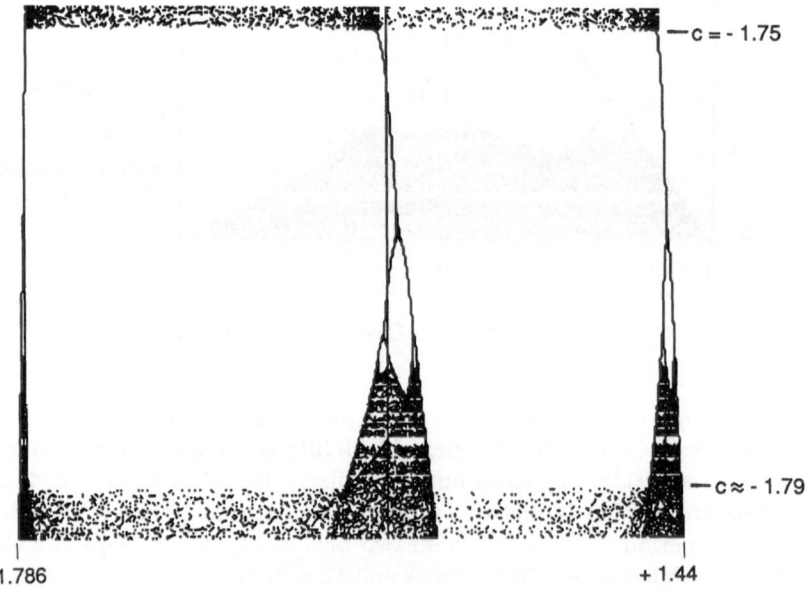

FIGURE 5.2.3. Blowup of cascade of bifurcations across window near $c = -1.75$.

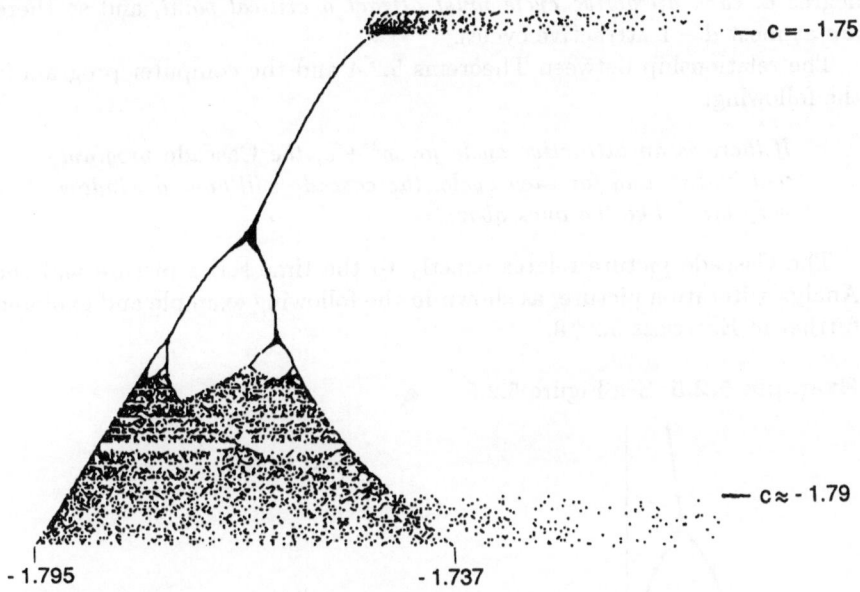

$$c = -1.75$$

$$c \approx -1.79$$

-1.795   -1.737

FIGURE 5.2.4. Further blowup of leftmost portion of Figure 5.2.3.

Figure 5.2.4 is a blowup of the left-hand branch of Figure 5.2.3. Note that this blowup has a cascade of bifurcations, and a window of period 3 (which actually is part of a cycle of period 9), exactly like the total picture of Figure 5.2.2. The horizontal scales are different in Figures 5.2.2 and 5.2.4, but one can show that in fact the structures (windows and their associated periods, as well as the order in which they appear) coincide down to the finest detail. Even if you scale them correctly, they will still not quite superpose (but they come amazingly close).

The real justification for the *Cascade* program described above is the following theorem:

Theorem 5.2.4. *If the polynomial $x^2 + c$ has an attractive cycle, then the cycle attracts 0. In particular, $x^2 + c$ can have at most one attractive cycle.*

There does not seem to be an easy proof of Theorem 5.2.4. It was found at the beginning of this century by a Frenchman, Pierre Fatou, who proceeded to lay the foundations, from the point of view of complex analysis, of the subject which we will go into in Section 5.6, more particularly in connection with Theorem 5.6.6. So we shall accept it without proof and discuss what it tells us.

The significance of the point 0 in Theorem 5.2.4 is that it is a *critical point* (point where the derivative vanishes) of $x^2 + c$. In fact, 0 is the only critical point of $x^2 + c$. More generally, when iterating polynomials of higher

degree *d, each attractive cycle must attract a critical point,* and so there
are at most $d - 1$ attractive cycles.

The relationship between Theorems 5.2.4 and the computer program is
the following:

> *If there is an attractive cycle for $x^2 + c$, the* Cascade *program
> will find it, and for each cycle, the cascade will have a window
> very much like the ones above.*

The Cascade picture relates exactly to the time series picture and the
Analyzer iteration picture, as shown in the following example and explored
further in Exercises 5.2#8.

Example 5.2.5. See Figure 5.2.5.   ▲

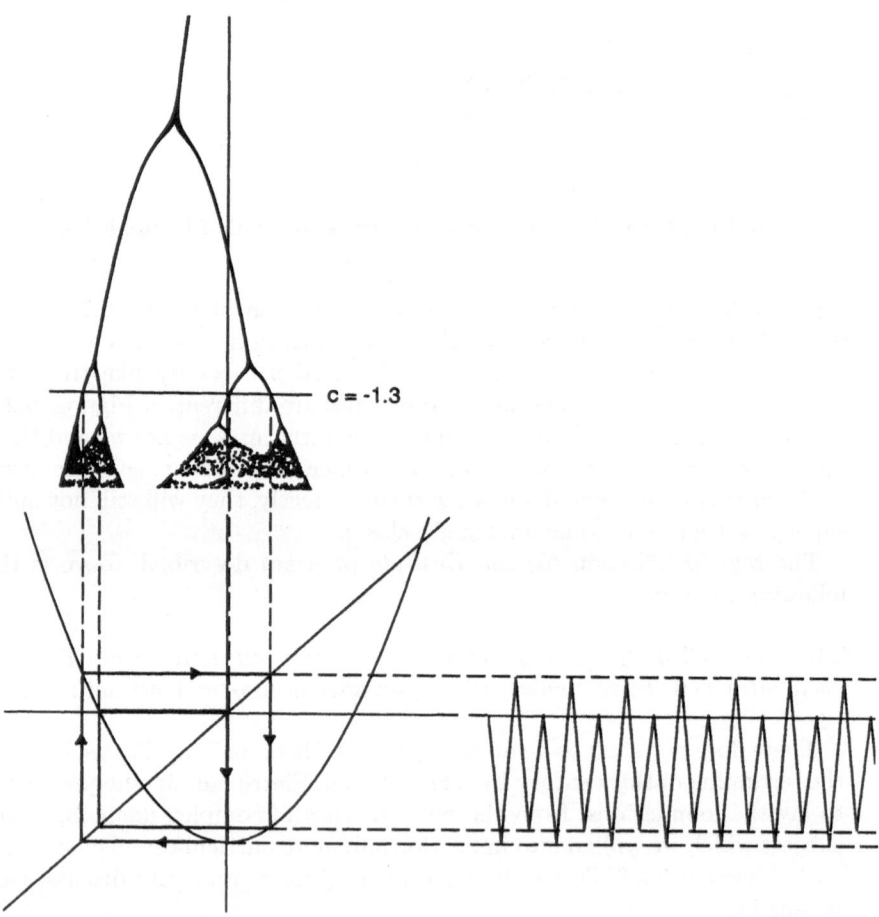

FIGURE 5.2.5. Iterating $x^2 - 1.30$; cycle order four. Iteration matched with
cascade of bifurcations and time series.

In the twentieth century, especially since 1980 when computer graphics became easily accessible, a great deal of effort has gone into understanding the iteration of quadratics. The subject is now fairly well understood. The detailed description of the "islands of stability," in which orbits are attracted to attractive cycles, has been explored and described in great detail by such eminent contemporary researchers as Feigenbaum, John Milnor, and Dennis Sullivan, as well as others to be mentioned later. Section 5.6 discusses further the iteration of quadratic polynomials, for the case where a real number x is generalized to a complex number z. A list of references is provided at the end of the this volume.

5.3 Newton's Method

Solving equations is something you have to do all the time, and anything more than a quadratic equation is apt to be impossible to solve "exactly." So "solving an equation" usually means finding an algorithm which *approximates* the roots. The most common algorithm is an iterative scheme: *Newton's method*. Newton's method is used as a subroutine in practically every computational algorithm—it is one of the real nuts and bolts of applied mathematics.

Our purpose in this section is first to show the complications that are inherent in an ostensibly simple method, and second to show a surprising relation to ideas discussed in the previous section. We shall first discuss, with theorems, the orderly results of Newton's method; then we shall examine the global disorder that also results.

If f is a function (say of one real variable, though we will be able to generalize in Volume II, Ch. 13), then the iterative system for Newton's method is

$$N_f(x) = x - \frac{f(x)}{f'(x)}. \tag{11}$$

The main point to observe about the defining equation (11) for Newton's method is that x_0 is a root of the equation $f(x) = 0$ if and only if x_0 is a *fixed point* of N_f.

Moreover, if $f'(x_0) \neq 0$ (the general case), then $N_f'(x_0) = 0$, which means that N_f has the roots of f as *superattracting* fixed points. This means that N_f converges extremely fast when it converges: the number of correct decimals doubles at each iteration.

In the subsections that follow, we shall examine various aspects of Newton's method.

Geometric Interpretation of Newton's Method

It is quite easy to see geometrically what Newton's method does. At the
point x_0, substitute for f its *linear approximation*

$$L_{x_0}(x) = f(x_0) + f'(x_0)(x - x_0), \tag{12}$$

and look for a root of L_{x_0}, namely

$$x = x_0 - \frac{f(x_0)}{f'(x_0)}.$$

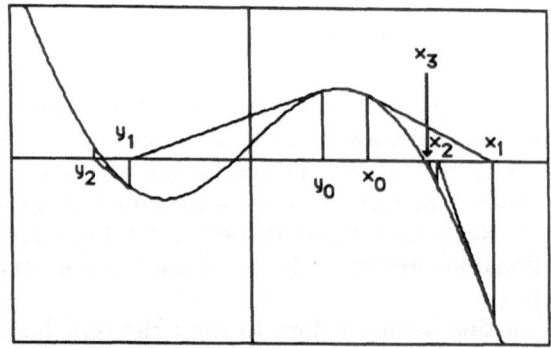

FIGURE 5.3.1. Newton's method, starting at x_0 and at y_0.

This is illustrated by Figure 5.3.1 (which you probably have seen before).
Given an initial guess x_0, Newton's method draws a tangent line to $f(x)$,
intersecting the x-axis at x_1; the process is repeated from x_1, converging
on the right-hand root of $f(x)$. A second example is provided in the same
drawing, starting at y_0; in this case Newton's method converges to the left-
hand root (rather than to the closest root, because of the shallow slope of
the initial tangent line).

Notice that this geometric idea goes a long way towards justifying the
principle that if you start near a root then Newton's method should converge to the root, since the linear approximation L_{x_0} is close to f near
x_0.

Local Behavior of Newton's Method

The following major theorem provides most of the theory concerning Newton's method. It is a fairly hard theorem, and actually hardly worth the
work, if we were only interested in one variable. Indeed, in one real variable,
for roots where the graph of the function is not tangent to the x-axis, there
is the *bisection method* (as in the *Analyzer* program), which works more

slowly and more surely than Newton's method. But the bisection method does not generalize to higher dimensions, whereas both Newton's method and Theorem 5.3.1 below *do*.

Theorem 5.3.1. *Let f be a function of a single variable and suppose that $f'(x_0) \neq 0$. Define*

$$h_0 = \frac{-f(x_0)}{f'(x_0)},$$
$$x_1 = x_0 + h_0,$$
$$J_0 = [x_1 - |h_0|, x_1 + |h_0|],$$
$$M = \sup_{x \in J_0} |f''(x)|.$$

If

$$2\left| \frac{f(x_0)M}{(f'(x_0))^2} \right| < 1,$$

then the equation $f(x) = 0$ has a unique solution in J_0, and Newton's method with initial guess x_0 converges to it.

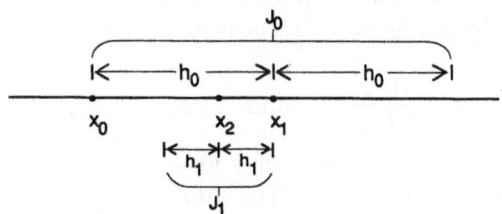

FIGURE 5.3.2.

Proof. Let $h_1 = -\dfrac{f(x_1)}{f'(x_1)}$. We need to estimate h_1, and $f(x_1)$ and $f'(x_1)$ along the way. First, using the triangle inequality,

$$|f'(x_1)| \geq |f'(x_0)| - |f'(x_1) - f'(x_0)|$$
$$\geq |f'(x_0)| - M|x_1 - x_0|$$
$$\geq |f'(x_0)| - \frac{|f'(x_0)|}{2}$$
$$\geq \frac{1}{2}|f'(x_0)|. \tag{13}$$

Estimating $f(x_1)$ is trickier, and involves an integration by parts:

$$f(x_1) = f(x_0) + \int_{x_0}^{x_1} f'(x)dx$$

$$= f(x_0) - \left[(x_1 - x)f'(x)\right]_{x_0}^{x_1} + \int_{x_0}^{x_1} (x_1 - x)f''(x)\,dx$$

$$= \int_{x_0}^{x_1} (x_1 - x)f''(x)\,dx.$$

Make the change of variables $th_0 = x_1 - x$, and the formula above gives

$$|f(x_1)| \le M|h_0|^2 \int_0^1 t\,dt = \frac{M|h_0|^2}{2}. \tag{14}$$

These two inequalities (13) and (14) tell us what we need, namely

$$|h_1| = \left|\frac{f(x_1)}{f'(x_1)}\right| \le \frac{M|h_0|^2}{2}\left|\frac{2}{f'(x_0)}\right| < \frac{|h_0|}{2}$$

and

$$\left|\frac{f(x_1)M}{f'(x_1)^2}\right| \le M\frac{|h_0|}{2}\frac{2}{|f'(x_0)|} = M\left|\frac{f(x_0)}{f'(x_0)^2}\right|.$$

If we define $x_2 = x_1 + h_1$ and $J_1 = [x_2 - |h_1|, x_2 + |h_1|]$, we see that $J_1 \subset J_0$, and that the same requirements are satisfied by x_1, h_1 and J_1 as were satisfied by x_0, h_0 and J_0. As such, we can continue by induction, defining

$$h_k = -\frac{f(x_k)}{f'(x_k)} \qquad \text{and} \qquad x_{k+1} = x_k + h_k.$$

All the points x_k lie in J_0, and they form a convergent sequence, since the series $h_0 + h_1 + \ldots$ converges. The limit

$$x_f = \lim x_k$$

is a root of $f(x) = 0$, since

$$\underbrace{f(x_k)}_{\text{continuous}} + \underbrace{f'(x_k)}_{\text{bounded}} \underbrace{(x_{k+1} - x_k)}_{\to 0} \to 0. \qquad \square$$

Theorem 5.3.1 apparently gives only geometric convergence for Newton's method, from $h_{k+1} \le h_k/2$. This would mean that you get one more correct digit in base 2 for each iteration, or that it takes a bit more than 3 iterations to get an extra digit in base 10. The truth is much better, and we very nearly have it proved.

Theorem 5.3.2. *Under the conditions of Theorem 5.3.1, Newton's method leads to quadratic convergence:*

$$|h_{k+1}| \le \frac{M}{|f'(x_k)|}|h_k|^2.$$

Proof. This is obtained by dividing inequality (14) by $|f'(x_1)|$, then using inequality (13). Generalize to get statement for k^{th} step. $\qquad\square$

Example 5.3.3. Consider the equation $x^3 - 2x - 5 = 0$.

Taking $x_0 = 2$, we find $f(x_0) = -1$, $f'(x_0) = 10$, $h_0 = .1$, and $J_0 = [2, 2.2]$. The second derivative of the polynomial is $6x$, whose supremum on J_0 is $M = 13.2$.

Since $|Mf(x_0)/(f'(x_0)^2| = .132 < .5$, Theorem 5.3.1 guarantees that there is a root in the interval $[2, 2.2]$, and that Newton's method starting at $x_0 = 2$ will converge to it. $\qquad\blacktriangle$

GLOBAL BEHAVIOR OF NEWTON'S METHOD

You pay for the speed at which Newton's method converges by the uncertainty of what the method will converge to: if you do not start sufficiently close to the desired root (whatever that means), the method may converge to any root, or fail to converge at all. We shall illustrate this statement in the next subsection, especially Examples 5.3.6–5.3.8, but we begin with a simpler case, in fact very nearly the only case in which Newton's method is actually understood as a dynamical system.

Example 5.3.4. *Quadratic equations and square roots.* Suppose we want to solve the quadratic equation

$$f(x) = (x - a)(x - b) = 0$$

by Newton's method. Of course, this is a ridiculous thing to do, since we already know the roots, namely a and b. However, if we want to understand how Newton's method works, it is best to start with a simple example.

In this case, Newton's method as defined by equation (11) gives

$$N_f(x) = x - \frac{x^2 - (a+b)x + ab}{2x - (a+b)} = \frac{x^2 - ab}{2x - (a+b)}.$$

There is no trouble in seeing that a and b are indeed superattracting fixed points, but in this case it is possible to analyze the system completely, by a clever change of variables.

$$x = \varphi(y) = \frac{by - a}{y - 1}.$$

Then we can show (below) that

$$\varphi^{-1} \circ N_f \circ \varphi(y) = y^2 = g(y) \qquad (15)$$

Remark. The mapping φ sends $y = 0$ and $y = \infty$ to $x = a$ and $x = b$.

The proof of equation (15) is a "simple" computation (Exercise 5.3#8a). Note that

$$\varphi^{-1}(x) = \frac{x - a}{x - b}.$$

and we have

$$g(y) \xrightarrow{} y^2$$
$$\varphi \downarrow \qquad\qquad \downarrow \varphi$$
$$\xrightarrow{N_f}$$

Computing away, we find

$$y \xrightarrow{\varphi} \frac{by - a}{y - 1} \xrightarrow{N_f} \frac{\frac{(by-a)^2}{(y-1)^2} - ab}{\frac{2(by-a)}{(y-1)} - (a+b)} = \frac{by^2 - a}{y^2 - 1} \xrightarrow{\varphi^{-1}} \frac{\frac{by^2-a}{y^2-1} - a}{\frac{by^2-a}{y^2-1} - b} = y^2,$$

which indeed proves equation (15). This fact allows us to understand everything about Newton's method for quadratic polynomials, by simply changing to the variable y where everything is known because $g(y) = y^2$. We leave it to the reader (Exercise 5.3#8b) to verify the following:

> If you start Newton's method to the right of $(a+b)/2$, you will converge to the larger root, and if you start to the left, you will converge to the smaller root. If you attempt to start at $(a+b)/2$, Newton's method does not converge at all.

To find the square root of a real number a we need to solve the equation $x^2 - a = 0$, for $a > 0$. Newton's method (11) in this case consists of iterating

$$N_a(x) = x - \frac{(x^2 - a)}{2x} = \frac{1}{2}\left(x + \frac{a}{x}\right).$$

By Exercise 5.3#8b you have shown that if the initial x_0 is chosen positive, Newton's method will converge to the positive square root of a; and if x_0 is chosen negative, the method will lead to the negative square root.

▲

DISORDER IN NEWTON'S METHOD

Various unpleasant things may cause difficulty when applying Newton's method. For instance,

(a) Newton's method does not work if the *tangent* at $f(x_0)$ is *horizontal*.

Example 5.3.5. See Figure 5.3.3. ▲

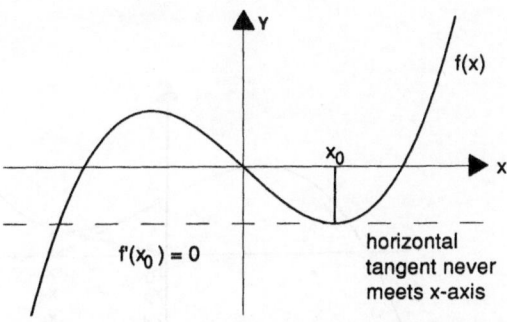

FIGURE 5.3.3. Horizontal tangent at x_0.

(b) Newton's method may converge, but to a different root than expected.

Example 5.3.6. Figure 5.3.4 shows an orbit which converges not to the root closest to the initial point, but to the farthest. ▲

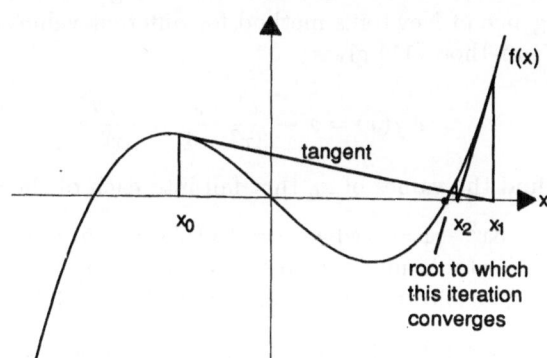

FIGURE 5.3.4. Convergence to a further root.

(c) Newton's method does not work, because the iteration does something else, such as being cyclic of order > 1, or being attached to a cycle of period > 1, or being "chaotic." ▲

Example 5.3.7. For $x^3 - x + \sqrt{2}/2$, 0 is part of an attracting cycle period 2.

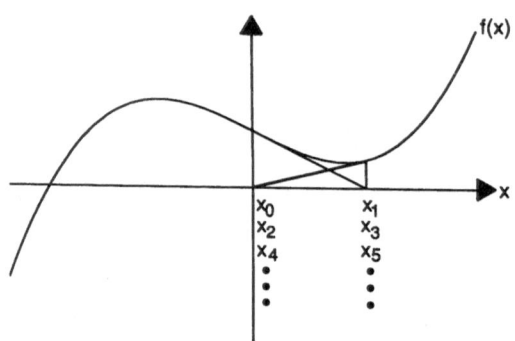

FIGURE 5.3.5. Newton's method in a cycle.

For quadratic polynomials, only difficulty (a) can occur. Cubic polynomials are a much richer hunting ground for strange behavior.

Example 5.3.8. Consider $f(x) = x^3 - x + c$, for which we already *know* the roots are 0 and ± 1. Nevertheless this is a good example for analyzing the convergence of Newton's method for different values of x_0.

Newton's method (11) gives

$$N_f(x) = x - \frac{x^3 - x}{3x^2 - 1} = \frac{2x^3}{3x^2 - 1}.$$

We will find the values of x_0 that fall into each of the above classes:

(a) $f'(x) = 3x^2 - 1 = 0$ when $x = \pm 1/\sqrt{3} \approx \pm.57735$, so for these values of x_0 Newton's method fails to converge because of the horizontal tangent, as shown in the example of Figure 5.3.3.

(b) A cycle order 2 will occur, by symmetry, where $N_f(x) = -x$.

$$\frac{2x^3}{3x^2 - 1} = -x \quad \text{when} \quad x = 0, \quad x = \pm\frac{1}{\sqrt{5}} \approx \pm.44721.$$

The point $x_0 = 0$ is a fixed point, which *does* converge under Newton's method, but the others form a cycle of order 2 that does not converge to a root, similar to that shown in the example of Figure 5.3.5.

(c) As shown in the example of Figure 5.3.4, if $x_0 = -.5$, Newton's method iterates to neither of the closer roots, but to $+1$. There are, in fact, whole intervals of values of x_0 for which Newton's method behaves like this, and we can further analyze this phenomenon.

For an initial guess x_0 chosen from the intervals $(-1/\sqrt{3}, -1/\sqrt{5})$ and $(1/\sqrt{5}, 1/\sqrt{3})$, the picture of what happens under Newton's method is complicated: In Figure 5.3.6 all those initial values that eventually iterate to the right-hand root $(r = +1)$ are "colored" with a thick line; those that converge to the left-hand root $(r = -1)$ are "colored" in normal thickness; those that converge to the center root $(r = 0)$ are "colored" in white. Those points x_0 for which Newton's method does not converge to any root at all are indicated by the tic marks; there are more and more of them as x_0 approaches the inside endpoint of the complicated intervals.   ▲

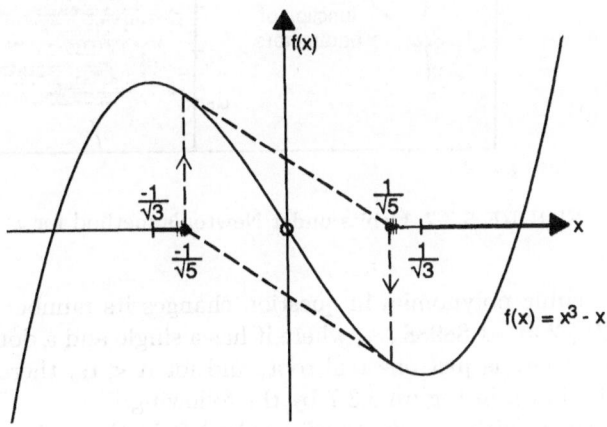

FIGURE 5.3.6. Convergence of Newton's method for different x_0.

Example 5.3.8 is just the tip of the iceberg as to the delicacy in certain regions about how close x_0 is to the root being sought by Newton's method. If you have a picture of the graph of the function to which you wish to apply Newton's method, you may be able to choose x_0 wisely; if not, try to get as close as you can—a good choice would be a place where $f(x)$ changes sign. If your result is somewhere far from what you expected, the picture may be like Figure 5.3.6.

Let us now study from a different point of view the behavior under iteration of cubic polynomials. The program *Cascade* will plot for you the orbit of 0 under iteration by Newton's method for the polynomial $x^3 + \alpha x + 1$. (Exercise 5.3#4b asks you about the generality of this expression, to show that any cubic polynomial can be written this way with just one parameter if you make an appropriate change of variables.) Just as in the last section for Figure 5.2.2, the cascade of bifurcations for quadratics, you can choose a vertical range for α, how many (ℓ) points to mark and how many (k) not to mark; only the function to iterate has changed. The picture you get if you let α vary from -2 to -1, for $k = 20$ and $\ell = 100$, is shown in Figure 5.3.7.

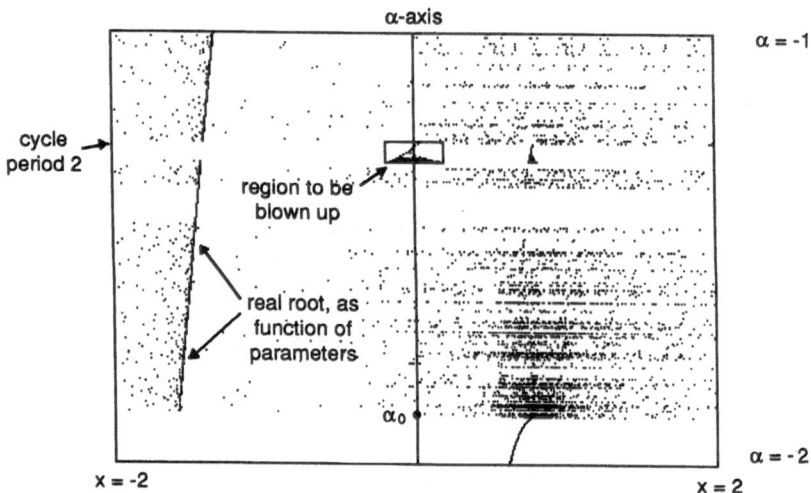

FIGURE 5.3.7. Orbits under Newton's method for $x^3 + \alpha x + 1$.

The cubic polynomial in question changes its number of roots at $\alpha_0 = -(3/2)\sqrt[3]{2} \approx -1.88988\ldots$, where it has a single and a double real root. For $\alpha > \alpha_0$ there is just one real root, and for $\alpha < \alpha_0$ there are 3 real roots. This is shown in Figure 5.3.7 by the following:

We see a fairly steady "line" to the left in the region $\alpha > \alpha_0$, which is the single real root, and which frequently does attract 0 (one of the critical points and the starting point for iteration on each line of this picture). Gaps in this "line" represent values of α for which 0 is not being attracted to this root. For $\alpha < \alpha_0$, there is another perfectly steady "line", representing the fact that one of the two new roots (that appear when $\alpha < \alpha_0$) consistently attracts 0. However, in the middle of the picture (where there are gaps in these "lines") there appear to be more complicated regions, where 0 does something other than being attracted to the root, and we see a blowup of such a region below in Figure 5.3.8.

Figure 5.3.8 is strikingly similar to Figure 5.2.2 for straight (nonNewton) iteration of quadratics, and in fact reproduces it in full detail. This appearance of the "cascade picture" in settings other than quadratic polynomials is a large part of why the cascade of bifurcations has attracted so much attention recently. It appears to be a "universal drawing," an archetypal figure which can be expected to show up in many different settings. This is an instance of the universality of the cascade picture investigated by Feigenbaum.

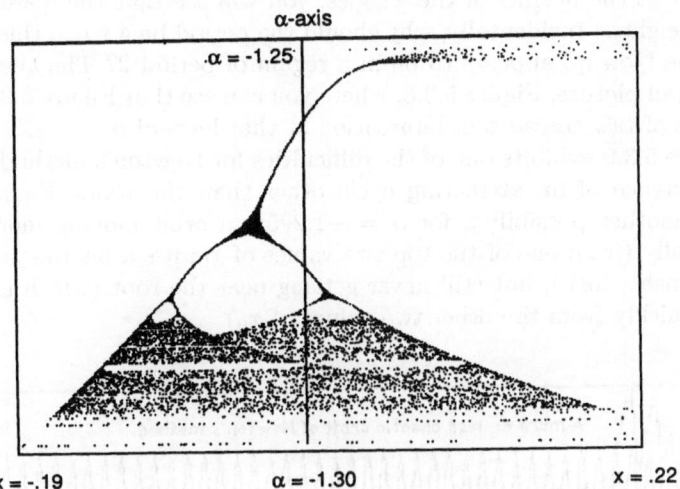

FIGURE 5.3.8. A cascade of bifurcations for $x^3 + \alpha x + 1$.

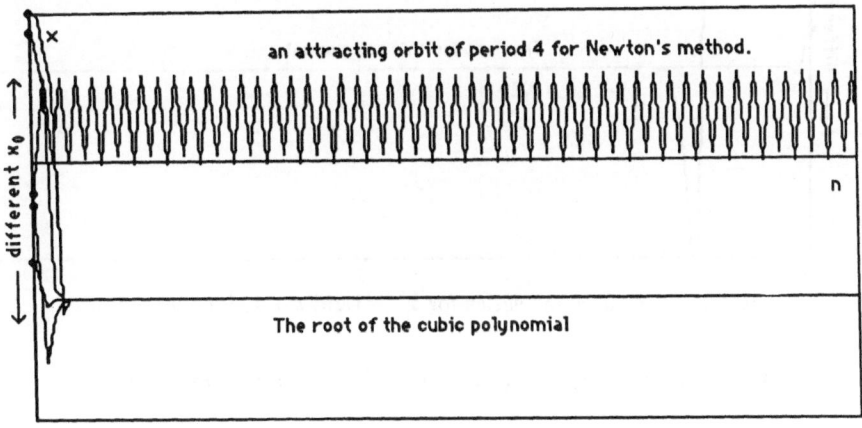

FIGURE 5.3.9. Time series for $x^3 - 1.28x + 1$, for different x_0.

There are actually infinitely many other blowups of Figure 5.3.7 which will show the same picture (in fact infinitely many blowups even within the region already blown up).

Figure 5.3.9 is a superposition of several time series pictures for five different values of x_0, for $\alpha = -1.28$, which occurs in the middle of Figure 5.3.7. We see some orbits being attracted (very fast) to the root, and another (the middle value of x_0) being attracted to an attracting cycle of period 4. (You can tell that it is period 4 rather than period 2 by looking carefully at the heights of the wiggles; you will see that there are two different heights.) Incidentally, why should the period be 4 when this value of α, in the blowup, appears to be in a region of period 2? The answer is in the overall picture, Figure 5.3.8, where you can see that Figure 5.3.9 covers just one of *two* cascades of bifurcation at that level of α.

Figure 5.3.9 exhibits one of the difficulties for Newton's method, namely the existence of an attracting cycle other than the roots. Figure 5.3.10 shows another possibility, for $\alpha = -1.295$: an orbit moving more or less chaotically (from one of the top two values of x_0; it's a bit too jumbled to distinguish which), but still never getting near the root (which is reached fairly quickly from the other two values of x_0).

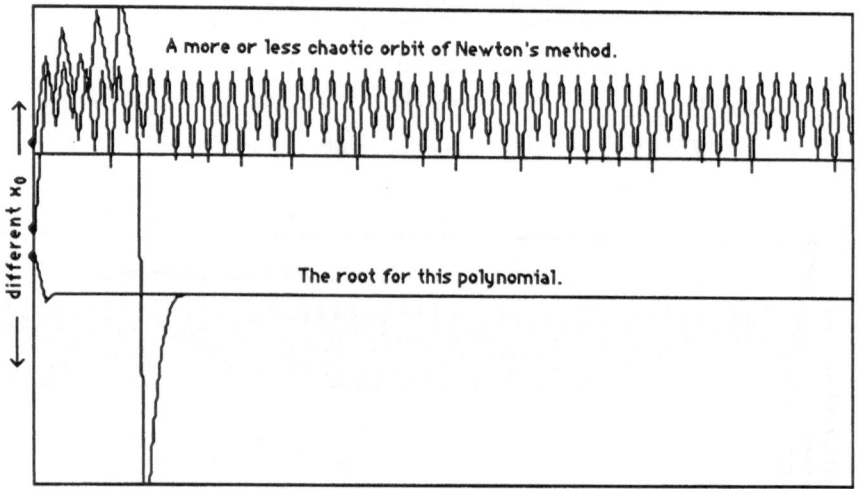

FIGURE 5.3.10. Time series for $x^3 - 1.295x + 1$, for different x_0.

GENERALIZATION OF NEWTON'S METHOD TO MORE VARIABLES

Newton's method in several variables is an even more essential tool in several variables than it is in one variable, and also far more poorly understood. The bare bones of this generalization are as follows:

Newton's method as expressed by equation (11) at the beginning of this Section 5.3 transfers exactly to higher dimensions, where both $\mathbf{f}$ and $\mathbf{x}$ are vector quantities. If $\mathbf{f}: \mathbb{R}^n \to \mathbb{R}^n$ is a function and we look for a solution of $\mathbf{f}(0) = 0$ near x_0, we may replace $\mathbf{f}$ by its linear approximation

$$L_{\mathbf{x}_0}(\mathbf{x}) = \mathbf{f}(\mathbf{x}_0) + (d_{\mathbf{x}_0}\mathbf{f})(\mathbf{x} - \mathbf{x}_0) \tag{16}$$

where $\mathbf{d}_{\mathbf{x}_0}\mathbf{f}$ is the $(n \times n)$-matrix of partial derivatives of the components of $\mathbf{f}$ evaluated at $\mathbf{x}_0$.

Example 5.3.9. In Volume II, Section 8.1, we shall use this scheme in $\mathbb{R}^2$, with

$$\mathbf{x} = \begin{bmatrix} x \\ y \end{bmatrix} \quad \text{and} \quad \mathbf{f} = \begin{bmatrix} f(x,y) \\ g(x,y) \end{bmatrix};$$

this equation (16) would become

$$L_{\mathbf{x}_0} \begin{bmatrix} x \\ y \end{bmatrix} = \begin{bmatrix} f(x_0, y_0) \\ g(x_0, y_0) \end{bmatrix} + \underbrace{\begin{bmatrix} \dfrac{\partial f(x,y)}{\partial x} & \dfrac{\partial f(x,y)}{\partial y} \\[2ex] \dfrac{\partial g(x,y)}{\partial x} & \dfrac{\partial g(x,y)}{\partial x} \end{bmatrix}_{\mathbf{x}_0}}_{d_{\mathbf{x}_0}} \begin{bmatrix} x - x_0 \\ y - y_0 \end{bmatrix}. \quad \blacktriangle$$

If $d_{\mathbf{x}_0}\mathbf{f}$ is invertible, the equation

$$L_{\mathbf{x}_0}(\mathbf{x}) = 0$$

has the unique root

$$\mathbf{x}_1 = \mathbf{x}_0 - (d_{\mathbf{x}_0}\mathbf{f})^{-1}(\mathbf{f}(\mathbf{x}_0)) = N_f(\mathbf{x}_0).$$

Exactly as in the single variable case, you may expect that if $\mathbf{x}_0$ is chosen sufficiently near a root of $\mathbf{f}$, then the sequence of iterates of N_f starting at $\mathbf{x}_0$ will converge to the root.

5.4 Numerical Methods as Iterative Systems

In Section 3.3–3.5, we discussed at some length the question of what happens to numerical methods if we *fix the independent variable t and decrease stepsize $h = \Delta t$*? Now we want to discuss a quite different aspect: what happens if we *fix stepsize h and increase the independent variable t*? It is really in order to study this second question that the present chapter is included.

At heart, the kind of thing we want to discuss is illustrated by the two pictures on the next page as Figures 5.4.1 and 5.4.2.

Figure 5.4.1 was obtained by "solving" $x' = x^2 - t^2$, using *DiffEq* with Runge–Kutta, $-10 \le t \le 10$, and stepsize $h = 0.3$. The middle of the picture is reasonable, but the jagged "solutions curves" to the left and right are wrong; the solutions of the differential equation do not look like that. In fact, they look like the picture in Figure 5.4.2.

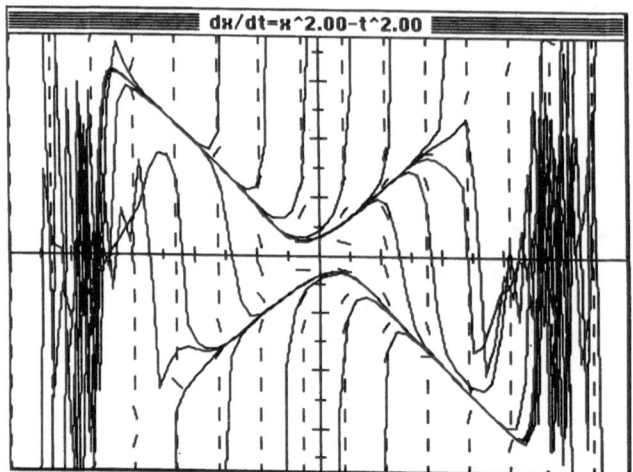

FIGURE 5.4.1. Some solutions for $x' = x^2 - t^2$ by Runge–Kutta with $h = 0.3$.

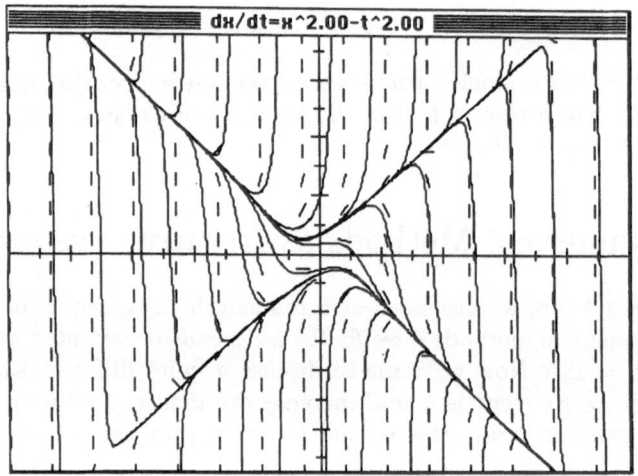

FIGURE 5.4.2. Some solutions for $x' = x^2 - t^2$ by Runge–Kutta with $h = 0.03$.

Figure 5.4.2 is obtained from the same program (*DiffEq* with Runge–Kutta and $-10 \leq t \leq 10$), changing only the stepsize, to $h = 0.03$. It is easy to show that the funnel and antifunnel are really there, so that at least these properties of the actual differential equation are accurately reflected by the drawing. In reality, if solutions are drawn for larger t's, even with this stepsize the solutions in the funnel will eventually become jagged junk. You can right now experiment, to find for which t this occurs; after having read the section you should be able to predict it without the computer. Exercise 5.4#6 will ask you to do so.

Where did the jaggles come from? We will explore this question in this section and see that it comes from a lack of *"numerical stability"* of the computation (not a lack of stability of the solution itself, but of the numerical scheme for approximating solutions of differential equations).

A first observation is that *each of the numerical methods we have discussed may be interpreted as an iterative system.* Consider for instance *Euler's method* applied to $x' = f(t, x)$:

$$t_{n+1} = t_n + h$$
$$x_{n+1} = x_n + hf(t_n, x_n).$$

The successive points (t_n, x_n) are the orbit of (t_0, x_0) under iteration of the mapping $\tilde{F}_h \colon \mathbb{R}^2 \to \mathbb{R}^2$ given by

$$\tilde{F}_h \begin{bmatrix} t \\ x \end{bmatrix} = \begin{bmatrix} t + h \\ F_h \end{bmatrix}, \quad \text{where} \quad F_h \equiv x + hf(t, x).$$

We shall study the behavior of F_h, the second coordinate of $\tilde{F}_h$.

We definitely do want to study the nonautonomous case, but first we will see what happens for autonomous equations. If the differential equation is *autonomous,* of the form

$$x' = f(x),$$

then, for Euler,

$$F_h(x) = x + hf(x). \tag{17}$$

Actually, for autonomous systems it is also true of the other methods under discussion that there exists an analogous function F_h for the second coordinate of $\tilde{F}_h$ that does not depend on t, although the formulas corresponding to equation (17) are more complicated (Exercises 5.4#1):

Midpoint Euler gives $F_h(x) = x + hf\left(x + \frac{h}{2}f(x)\right)$;

Runge–Kutta gives the rather awesome formula

$$F_h(x) = x + \frac{h}{6}\left(f(x) + 2f\left(x + \frac{h}{2}f(x)\right) + 2f\left(x + \frac{h}{2}f\left(x + \frac{h}{2}f(x)\right)\right) +$$
$$f\left(x + hf\left(x + \frac{h}{2}f\left(x + \frac{h}{2}f(x)\right)\right)\right)\right).$$

The following example, although simple and completely analyzable, is still the most important example in understanding numerical stability.

Example 5.4.1. We will try various numerical methods for the archetypical linear differential equation $x' = -\alpha x$, with $\alpha > 0$. Since the solutions of this equation are $x(t) = e^{-\alpha t}$, all solutions go monotonically to 0, faster and faster as α gets larger.

A perfect numerical method with step h would give

$$F_h(x) = e^{-\alpha h}x.$$

Now let us examine what in fact our actual numerical methods yield:

(a) *Euler's method* with step h comes down to iterating

$$F_h(x) = x - \alpha h x = (1 - \alpha h)x.$$

If h is very small, then $1 - \alpha h$ is quite close to $e^{-\alpha h}$, in fact it is the value of the Taylor polynomial approximation of degree 1 (Appendix A3). If h is larger, things are not so good.

Since the nth iterate $F^{\circ n}(x) = (1 - \alpha h)^n x$, the orbit of x will go to zero if and only if $|(1 - \alpha h)| < 1$, which will occur if $h < 2/\alpha$. Moreover, the orbit of x under F_h

will go monotonically to 0 if $h < 1/\alpha$,

will land on 0 in one move and stay there if $h = 1/\alpha$,

will tend to zero, oscillating above and beneath zero if $1/\alpha < h < 2/\alpha$.

So we see that only if the step h is chosen smaller than $1/\alpha$ does the numerical method reproduce even the qualitative behavior of the differential equation; the numerical method does not even give solutions which tend to 0 at all if $h > 2/\alpha$. ▲

Euler's method applied to $x' = -\alpha x$, as in Example 5.4.1, exhibits "*conditional stability*," i.e., the stability of approximate solutions depends on the stepsize. Note the peculiar feature that *the more strongly 0 attracts solutions, the shorter the step must be* for the numerical method to reflect this attraction. This phenomenon, known as "*stiffness*," is at the root of many difficulties with numerical methods. We shall see this particularly in Volume III when we try to solve analogs of the heat equation. These are the situations in which implicit methods are particularly useful, as shown below in part (d).

Example 5.4.2. Let us try other methods on $x' = -\alpha x$.

(b) *Midpoint Euler,* applied to the same equation with stepsize h, amounts to iterating

$$F_h(x) = x + h\left(-\alpha\left(x + \frac{h}{2}(-\alpha x)\right)\right) = \left(1 - \alpha h + \frac{\alpha^2 h^2}{2}\right)x.$$

We can analyze this system, exactly as above. For h small, we now see the second degree Taylor polynomial (again, see Appendix A3), and can expect the numerical approximation to be quite good. As you can confirm in Exercise 5.4#2a, we find that all solutions tend monotonically to 0 if $0 < h < 2/\alpha$, but if $h > 2/\alpha$, the orbit goes to ∞.

(c) *Runge–Kutta* applied to the same equation with stepsize h, amounts to iterating

$$F_h(x) = \left(1 - \alpha h + \frac{h^2\alpha^2}{2} - \frac{h^3\alpha^3}{6} + \frac{h^4\alpha^4}{24}\right)x.$$

Exercise 5.4#2b asks you to analyze the iteration of this function and find for what h do orbits tend to 0? You will find a wider region of conditional stability.

(d) *Implicit Euler* applies the equation

$$x_{i+1} = x_i + hf(t_{i+1}, x_{i+1}) = x_i - h\alpha x_{i+1},$$

which gives

$$x_{i+1} = x_i/(1 + \alpha h).$$

The function $1/(1 + \alpha h)$ has the Taylor series

$$1 - \alpha h + \alpha^2 h^2 - \alpha^3 h^3 + \dots ,$$

and for small h is close to $1 - \alpha h$. *Since the quadratic term is different from the quadratic term of the Taylor series, one expects that this method will have the same order as Euler's method, i.e., order 1.* The implicit method's great advantage comes from the fact that for all positive h, we have

$$0 < 1/(1 + \alpha h) < 1.$$

Thus for all step lengths, all solutions tend monotonically to 0, as they should. So for this equation, this implicit method is much better than the explicit ones; there will never be any jagged "solutions."   ▲

Examples 5.4.1 and 5.4.2 are of course exceptionally simple; in general we cannot hope to analyze iterative numerical methods so simply.

Example 5.4.3. Now we will examine how the simplest numerical method deals with

$$x' = x - x^2. \tag{18}$$

Euler's method leads to iterating

$$F_h(x) = x + h(x - x^2), \tag{19}$$

a second degree polynomial. It is not too hard to show (Exercise 5.4#3a) that by choosing h appropriately and making a change of variables, you can get any quadratic polynomial this way.

So studying the solutions of equation (18) by Euler's method is just another way of studying iteration of quadratic polynomials. We will not repeat Section 5.2, but will apply the insight gained there to study the differential equation (18).

First, observe that $x = 1$ is a solution of the differential equation, which in fact attracts nearby solutions. Also observe, from equation (19), that $x = 1$ is a fixed point of F_h. If we set $x = 1 + u$ and write F_h in terms of u, we find

$$F_h(1 + u) = 1 + ((1 - h)u - hu^2). \qquad (20)$$

From the form of equation (20) we can see that $x = 1$ (or $u = 0$) is an *attracting* fixed point if and only if $|1 - h| < 1$, i.e., if and only if $0 < h < 2$.

Figures 5.4.3 and 5.4.4 represents solutions, through several different starting points each, of equation (18) under Euler's method (for steps $h = 1.95$ above and $h = 2.1$ below). Please observe that in the upper drawing the approximate solutions are all attracted to the actual solution $x = 1$, whereas in the lower drawing they are not, but instead are all attracted to a cycle of order 2, which represents some sort of "spurious solution."   ▲

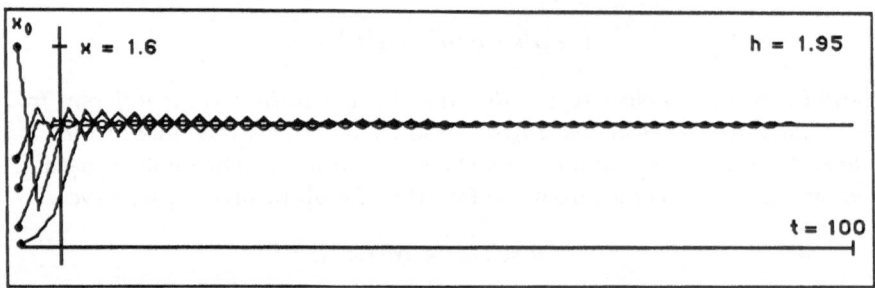

FIGURE 5.4.3. Iterating $x' = x - x^2$ by Euler's method, for stepsize 1.95.

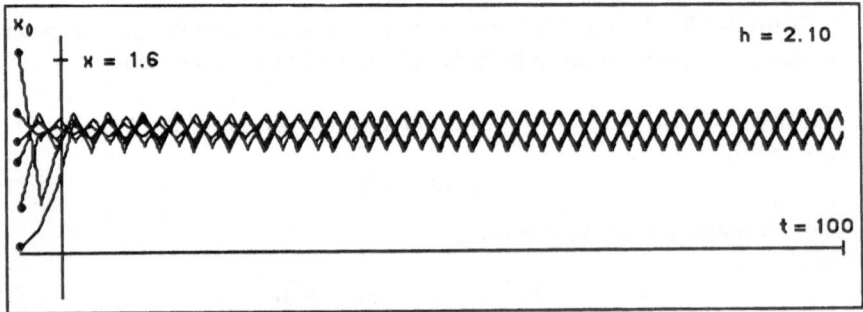

FIGURE 5.4.4. Iterating $x' = x - x^2$ by Euler's method, for stepsize 2.10.

Example 5.4.4. Now we will try a *nonautonomous* equation:

$$x' = -tx.$$

This is the equation of Examples 1.1.1 and 1.1.2, and the exact solutions are

$$x = Ce^{-t^2/2}.$$

Figure 5.4.5 shows various approximate solutions, all for stepsize $h = 0.3$, for the three numerical methods we have been using. We observe that all three methods appear to behave properly, giving solutions which tend to 0 as they should, for small t. But suddenly, as t becomes large (about 5 for Euler, 6.6 for Midpoint Euler, 9.3 for Runge–Kutta), solutions stop tending to 0; in fact, the solution 0 loses its stability, and solutions fly away from it.

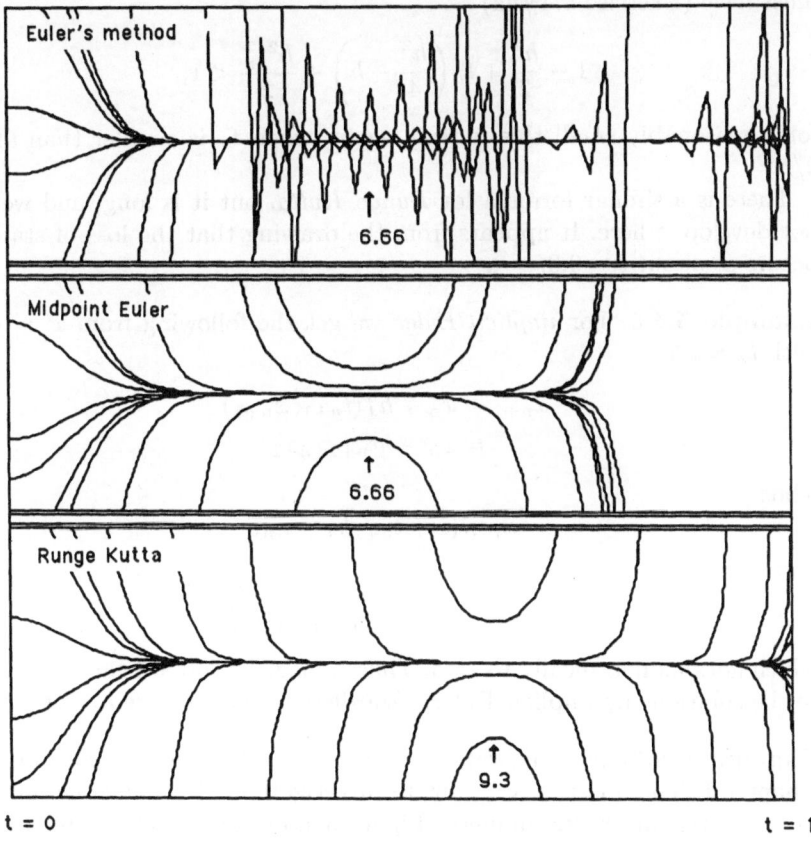

FIGURE 5.4.5. Solving $x' = -tx$ by different methods, for fixed stepsize.

Let us see what the theory predicts. For *Euler's method,* we are iterating

$$x_{n+1} = x_n + h(-t_n x_n) = x_n(1 - ht_n).$$

In order for the sequence (x_n) to be tending to 0, we must have

$$|1 - ht_n| < 1,$$

i.e., $t_n < 2/h$. For $h = .3$, this means that the solution 0 should lose its stability at $t = 6.666\ldots$.

For *Midpoint Euler*, the formula for x_n is a bit more formidable:

$$x_{n+1} = x_n\left(1 - \frac{h^2}{2} + t_n\left(\frac{h^3}{4} - h\right) + \frac{h^2 t_n^2}{2}\right).$$

Unlike the case of Euler's method, this does not become negative. This explains the different kinds of loss of stability: Euler's method oscillates around the correct solution, but midpoint Euler pulls away on one side. As soon as t_n becomes large, the quadratic term dominates. For stability we must have (Exercise 5.4#4a)

$$1 - \frac{h^2}{2} + t_n\left(\frac{h^3}{4} - h\right) + \frac{h^2 t_n^2}{2} < 1;$$

for h reasonably small this is true, again, when t_n is smaller than about $2/h$.

There is a similar formula for *Runge–Kutta*, but it is long, and we will not develop it here. It appears from the drawing that the loss of stability occurs at about $t = 9.3$. ▲

Example 5.4.5. For *implicit Euler* we get the following from $x' = -tx$, with $t_n = nh$:

$$x_{n+1} = x_n + hf(t_{n+1}, x_{n+1})$$
$$= x_n - ht_{n+1}x_{n+1}.$$

Thus

$$x_{n+1}(1 + t_{n+1}h) = x_n,$$

and

$$x_{n+1} = x_n\left(\frac{1}{1 + (n+1)h^2}\right),$$

which goes monotonically to zero. Thus *no* jaggies will appear in drawings of the solutions by implicit Euler, regardless of the stepsize h. ▲

Example 5.4.6. We will return to $x' = x^2 - t$, the ubiquitous equation of Chapter 1. The question we want to investigate is "Do numerical approximations remain in the funnel?" Figure 5.4.6 shows much of the answer. All three pictures are for the region $0 < t < 15$, and for Euler's method, Midpoint Euler and Runge–Kutta with stepsize $h = 0.4$.

The computer picture clearly shows that the funnel loses its stability, at approximately $t = 6$ for Euler's method and midpoint Euler, and somewhat later for Runge–Kutta.

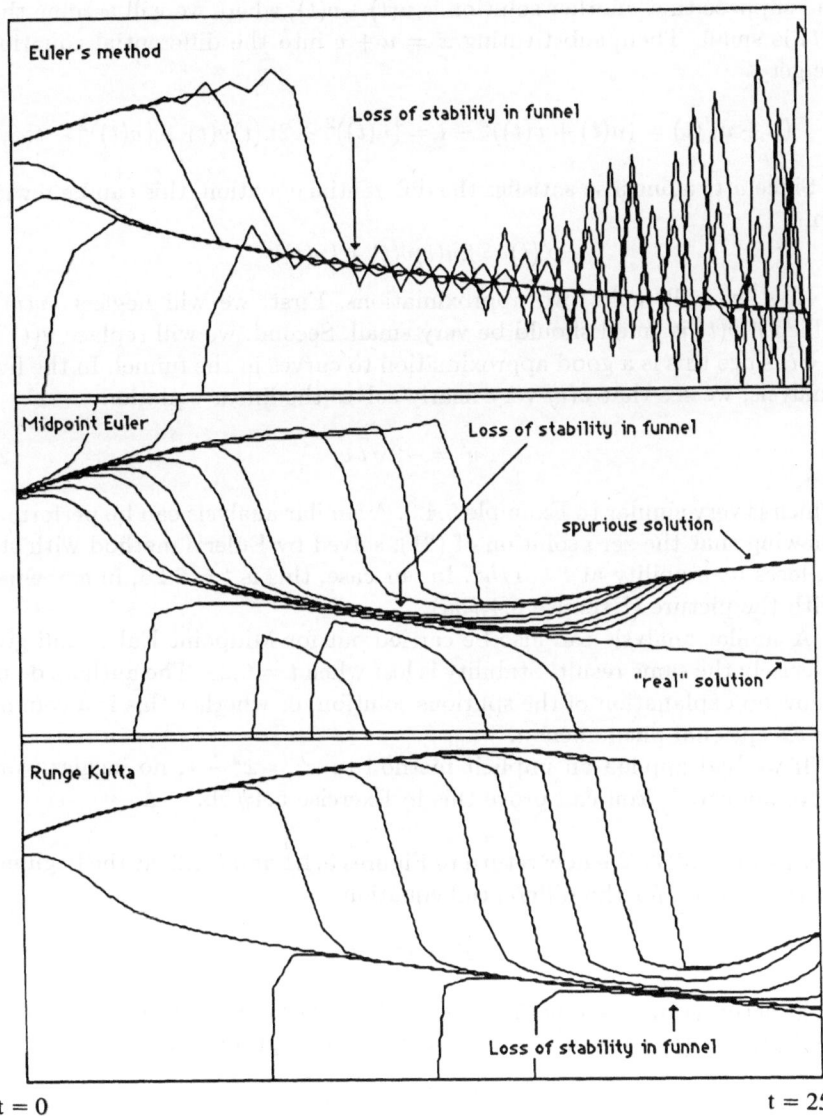

FIGURE 5.4.6. Iterating $x' = x^2 - t$ by different methods, for fixed stepsize.

We will analyze this phenomenon in a moment, but first observe that Midpoint Euler develops a really unpleasant behavior: a perfectly reasonable looking but spurious solution appears, designed to fool people who trust computers. (*We* know it is spurious from our analysis in Chapter 1.)

It is not quite obvious how to analyze Figure 5.4.6; the answer is to *linearize around a real solution.* Suppose $u(t)$ is a solution in the funnel, and suppose that another solution is $u(t) + v(t)$, where we will assume that $v(t)$ is small. Then, substituting $x = u + v$ into the differential equation, we get

$$(u + v)'(t) = (u(t) + v(t))^2 - t = (u(t))^2 + 2u(t)v(t) + (v(t))^2 - t.$$

Since $u(t)$ alone also satisfies the differential equation, this can be rewritten

$$v'(t) = 2u(t)v(t) + (v(t))^2.$$

Now we will make two approximations. First, we will neglect $(v(t))^2$, which if $v(t)$ is small should be very small. Second, we will replace $u(t)$ by $-\sqrt{t}$, since this is a good approximation to curves in the funnel. In the final analysis, we see that $v(t)$ very nearly solves the linear equation

$$v' = -2\sqrt{t}\, v, \tag{21}$$

which is very similar to Example 5.4.3. A similar analysis can be performed, showing that the zero solution of (21), solved by Euler's method with step h, loses its stability at $t = 1/h^2$. In our case, this is $t = 6.25$, in agreement with the picture (Exercise 5.4#5a).

A similar analysis can also be carried out for Midpoint Euler, and gives precisely the same result. Stability is lost when $t = 6.25$. The authors do not know an explanation of the spurious solution, or whether this is a common or exceptional phenomenon.

If we had applied an implicit method to $x' = x^2 - t$, no jaggies would have appeared. You can prove this in Exercise 5.4#5b. ▲

Example 5.4.7. We now return to Figures 5.4.1 and 5.4.2, at the beginning of the section, for the differential equation

$$x' = x^2 - t^2.$$

Exactly as in Example 5.4.6, if we let $u(t)$ be a solution in the funnel and $u(t) + v(t)$ a perturbation of it, then $v(t)$ approximately solves

$$v' = 2u(t)v(t),$$

and since $u(t) \approx -t$, we find that $v(t)$ approximately solves

$$v' = -2tv. \tag{22}$$

Equation (22) is almost that of Example 5.4.4, up to the factor of 2. So if we use a change of variables $s = \sqrt{2}t$, with $w(s) = v(t)$, then

$$w'(s) = -sw, \tag{23}$$

which is exactly the equation of Example 5.4.4. Since under Runge–Kutta with stepsize $h = 0.3$ solutions of (23) lose their stability at $s \approx 9.3$ and $t = s/\sqrt{2}$, we can expect solutions of

$$x' = x^2 - t^2$$

close to $u(t) = -t$, solved by Runge–Kutta with stepsize $h = 0.3$, to lose their stability at $\approx 9.3/\sqrt{2} \approx 6.6$.

This agrees with Figure 5.4.1, its t-coordinate runs from -7.5 to 7.5.

Exercise 5.4#6 asks you to perform a similar analysis for Figure 5.4.2 where $h = 0.03$, and to confirm your calculation by a computer drawing over an appropriate t domain. ▲

We conclude this section on conditional stability of numerical approximation with the remark that there exist higher order implicit methods of numerical approximation that have the precision of higher order methods like Runge–Kutta, but still have the stability of implicit Euler. However these methods are far more complicated and beyond the scope of this book. Consult the references listed at the end of this volume.

5.5   Periodic Differential Equations

This section is much closer to differential equations than the previous sections of this chapter. It reduces a surprisingly large class of differential equations to straight and simple iteration as described in Section 5.1. Periodic differential equations provide a first instance of the main way in which iterations arise when studying differential equations: *Poincaré sections*. This technique was introduced into the subject by Henri Poincaré, about 1890, and has remained an essential tool. The particular case we will study is very simple, but still brings out some of the main ideas. This will be investigated in a more substantial setting in Volume II, Chapter 13 on iteration in two dimensions.

By a *periodic differential equation* we shall mean throughout this section

$$x' = f(t, x) \tag{24}$$

where $f(t, x)$ will be assumed to be locally Lipschitz with respect to x, and, most importantly, *periodic of period T with respect to t*.

The periodicity of f means that $f(t, x) = f(t+T, x)$ for all t, x. Examples of such functions are

$x' = \cos(x^2 + t)$,   which is periodic of period 2π, or

$x' = x \operatorname{frac}(t)$,   where $\operatorname{frac}(t)$ is defined to be the fractional part of t, which is periodic of period 1.

The term "periodic differential equations" *sounds* more specialized than it is, as you can see from additional examples like

$$\left.\begin{array}{l} x' = 1 \\ x' = g(x) \end{array}\right\} \text{ both of which are periodic with any finite period.}$$

Remark. It is important to note that the hypothesis that f is periodic does *not* mean that the *solutions* of $x' = f(t, x)$ are periodic. It does mean that if you translate the whole field of slopes in the (t, x) plane to the right by T (or nT, where n is any integer), you find the same field as what you started with. Equivalently, if $\gamma(t)$ is a solution, then the translate $\gamma(t - nT)$ is also a solution for any integer n (but not in general the same solution).

Example 5.5.1. The differential equations studied in Example 2.5.2 are periodic. (See Figure 5.5.1.)

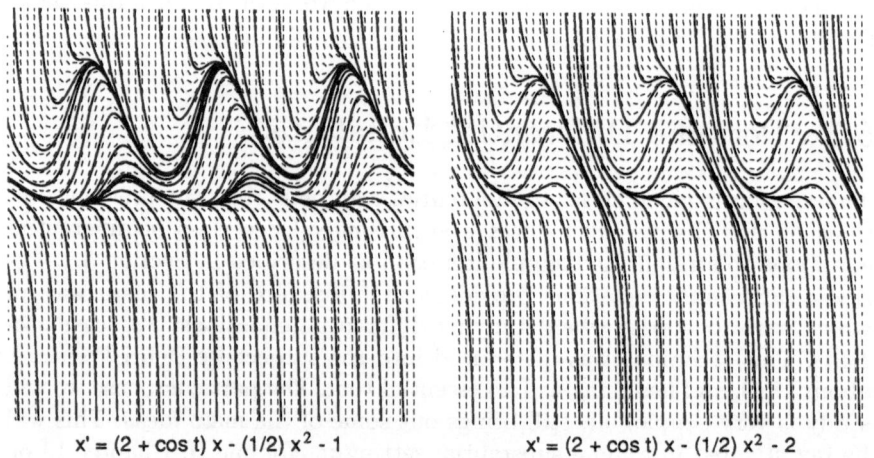

$$x' = (2 + \cos t)\, x - (1/2)\, x^2 - 1 \qquad\qquad x' = (2 + \cos t)\, x - (1/2)\, x^2 - 2$$

FIGURE 5.5.1. Periodic differential equations.

Particularly in the second drawing it is easy to see that while the direction field repeats with period 2π, the individual solutions $\gamma(t)$ are not periodic. However $\gamma(t - 2n\pi)$ is indeed also a solution, and one that has the same shape as $\gamma(t)$. ▲

To make the connection of periodic differential equations with iteration, we need to define a new function, called a *one period later mapping*, the simplest, one-dimensional Poincaré section.

Definition 5.5.2. Consider the differential equation $x' = f(t, x)$, with solutions $x = \gamma(t)$. Let $\gamma_{x_0}(t)$ be the solution with initial condition x_0 at

t_0, i.e., so that $\gamma_{x_0}(t_0) = x_0$. Then define the "*one period later mapping*" P_{t_0} as that which associates to x_0 the value of $\gamma_{x_0}(t)$ after one period, so that

$$P_{t_0}(x_0) = \gamma_{x_0}(t_0 + T).$$

The map P_{t_0} depends on the choice of t_0 and tells what happens to the solutions $\gamma(t)$ for each x_0.

P_{t_0} is constructed as follows, as illustrated in Figure 5.5.2:

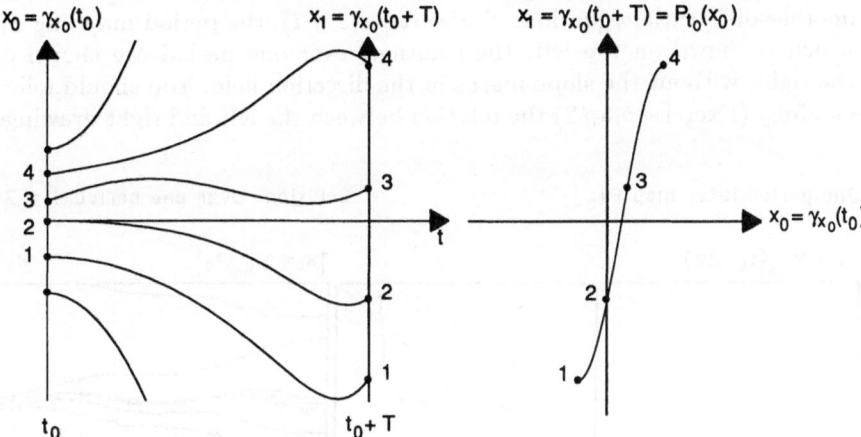

FIGURE 5.5.2. Construction of $P_{t_0}(x_0)$ for periodic differential equation $x' = f(t, x)$ with solutions $x = \gamma(t)$.

(i) choose t_0;

(ii) move vertically to x_0 at t_0;

(iii) for every value x_0 for which the solution $\gamma_{x_0}(t)$ makes it all the way across to $t_0 + T$, the *endpoints* $x_0 = \gamma_{x_0}(t_0)$ and $x_1 = \gamma_{x_0}(t_0 + T)$ *give a point* (x_0, x_1) of the graph of P_{t_0}.

In Figure 5.5.2 points 1, 2, 3, and 4 belong to the domain of P_{t_0}; the other values of x_0 do not. Also, as a consequence of the construction, you can note that

$$P_{t_0+T} = P_{t_0},$$

because the direction field at $t_0 + T$ simply repeats that at t_0. Exercise 5.5#1 asks you to show this fact directly from the definition, as a good way to master the role of all the subscripts therein.

For *linear* equations, P_{t_0} is a linear function $P_{t_0}(x) = ax + b$, for appropriate a and b (Exercise 5.5#8). For *separable* differential equations, the graph of P_{t_0} is sometimes a segment of the diagonal (Exercises 5.5#3c and d, 5, and 6).

The computer program *1-D Periodic Equations* will compute and draw these maps P_{t_0}, but we warn you that it is quite slow because each individual point on P_{t_0} is computed by following one entire solution on the direction field, which you already know takes several seconds. You can minimize this time by choosing a window for the x-range that only includes points in the domain of P_{t_0}, i.e., those points for which the solutions make it all the way across. Exercises 5.5#3 and 4 ask you to do this for the functions of Example 5.5.1.

For Figure 5.5.3 the program *1-D Periodic Equations* has plotted for another differential equation, $x' = \sin t \cos(x + t)$, the period mapping P_{t_0} which is shown on the left; the solutions over one period are shown on the right, without the slope marks in the direction field. You should follow carefully (Exercise 5.5#2) the relation between the left and right drawings.

One period later map P_{t_0} Solutions over one period T = 2π.

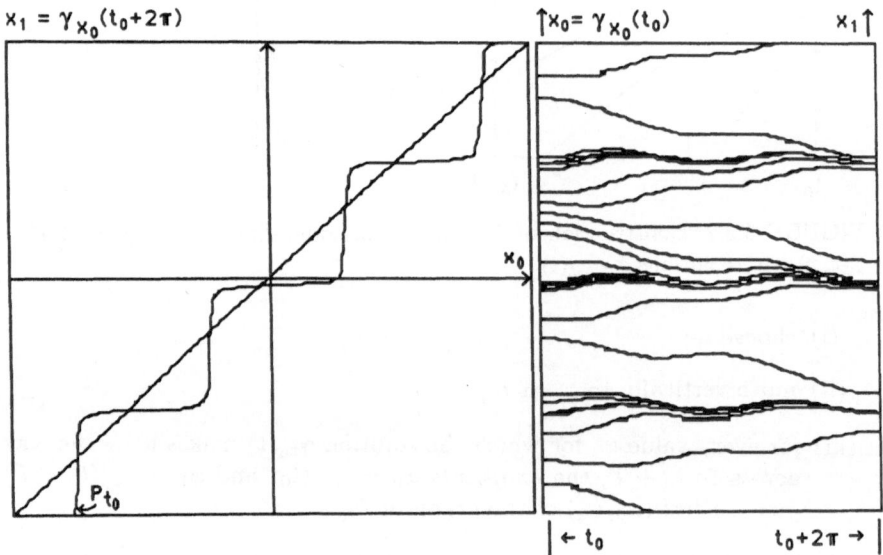

FIGURE 5.5.3. P_{t_0} compared with solutions $\gamma(t)$ for $x' = \sin t \cos(x + t)$.

Many features can be seen to correspond on the two drawings in Figure 5.5.3, more particularly attracting fixed points of P_{t_0} and attracting periodic solutions of the differential equation. Actually, the correspondence is very close; the point we want to make is that

> Iterating P_{t_0} is much the same as solving the differential equation $x' = f(t, x)$.

At least any information about "what happens in the long run" can be discovered by iterating P_{t_0}.

The reason is straightforward. Pick any x_0, and set

$$x_1 = P_{t_0}(x_0), \qquad x_2 = P_{t_0}(x_1), \ldots.$$

Then the solution $\gamma_{x_0}(t)$ of $x' = f(t, x)$ gives

$$\gamma_{x_0}(t_0 + T) = x_1, \quad \gamma_{x_0}(t_0 + 2T) = x_2, \ldots, \gamma_{x_0}(t_0 + nT) = x_n.$$

In other words, iterating P_{t_0} is just like solving the equation one period further.

Example 5.5.3. Consider the differential equation

$$x' = (x^2 - 1)(\cos t + 1). \tag{25}$$

Again, this equation (25) can be solved, to give

$$\frac{1}{2} \ln\left(\frac{x-1}{x+1}\right) = \sin t + t + C, \tag{26}$$

and the solution (26) can be solved explicitly for x in terms of t:

$$x(t) = \frac{1 + C^2 e^{2(t+\sin t)}}{1 - C^2 e^{2(t+\sin t)}}. \tag{27}$$

The solution $\gamma_{x_0}(t)$ with $\gamma_{x_0}(0) = x_0$ is then

$$\gamma_{x_0}(t) = \frac{(x_0 + 1) + (x_0 - 1)e^{2(t+\sin t)}}{(x_0 + 1) - (x_0 - 1)e^{2(t+\sin t)}}, \tag{28}$$

so that

$$P_{t_0}(x_0) = \gamma_{x_0}(2\pi) = \frac{x_0(1 + e^{4\pi}) + (1 - e^{4\pi})}{x_0(1 - e^{4\pi}) + (1 + e^{4\pi})}. \tag{29}$$

The graph of P_{t_0} is a hyperbola. Actually, it is only the left-hand branch of the hyperbola, the right-hand branch corresponds to solutions which "go through infinity"; you can check (Exercise 5.5#9) that if

$$x_0 > \frac{e^{4\pi+1}}{e^{4\pi-1}}, \tag{30}$$

then the formula (28) for γ_{x_0} does not correspond to a genuine solution of the differential equation, but goes to infinity at some t, $0 < t < 2\pi$. Figure 5.5.4 shows this graph, together with several orbits. Observe that P_{t_0} has two fixed points, ± 1, corresponding to the two constant solutions of the differential equation. Every point $x < 1$ iterates to -1, and every point

$x > 1$ leads to an increasing sequence which eventually lands in the region (30) where P_{t_0}, as given by equation (29), is not defined. ▲

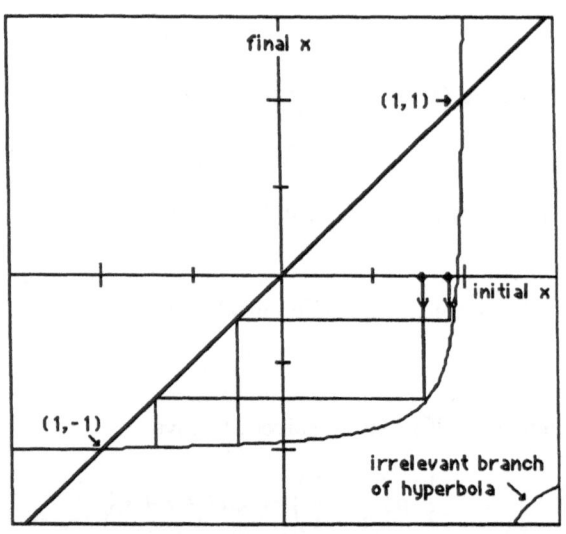

FIGURE 5.5.4. Iterating P_{t_0} for $x' = (x^2 - 1)(\cos t + 1)$.

Example 5.5.4. Consider the following differential equation periodic with period 2π:

$$x' = (\sin t + \alpha)x - \beta\, e^{\cos t}x^2. \tag{31}$$

This is a "Bernoulli equation" (Exercise 2.3#9). The Bernoulli "trick" involves setting $x = 1/u$, transforming equation (31) to

$$-\frac{u'}{u^2} = \frac{\sin t + \alpha}{u} - \frac{\beta e^{\cos t}}{u^2},$$

whereby multiplying by $-u^2$ gives

$$u' = -(\sin t + \alpha)u + \beta\, e^{\cos t}. \tag{32}$$

Equation (32) is a linear equation, which can be solved by variation of parameters, to give

$$u(t) = u(0)e^{\cos t - 1 - \alpha t} + \beta \int_0^t e^{\cos t - \cos s - \alpha(t-s)}e^{\cos s}ds$$

$$= u(0)e^{\cos t - 1 - \alpha t} + \frac{\beta}{\alpha}(e^{\alpha t} - 1)e^{\cos t - \alpha t}.$$

In particular, we find

$$u(2\pi) = e^{-2\pi\alpha}u(0) + \frac{\beta e}{\alpha}(1 - e^{-2\pi\alpha}) = Au(0) + B,$$

where A and B are the constants

$$A = e^{-2\pi\alpha}, \qquad B = \frac{\beta e}{\alpha}(1 - e^{-2\pi\alpha}).$$

Now going back to the variable $x = 1/u$, and as in Definition 5.5.2 calling $\gamma_{x_0}(t)$ the solution with $\gamma_{x_0}(0) = x_0$, we find

$$P_{t_0}(x_0) = \gamma_{x_0}(2\pi) = \frac{1}{(A/x_0) + B} = \frac{x_0}{A + Bx_0}. \tag{33}$$

Thus $P_{t_0}(x_0) = x_0/(A + Bx_0)$. *Actually again this is not quite true.* Only the right-hand branch of the hyperbola (33) belongs to the graph of P_{t_0}; $P_{t_0}(x_0)$ is undefined for $x_0 \leq -A/B$. Indeed, in those cases the solution we have written down goes through ∞, and is not really a solution of the differential equation on the entire interval $0 \leq t \leq 2\pi$.

From the point of view of iteration, there are three very different cases to consider: $A > 1$, $A = 1$, and $A < 1$; which correspond to α positive, zero or negative. We will only consider the case B positive, which corresponds to $\beta > 0$.

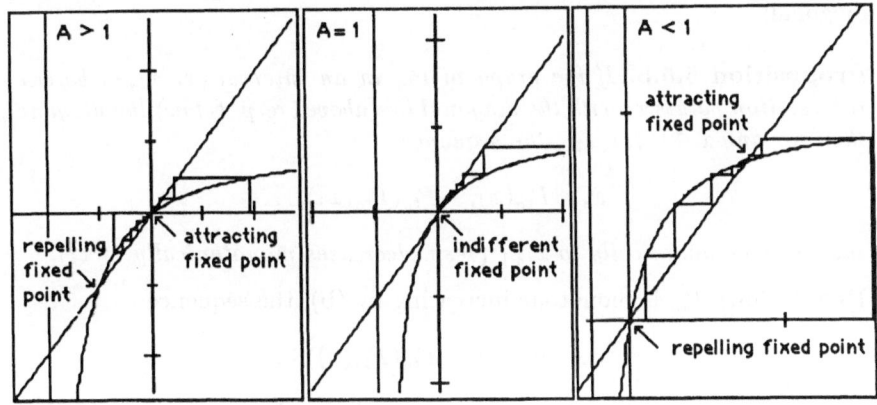

FIGURE 5.5.5. For $x' = (\sin t + \alpha)x - \beta e^{\cos t}x^2$, graphs of P_{t_0} for different values of $A = e^{-2\pi\alpha}$.

Graphs of P_{t_0} in these three cases are shown in Figure 5.5.5. We see (as you can verify in Exercise 5.5#10) that

If $A > 1$, 0 is an attracting fixed point and there is a negative repelling fixed point;

For $A = 1$, 0 is an indifferent fixed point, attracting points to the right and repelling points to the left;

If $A < 1$, 0 is a repelling fixed point and there is a positive attracting fixed point.

Exercise 5.5#11 asks you to show with graphs how these statements about P_{t_0} translate into statements about the solutions to the differential equation.   ▲

PROPERTIES OF THE FUNCTION P_{t_0}

(a) The function P_{t_0} is defined in some interval (a, b), where either end-point can be infinite.

(b) On this interval P_{t_0} is continuous and monotone increasing.

Any function g satisfying the two properties (a) and (b) is indeed the period-mapping function P_{t_0} for some differential equation, so there is not much more to say in general. In Exercise 5.5#12 you are asked to prove this if g is defined on all of $\mathbb{R}$ and is onto. The general case is not much harder, but a lot messier.

Functions satisfying (a) and (b) are especially simple under iteration. In general, such a function (in fact any differentiable function) will have a graph which will intersect the diagonal in points $\ldots, x_{-1}, x_0, x_1, \ldots$ and in each interval (x_i, x_{i+1}), the graph will lie either above or below the diagonal.

Proposition 5.5.5. *If the graph of P_{t_0} in an interval (x_i, x_{i+1}) between intersection points x_i with the diagonal lies above (resp. below) the diagonal, then for any $x \in [x_i, x_{i+1}]$, the sequence*

$$x, \quad P_{t_0}(x), \quad P_{t_0}(P_{t_0}(x)), \ldots$$

increases monotonically to x_{i+1} (resp. decreases monotonically to x_i).

Proof. Since P_{t_0} is monotone increasing by (b), the sequence

$$x, \quad P_{t_0}(x), \quad P_{t_0}(P_{t_0}(x)), \ldots$$

is either monotone increasing, monotone decreasing or constant; indeed, P_{t_0} being monotone increasing means that

$$x < y \Rightarrow P_{t_0}(x) < P_{t_0}(y). \tag{34}$$

Apply inequality (34) to the first two terms of the sequence:

$$\text{if } x < P_{t_0}(x) \quad \text{then } P_{t_0}(x) < P_{t_0}(P_{t_0}(x)),$$

$$\text{if } x > P_{t_0}(x) \quad \text{then } P_{t_0}(x) > P_{t_0}(P_{t_0}(x)).$$

The first of these alternatives occur if $x \in (x_i, x_{i+1})$ and the graph of P_{t_0} is above the diagonal in that interval; the second alternative occurs if the graph is below.

Similarly, if $x \in (x_i, x_{i+1})$, then $P_{t_0}(x) \in (x_i, x_{i+1})$ by applying (34) to x and x_i or x_{i+1}, so the entire sequence is contained in the interval, and in particular remains in the domain of P_{t_0}.

Thus the sequence is monotone and bounded, so it must tend to some limit. The limit must be in the interval $[x_i, x_{i+1}]$, and it must be a fixed point of P_{t_0} (Exercise 5.5#13). So it must be x_i (if the sequence is decreasing) or x_{i+1} (if the sequence is increasing).   □

Note the phrase "in general" preceding Proposition 5.5.5. It is really justified; the only way it can be wrong is if the graph of P_{t_0} intersects the diagonal in something other than discrete points, most likely in intervals, as in Exercises 5.5#3c, 5 and 6. We leave the very easy description of what happens in that case to the reader.

> Thus to understand P_{t_0} under iteration, all you need to do is to find the fixed points of P_{t_0} and classify the intervals between them as above or beneath the diagonal.

Example 5.5.6. Consider the equation

$$x' = (2 + \cos(t))x - \left(\frac{1}{2}\right)x^2 - a,$$

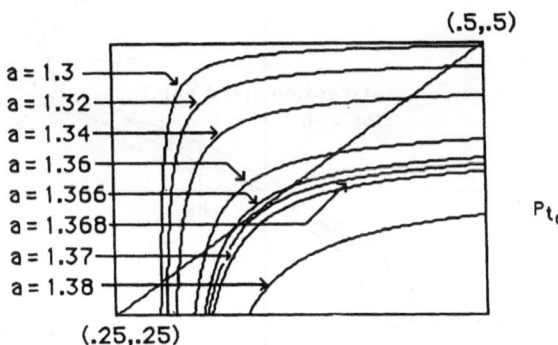

FIGURE 5.5.6. Graphs of P_{t_0} for $x' = (2 + \cos(t))x - (1/2)x^2 - a$.

which was introduced in Examples 2.5.2 and 5.5.1. In this case, no formula for P_{t_0} can be found in elementary terms. However, the graph of P_{t_0} can still be drawn from the graphs of solutions, and P_{t_0} helps to understand the strange change in the behavior of the solutions for different values of a. In Figure 5.5.6 we have plotted the function P_{t_0} for various values of a between 1.3 and 1.38.

The change of behavior for this differential equation occurs because the graph of P_{t_0} moves as follows: first there are two fixed points, then as a increases they coalesce into a single fixed point, then they disappear altogether. ▲

Example 5.5.6 illustrates again how changes in a parameter cause changes of behavior of an iteration. As in Section 5.2, this phenomenon is called *bifurcation*. In Volume II, Chapter 9 we will further discuss bifurcation behavior for differential equations. Meanwhile we give another example:

Example 5.5.7. Figure 5.5.7 illustrates how, as a result of small parameter changes and the resulting bifurcations of fixed points, the same initial condition can lead to an orbit attracted by an altogether different fixed point. It comes from the differential equation

$$x' = \cos(x^2 + \sin 2\pi t) - a$$

for various values of a as shown, and gives some idea of the sort of complication which may arise. However, there is nothing exceptional about this example; it is actually rather typical.

Exercise 5.5#14 will ask you to predict and verify the implications of Figure 5.5.7 for the solutions to this differential equation. ▲

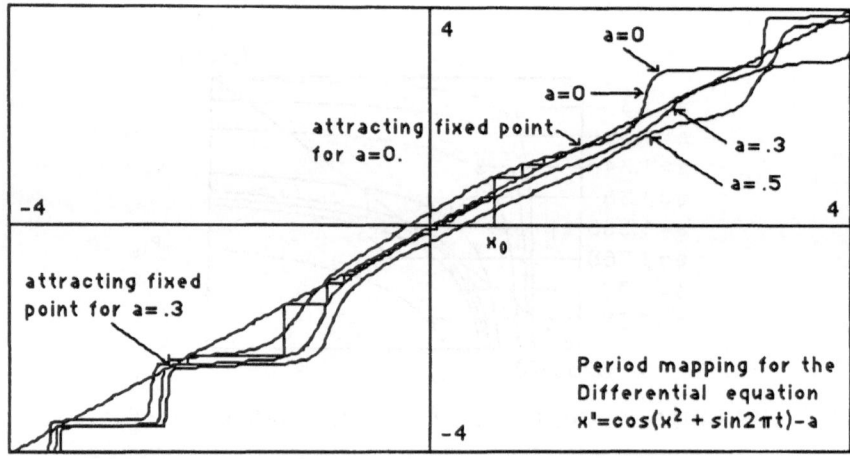

FIGURE 5.5.7. Graphs of P_{t_0} for $x' = \cos(x^2 + \sin 2\pi t) - a$.

The properties (a) and (b) make these iterative systems very simple for a one-dimensional differential equation. Unfortunately, when the dimension increases, there isn't anything so simple to say.

Furthermore, it is the *monotonicity* of P_{t_0} in the examples in this section that makes them simple. If you look back at Section 5.1, you will see that non-monotone functions, even in one dimension, can make for any number of complications.

5.6   Iterating in One Complex Dimension

Although this section is neither directly related to differential equations nor essential for understanding the basic iteration techniques underlying the applications to differential equations, it is easy to discuss complex iteration at this point. Iterating complex variables is at the heart of current research in dynamical systems. This research leads to the most beautiful pictures, which are no longer seen only by mathematicians. These computer-generated images currently appear on magazine covers and as art on walls. Because the explanation of how the pictures are made is so accessible, we cannot resist giving a peek at this topic, right here where we have already laid all the groundwork in Sections 5.1–5.3. We hope that this introduction will inspire you to experiment and read further.

Iteration of maps $f\colon \mathbb{R}^n \to \mathbb{R}^n$ is more complicated than iteration of maps $f\colon \mathbb{R} \to \mathbb{R}$. An important chapter in Volume II will be devoted to such things. However, iteration of appropriate maps $f\colon \mathbb{C} \to \mathbb{C}$ often behaves mathematically more like mappings in one than in two dimensions; this is particularly the case for polynomials. Further, there are many aspects of iteration which are simpler in the complex domain than in the real, and whose behavior sheds important light on the behavior of iteration of real numbers.

Not only does iteration of polynomials in the complex plane give rise to beautiful pictures; it is also often easier than iteration in the reals. Easy examples arise like the counting of periodic points. By the Fundamental Theorem of Algebra a polynomial of degree d always has d complex roots counted with multiplicity, whereas no such simple statement is true about real roots. For instance, every *complex* quadratic polynomial $z^2 + c$ has $2^p - 2$ complex periodic points of period exactly p, counting multiplicity, for every prime number p (Exercise 5.6#9). For a *real* polynomial $x^2 + c$, it can be quite difficult to see how many *real* periodic points of a given period it has: see Example 5.1.9 for the case $p = 2$. In some cases, questions about real polynomials cannot at the moment be answered without considering the polynomials as complex.

Example 5.6.1. Under iteration by $x^2 + c$, for how many real values of c is 0 periodic of period p, for some prime number p? Since the orbit of 0 is

$$x_0(c) = 0$$
$$x_1(c) = c$$
$$x_2(c) = c^2 + c$$
$$x_3(c) = (c^2 + c)^2 + c$$
$$x_4(c) = ((c^2 + c)^2 + c)^2 + c$$
$$\vdots$$

this question is simply: how many real roots does the equation $x_p(c) = 0$ have? This is an equation of degree 2^{p-1}, hence it has 2^{p-1} roots (counted with multiplicity). Note that $c = 0$ is one of the roots, but in that case $z = 0$ is a fixed point, hence not of exact period p. The point $z = 0$ is of period exactly p for the $2^{p-1} - 1$ other roots.

But how many of these roots are real? The answer is $(2^{p-1} - 1)/p$ for all primes $p \neq 2$. Two proofs of this fact are known; both rely on complex polynomials, and rather heavy duty complex analysis, far beyond the scope of this book. We simply ask you in Exercise 5.6#10 to verify this for $p = 3$, 5, 7, 11. ▲

For this entire section, we return to the basic class of functions studied for real variables in Sections 5.1 and 5.2—*quadratic polynomials of the form* $x^2 + c$. As has been shown (Exercises 5.1#19), studying $x^2 + c$ in effect allows us to study iteration of *all* quadratic polynomials.

If c is complex, then x must also be complex in order for iteration to make sense. Since the letter z is usually used to denote a complex variable, we will henceforth write *quadratic polynomials* as

$z^2 + c$, where both z and c may be complex numbers.

All the previous real examples are of course special cases.

BOUNDED VERSUS UNBOUNDED ORBITS

Experience has shown that the right question to ask about a polynomial (viewed as a dynamical system) is: *Which points have bounded orbits?*

Example 5.6.2. If c is *real* and you ask which *real* values of z have bounded orbits under $z^2 + c$, the answer is as follows (shown in Exercise 5.1#10):

(a) If $c > 1/4$ there are no points with bounded orbits;

(b) If $-2 \leq c \leq 1/4$, the points of the interval

$$[-(1 + \sqrt{1 - 4c})/2, (1 + \sqrt{1 - 4c})/2]$$

have bounded orbits, and no others.

(c) If $c < -2$, there is a Cantor set of points with bounded orbits. (Cantor sets are elaborated in a later subsection.)

For illustrations of these cases, you can look back to some of the pictures for iterating real variables $x^2 + c$ in Section 5.1. Figure 5.1.10 has two examples of case (b); both of the orbits shown are bounded. In the first case, $c = -1.75$, the points with bounded orbits form the interval $[-1.9142\ldots, -1.9142\ldots]$. An example of an *unbounded orbit* is provided by the rightmost orbit of Figure 5.1.6.   ▲

In the complex plane, the sets of points with bounded orbits are usually much more complicated than intervals, but quite easy to draw using a computer. However, we will need a different type of picture, since the variable z requires two dimensions on a graph.

PICTURES IN THE z-PLANE

If we choose a particular value of c, we can make a picture showing what happens to the orbit of each point z as follows: Consider some grid of points in the complex plane, and for each point, take that value for z_0 and iterate it a certain number of times (like 100, or 300). If the orbit remains bounded for that number of iterates, color the point black; if the orbit is unbounded, leave the point white. (For $z^2 + c$ there is a practical test for knowing whether an orbit is unbounded: whenever an iterate exceeds $|c| + 1$ in absolute value, the successive iterates will go off toward infinity— Exercise 5.6#4.) The resulting black set will be an approximation to the set of points with bounded orbits, for that value of c.

The set of points with bounded orbits under iteration of $P_c = z^2 + c$ (i.e., the set of points approximated by the black points of the drawing described above) is denoted by

$$K_c \equiv \big\{ z \mid \underbrace{(z^2 + c)^{\circ n}}_{P_c} \not\to \infty \big\}.$$

The variable z is called the dynamical variable (i.e., the variable which is being iterated), so K_c, which lies in the z-plane, is a subset of the *dynamical plane* (to be distinguished from pictures in the c-plane or *parameter plane,* which will be discussed later). The name *Julia set* was given to the *boundary* of a set K_c, in honor of a French mathematician, Gaston Julia. Julia and his teacher, Pierre Fatou, did the seminal work in this field. Technically the sets of points with bounded orbits are called "filled-in" Julia sets, but we shall call them simply Julia sets, since we will not need to discuss the boundaries separately.

Examples 5.6.3. See the pictures shown in Figures 5.6.1 and 5.6.2 on the next page, all printed to scale inside a square -2 to 2 on a side.

If $c = 0$, then the Julia set K_0 is the unit disk. Confirm by iterating z^2.

If $c = -2$, then the Julia set K_{-2} is the interval $[-2, 2]$ (Exercise 5.6#11).

If $c = -1$, the Julia set K_{-1} is far more complicated. This is a good set to explore, as in Exercise 5.6#6, in order to learn about the dynamics of complex iteration—they are full of surprises for the newcomer. ▲

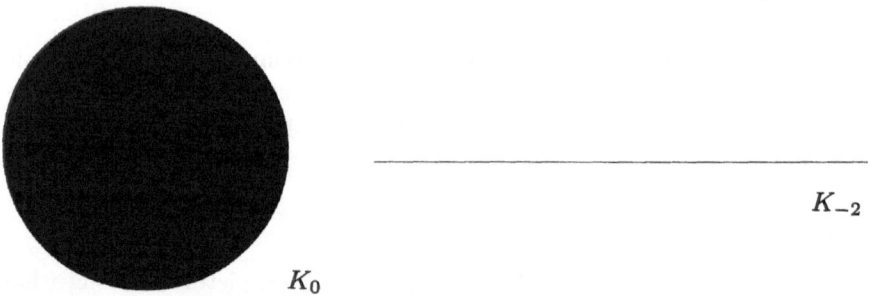

K_{-2}

K_0

FIGURE 5.6.1. The simplest Julia sets, in the z-plane, iterating $z^2 + c$ for $c = 0$, $c = -2$.

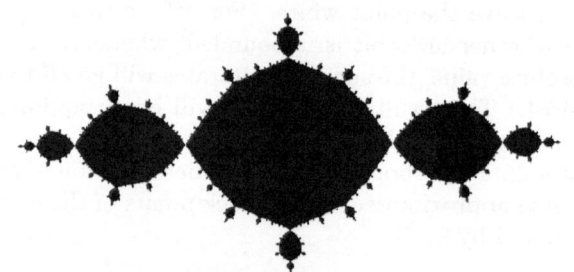

K_{-1}

FIGURE 5.6.2. The Julia set for $z^2 - 1$.

The only geometrically simple Julia sets are K_0 and K_{-2}. All the others are "fractal." This word does not appear to have a precise definition, but is meant to suggest that the sets have structure at all scales. Some other Julia sets are pictured in Examples 5.6.4.

Examples 5.6.4. (See Figure 5.6.3 on the next page.) ▲

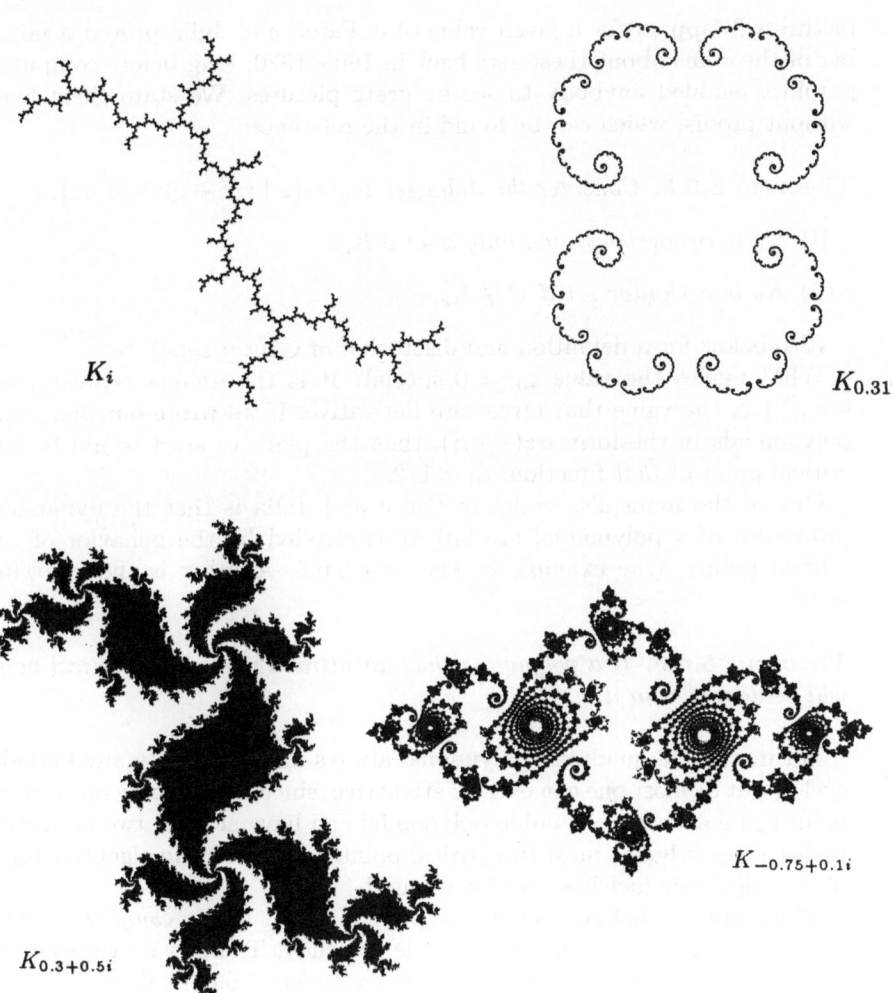

FIGURE 5.6.3. More typical Julia sets, in the z-plane.

Most Julia sets cannot be more than roughly sketched by hand, but we fortunately live in the age of computer graphics where there are several kinds of programs that draw them easily on personal computers. See the references at the end of this section for software to help you sample other K_c's.

With the infinite variety of patterns for Julia sets, it is not surprising that a great deal of study has gone into being able to predict what sort of picture will appear for a given value of c. Fatou and Julia proved a number of theorems about these sets back in 1905–1920, long before computer graphics enabled anybody to see accurate pictures. We state them here without proofs, which can be found in the references.

Theorem 5.6.5. *Consider the Julia set* $K_c = \{z \mid (z^2 + c)^{\circ n} \not\to \infty\}$.

   (i) K_c *is connected if and only if* $O \in K_c$.

   (ii) K_c *is a Cantor set if* $O \notin K_c$.

(See below for a definition and discussion of Cantor sets.)

What makes the value $z_0 = 0$ special? It is the unique *critical point* for $z^2 + c$, the value that gives zero derivative. If we wrote our quadratic polynomials in the form $\alpha z(1 - z)$, then the place to start would be the critical point of *that* function, $z_0 = 1/2$.

One of the main discoveries of Fatou and Julia is that the dynamical properties of a polynomial are largely controlled by the behavior of the critical points. One example is Theorem 5.6.5. Another is the following result:

Theorem 5.6.6. *If a polynomial has an attractive cycle, a critical point will be attracted to it.*

For instance a quadratic polynomial always has infinitely many periodic cycles, but at most one can ever be attractive, since there is only one critical point to be attracted. A cubic polynomial can have at most two attractive cycles, since it has at most two critical points, and so on. No algebraic proof of this algebraic fact has ever been found.

The dominant behavior of critical points is what makes *complex* analytic dynamics in one variable so amenable to study. There is no theorem for iteration in $\mathbb{R}^2$ corresponding to Theorems 5.6.5 or 5.6.6. As far as we know, there is no particularly interesting class of points to iterate in general in $\mathbb{R}^n$.

Consequences of Theorems 5.6.5 and 5.6.6 may be seen looking back at Examples 5.6.3 and 5.6.4; if you iterate $z^2 + c$ for any of the given values of c, starting at $z = 0$, and look at what happens to the orbit, you will find bounded orbits and corresponding connected K_c's, for all except $K_{0.3125}$ and $K_{-0.75+0.1i}$. Theorem 5.6.5 tells us that these last two Julia sets are Cantor sets, which we shall finally explain.

CANTOR SETS

Georg Cantor was a German mathematician of the late nineteenth century. In the course of investigations of Fourier series, Cantor developed world-shaking new theories of infinity and invented the famous set that bears his name. These results were considered most surprising and pathological at the time, though since then they have been found to occur naturally in many branches of mathematics.

A *Cantor set* $X \subset \mathbb{R}^n$ is a closed and bounded subset which is *totally disconnected* and simultaneously *without isolated points.*

The first property says that the only connected subsets are points; the second says that every neighborhood of every point of X contains other points of X. These two properties may seem contradictory: the first says that points are never together and the other says that they are never alone.

Benoit Mandelbrot of IBM's Thomas Watson Research Center, the father of "fractal geometry," has suggested calling these sets *Cantor dusts.* This name seems quite auspicious: it is tempting to think that a dust is made up of specks, but that no speck is ever isolated from other specks.

The compatibility of the requirements above is shown by the following example, illustrated by Figure 5.6.4.

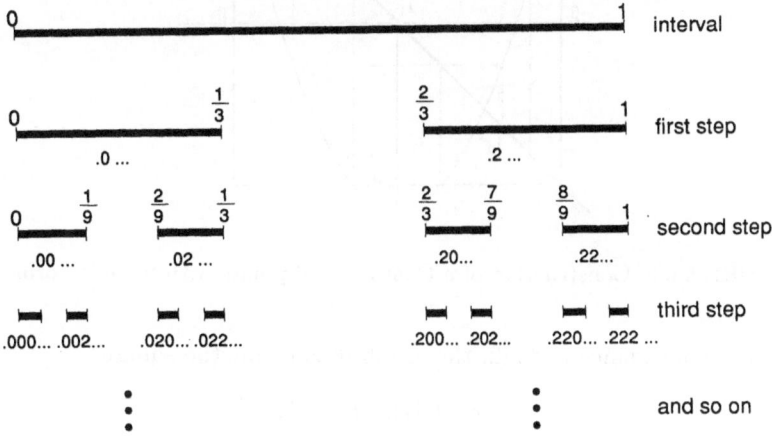

FIGURE 5.6.4. Construction of the Cantor set.

The Cantor set, the one originally considered by Cantor, is the set $X \subset \mathbb{R}$ made up of those numbers in $[0, 1]$ which can be written in base 3 without using the digit 1.

The set X can be constructed as follows: From $[0, 1]$ begin by removing the middle third $\left(\frac{1}{3}, \frac{2}{3}\right)$. All the numbers in the interval $\left(\frac{1}{3}, \frac{2}{3}\right)$ are not in X, since they will all use the digit "1" in the first place after the "decimal" point. Next remove the middle thirds of both intervals remaining, which means that the intervals $\left(\frac{1}{9}, \frac{2}{9}\right)$ and $\left(\frac{7}{9}, \frac{8}{9}\right)$ are also excluded, those numbers that will use the digit "1" in the second place. Then remove the middle thirds of the four intervals remaining, then remove the middle thirds of the eight intervals remaining, and so on.

You are asked in Exercise 5.6#13 to show that this set indeed satisfies the stated requirements for a Cantor set, and to consider the endpoints of the excluded intervals.

Example 5.6.7. The sets K_c can be constructed in a remarkably similar way when c is real and $c < -2$. Actually, we will construct only the real part of K_c; but it turns out that in this case the whole set is contained in the real axis. This fact appears to require some complex analysis in its proof, so the reader will have to take it on faith. The iteration graph of P_c looks like Figure 5.6.5.

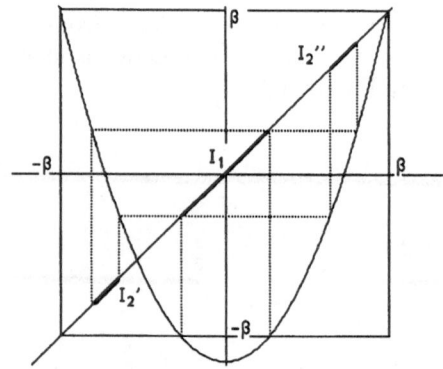

FIGURE 5.6.5. Construction of a Cantor set of points with bounded orbits.

In particular, since $c < -2$, the graph dips below the square

$$\{(x,y)\,|\,|x|,|y| \le \beta\},$$

where $\beta = (1 + \sqrt{1 - 4c})/2$ is the rightmost fixed point. Clearly the points $x \in \mathbb{R}$ with $|x| > \beta$ escape to ∞. But then so do the values of x in the "dip," which form an interval I_1 in the middle of $[-\beta, \beta]$. In particular, $O \in I_1$, hence does escape. Now I_1 has two inverse images I_2', I_2'', one in each of the two intervals of $[-\beta, \beta] - I_1$. Now each of the intervals I_2' and I_2'' has two inverse images, so there are four intervals I_3, one in each of the four pieces of $[-\beta, \beta] - I_1 - I_2' - I_2''$. This process of going back to two inverse images of each interval can continue indefinitely.

The intervals being removed are not in this case the middle thirds, and it is not quite clear that no intervals remain when they have all been removed. In fact, it is true but we omit the proof, which is rather hard. ▲

Cantor dusts need not lie on a line.

Example 5.6.8. Consider the polynomial $z^2 + 2$. In Exercise 5.6#5 you are asked to show that all points z with $|z| \geq 2$ escape to ∞. The inverse image of the circle $|z| = 2$ is a figure eight, touching the original circle at $\pm 2i$, and the polynomial is a 1–1 map of each lobe onto the disc of radius 2. Thus in each lobe you see another figure eight, the inside of which consists of the points which take two iterations to escape from the disc of radius 2. Now inside of each lobe of these there is another figure eight, and so on. Again, it is true but not easy to prove that the diameters of the figure eights tend to zero. If the reader will take that on faith, then the Julia set is clearly a Cantor dust: Any two points of the Julia set will eventually be in different lobes of some figure eight, hence can never be in the same connected part of the Julia set; on the other hand, any neighborhood of any point will necessarily contain whole figure eights, hence must contain infinitely many points of the Julia set. ▲

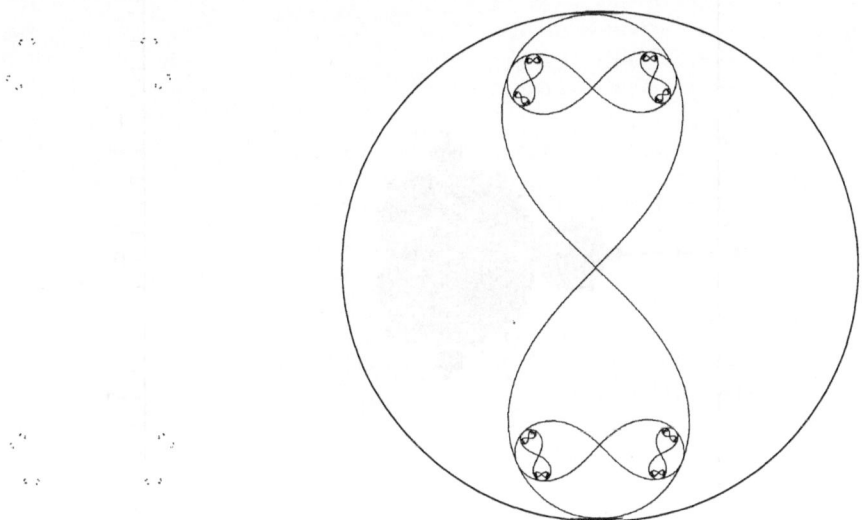

K_2 (Cantor Dust)   Construction of K_2

FIGURE 5.6.6. The Julia set K_2, showing the inverse image contours that bound and define it.

The Picture in the Complex c-Plane

In the late 1970's, Mandelbrot (among others) looked in the *parameter space* (that is, the complex c-plane) for quadratic polynomials of form $z^2 + c$ and plotted computer pictures of those values of c for which the orbit of 0 is bounded, after a large finite number of iterations. The computer algorithm is the same as described for the construction of Julia sets, to iterate $z^2 + c$ at each point of some grid, but this time it is c that is determined by the given point, and the starting value for z is what is fixed, at 0.

The set M that resulted from Mandelbrot's experiment, shown in Figure 5.6.7, is called the *Mandelbrot set*. Studying the Mandelbrot set is like studying all quadratic polynomials *at once*.

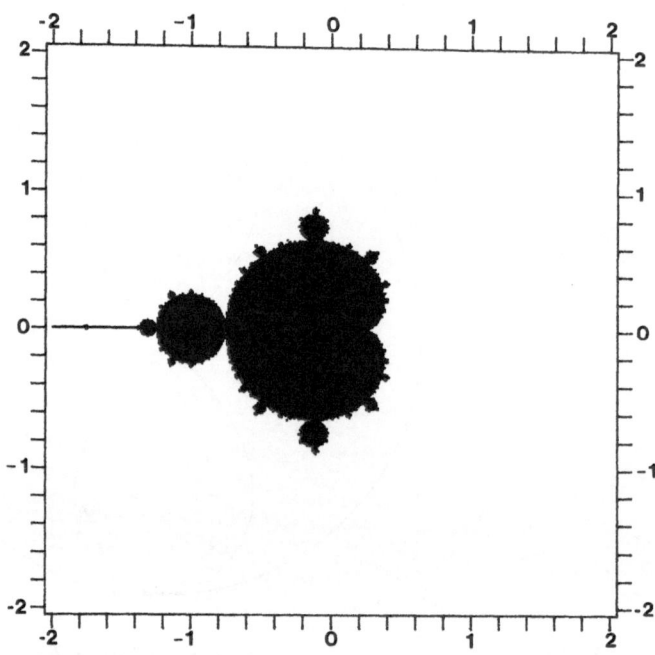

FIGURE 5.6.7. The Mandelbrot set, in the c-plane; the values of complex c such that iterating $z^2 + c$ from $z_0 = 0$ gives bounded orbits.

The Mandelbrot set is marvelously complicated. The closer we zoom in on the boundary, the more detail appears, as illustrated in Figure 5.6.8, on pp. 266–267. Notice how jumbled and intertwined are the regions of bounded orbits (black) and the regions of unbounded orbits (white). Yet with the advent of high speed and high resolution computer graphics, we are able, quite easily, to explore this set.

Iteration pictures such as these are made more detailed by using contour lines representing "escape times," telling how many iterations are necessary before the result exceeds the value that assures the orbit will not be bounded. The outer contour is a circle radius 2; outside this circle every point has "escaped" before the iteration even begins. The next contour in is the inverse image of the first contour circle; points between these contours escape in one iteration. The next contour is the inverse image of the second; points between this contour and the last will escape in two iterations, and so on. The pictures in Figure 5.6.8 in fact show *only* the contours for escape times; these contours get denser and denser as they get closer to the boundary of M, so they actually outline it.

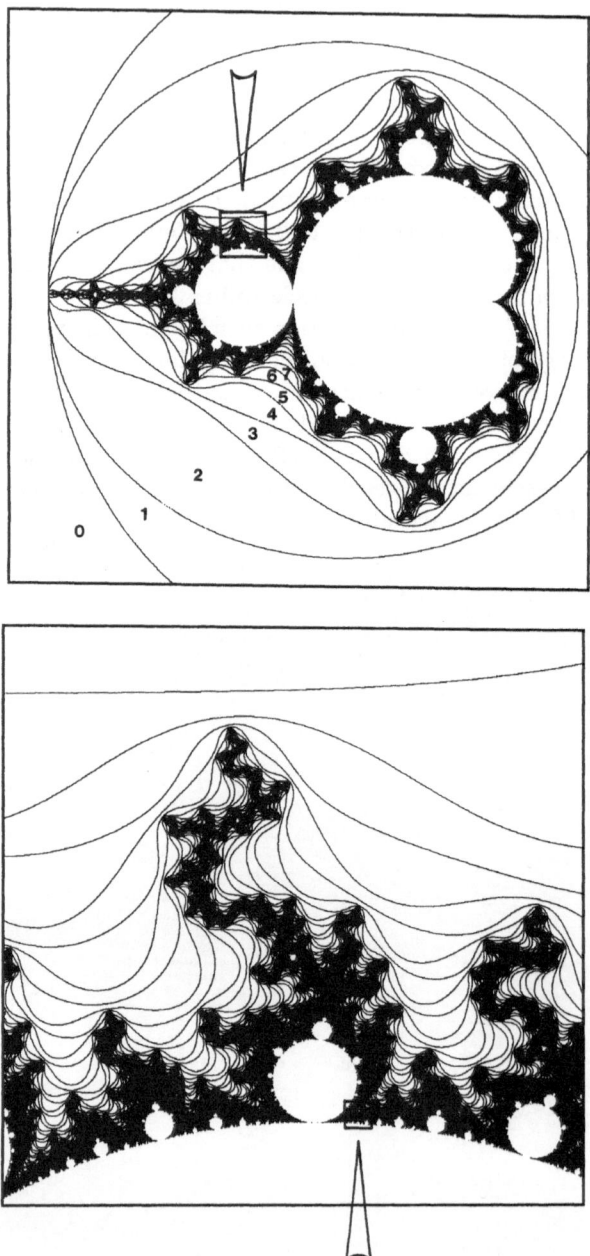

FIGURE 5.6.8. Zooming in on the boundary of the Mandelbrot set, in the c-plane. Each region that is blown up in the next picture is marked with a black box and arrow.

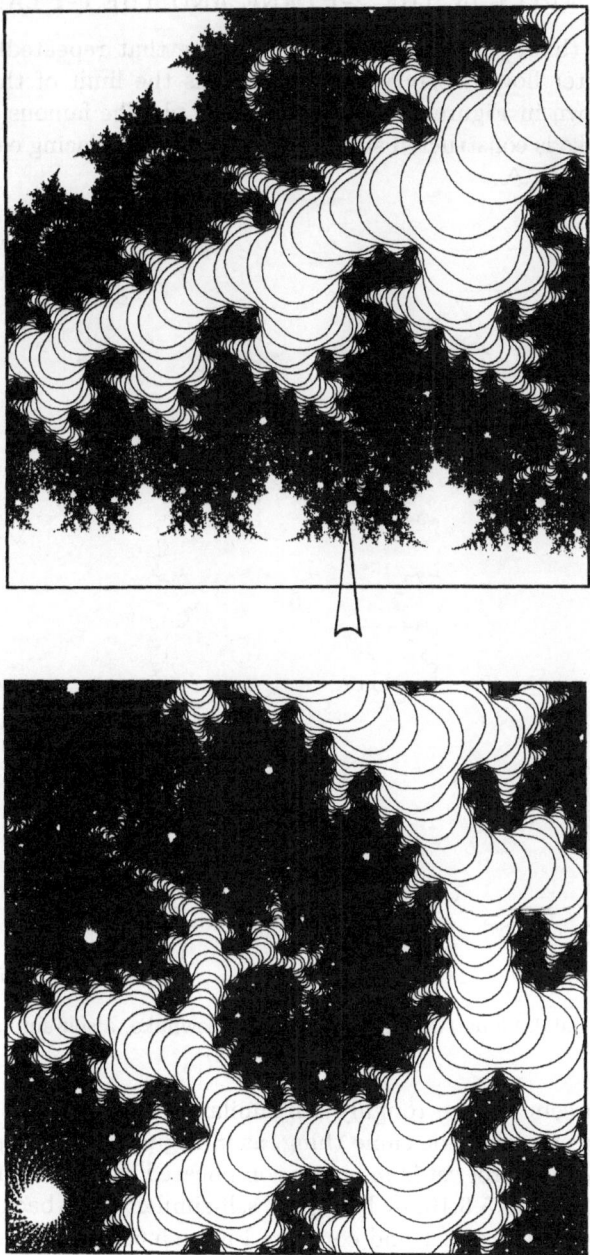

FIGURE 5.6.8. (cont.) The bands represent how many iterations are necessary for those points to escape beyond the circle radius 2, where they are sure to go to infinity. In the first frame they are numbered accordingly.

SELF-SIMILARITY IN THE z-PLANE AND THE c-PLANE

Some people think that self-similar sets (those that repeatedly look the same no matter how closely you look, such as the limit of the sequence of figures shown in Figure 5.6.9) are complicated. The famous self-similar Koch snowflake is constructed simply by repeatedly replacing each straight line segment with Λ.

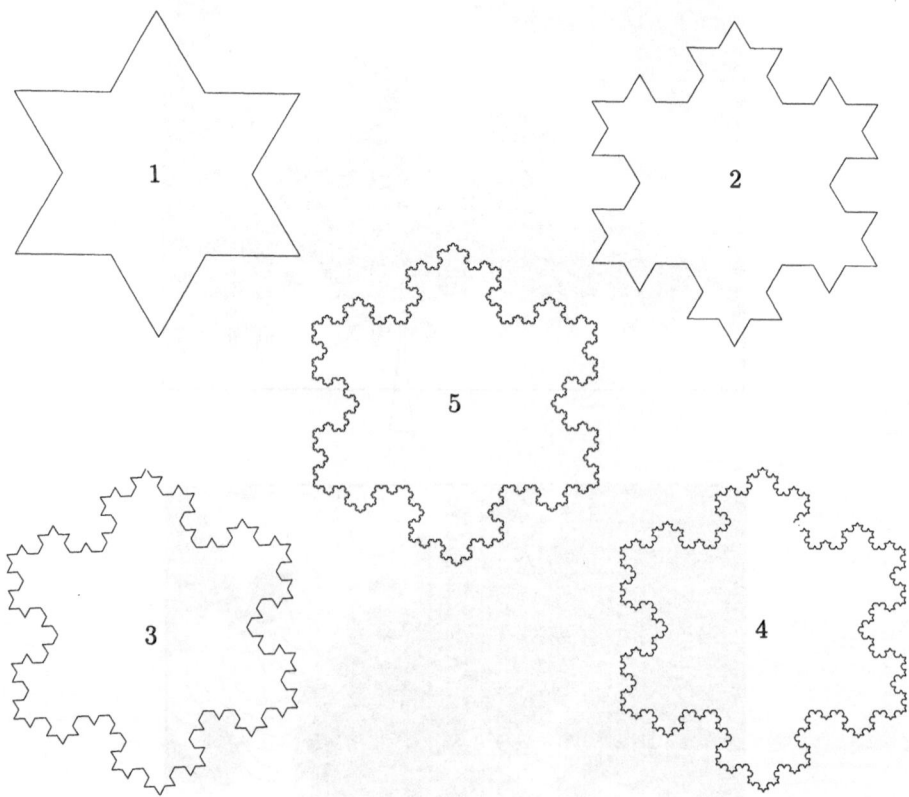

FIGURE 5.6.9. Construction of the Koch snowflake, a self-similar figure.

The Julia sets K_c are roughly self-similar. If you blow up around any point, you keep seeing the same thing, as in Figure 5.6.10 on pp. 270–271.

But the Mandelbrot set is *far* more complicated than the self-similar sets of Figures 5.6.9 and 5.6.10, by being *not* self-similar. Look back at the blow-ups of the Mandelbrot set and compare Figure 5.6.7 with Figure 5.6.8. As you zoom in on the boundary in the first picture, it gets more complicated rather than simpler, and although similar shapes may appear, they are different in detail at every point. In fact, the Mandelbrot set exhibits in various places all the shapes of all the possible associated K_c's.

UNIVERSALITY OF THE MANDELBROT SET

The Mandelbrot set is of intense mathematical interest, among other reasons because it is *universal*. Somewhat distorted copies of M show up in parameter space pictures for iteration of all sorts of nonquadratic functions—higher degree polynomials, rational functions (including Newton's method in the complex plane), and even transcendental functions (like $\lambda \sin z$).

Despite its complication, the Mandelbrot set is now quite well understood. The first result in this direction is that it is *connected* (those "islands" off the edges are actually connected to the whole set by thin "filaments"). Further, a complete description of how the "islands" are connected up by the "filaments" has been found, but the description is not easy. The interested reader can get a flavor of how this is done in the book by Peitgen and Richter. See the References.

To tie things together a bit, one should realize that the part of M on the real axis is essentially the cascade picture of Section 5.2.

Indeed, the cascade picture is also obtained by letting c vary (only on the real axis) and plotting for each value of c the orbit of 0. The windows of order correspond to regions in which the critical point is attracted to an attractive cycle, the order of which can be read off from the cascade picture. These windows correspond to islands of the Mandelbrot set along the real axis, as Figure 5.6.11 on p. 272 illustrates.

Note: This particular drawing was computed with *no* unmarked points, that is, with $K = 0$. You might want to contrast it with Figure 5.2.2. The white spaces are in the same places, but in Figure 5.6.11, the branches are greatly thickened.

The universality of the cascade picture exhibited in Sections 5.2 and 5.3 is simply a real form of the universality of M.

Further discussion of iteration in the complex plane is beyond the scope of this text. References for further reading are listed at the end of this volume.

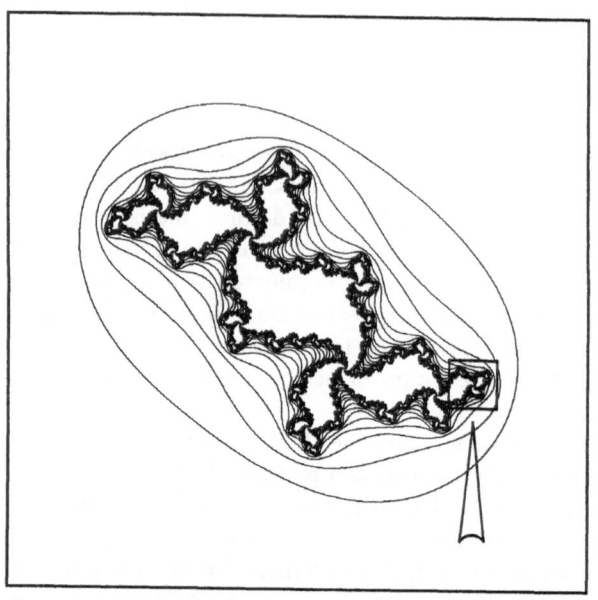

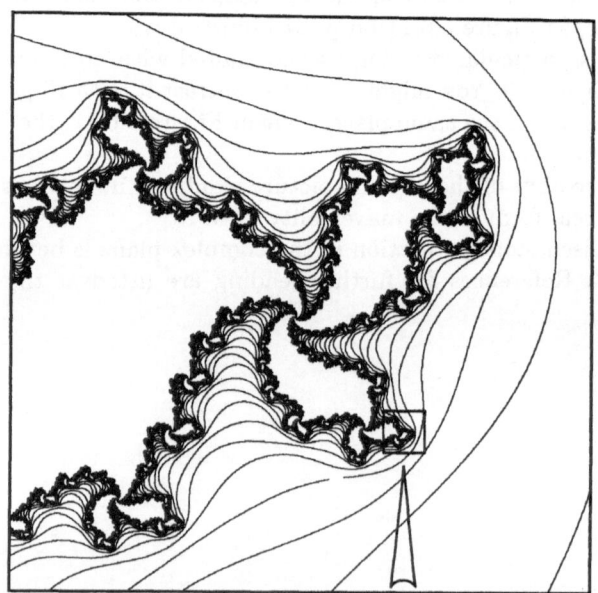

FIGURE 5.6.10. The Julia set for $c = -0.2 + 0.75i$ and successive blow-ups. Each region that is blown up in the next picture is marked with a black box and arrow.

FIGURE 5.6.10. (cont.)

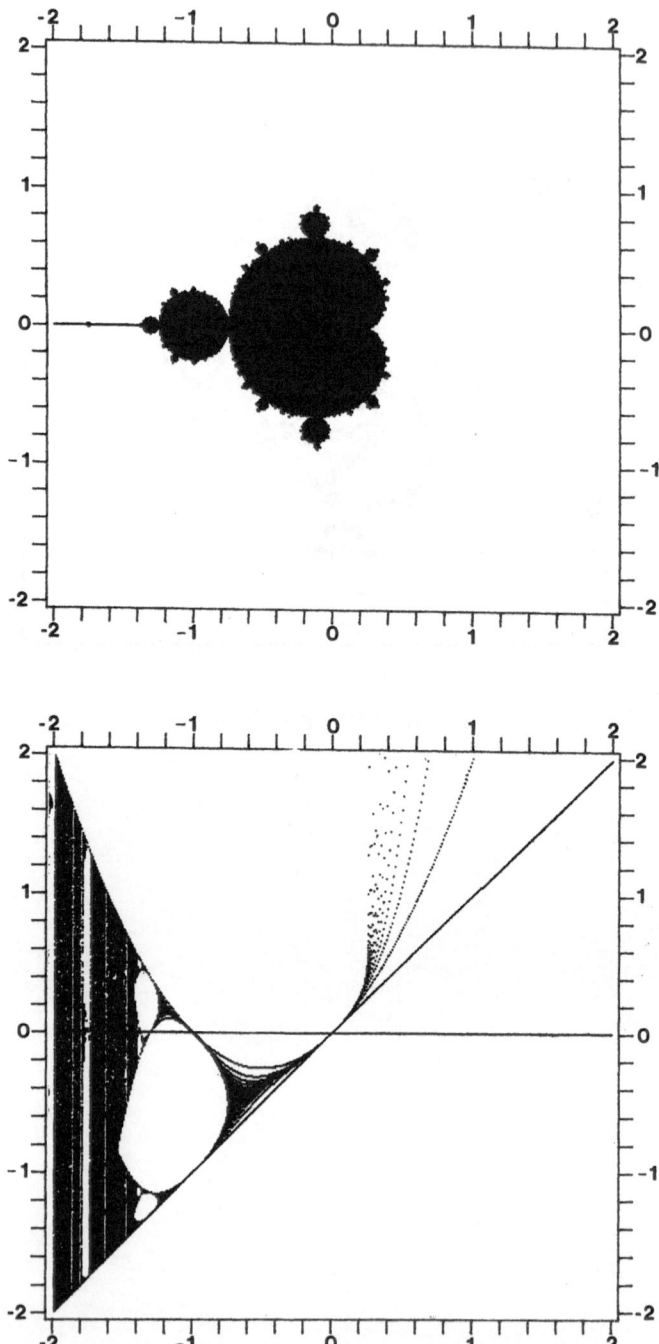

FIGURE 5.6.11. Relation between the Mandelbrot set and the cascade picture.

Exercises 5.1 Iterating Graphically

5.1#1. Analyze each of the following functions under iteration. Find the fixed points and determine the classification of each fixed point (attracting, superattracting, or repelling). Indicate all cycles of order two. Graph the equation and then iterate it by hand for a representative selection of x_0's. Finally, use *Analyzer* to confirm your handwritten iteration and to qualitatively determine and indicate regions on the x-axis for which all initial points converge to the same attracting fixed point.

$\qquad$ (a) $f(x) = -\frac{1}{2}x + 1$ $\qquad\qquad$ (e) $f(x) = 3$

$\qquad$ (b) $f(x) = \frac{1}{2}x$ $\qquad\qquad\quad$ (f) $f(x) = x^2$

$\qquad$ (c) $f(x) = 2x + 1$ $\qquad\qquad$ (g) $f(x) = x^3$

$\qquad$ (d) $f(x) = -2x + 1$ $\qquad\quad$ (h) $f(x) = 1/x$

5.1#2. To see that anything can happen with an indifferent fixed point, analyze as in 5.1#1 the behavior of the following equations under iteration:

$\qquad$ (a) $f(x) = e^{-x} - 1$ $\qquad$ (d) $f(x) = \ln(1 - x)$

$\qquad$ (b) $f(x) = -x$ $\qquad\qquad$ (e) $f(x) = x + e^{-x^2}$

$\qquad$ (c) $f(x) = -\tan x$ $\qquad$ (f) $f(x) = 0.5x + 0.5x^2 + 0.125$

5.1#3.

(a) Prove, using Taylor series expansion, that the difference between a superattracting fixed point, x_s, and an attracting fixed point, x_a, is the fact that, for sufficiently small $|u|$,

$$x_n = f(x_a + u) = x_a + Mu, \qquad 0 < |M| < 1,$$

while

$$x_n = f(x_s + u) = x_s \pm Nu^p, \qquad p \geq 2.$$

Hence, for sufficiently small $|u|$, solutions will converge faster (i.e., with fewer iterations) to x_s than to x_a.

(b) Compare Exercises 5.1#1e,f,g with Exercises 5.1#1a,b. How does this comparison confirm the preceding result?

5.1#4°.

(a) Use a Taylor series expansion to prove: If at some fixed point x_0 of a function to be iterated $f(x)$, $f'(x_0) = 1$ and $f''(x_0) > 0$, then in a sufficiently small neighborhood of x_0, points to the left of x_0 will be attracted to x_0 under iteration by $f(x)$, and points to the right will be repelled. Confirm your result with

$$f(x) = \frac{1}{3}x^3 + \frac{2}{3}.$$

(b) What will happen in the neighborhood of x_0 if $f''(x_0) < 0$? Confirm your result with

$$f(x) = \ln(1 + x).$$

(c) Show that anything can happen in the neighborhood of x_0 if $f''(x_0) = 0$ by iterating the following functions:

   (i) $f(x) = \sin x$

   (ii) $f(x) = e^x - 1 - \dfrac{x^2}{2}$

   (iii) $f(x) = e^x - 1 - \dfrac{x^2}{2} - \dfrac{x^3}{6} - x^4$

   (iv) $f(x) = e^x - 1 - \dfrac{x^2}{2} - \dfrac{x^3}{6} + x^4$

(d) (harder) What will happen in the neighborhood of x_0 if $f'(x) = -1$? Confirm your results with $f(x) = -g(x)$, where $g(x)$ is any of the functions from (a), (b) and (c). Hint: Write the Taylor polynomial of $f^{\circ 2}$ to degree 3.

5.1#5. Near a point x_0, we define the triangular regions formed by the lines $x - x_0 = \pm(y - f(x_0))$, as shown.

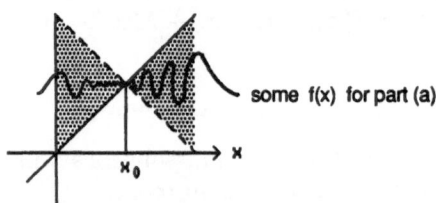

some f(x) for part (a)

Consider a function $f(x)$ with fixed point at x_0.

(a) If $f(x)$ lies entirely within the gray regions, justify with an $\varepsilon - \delta$ type argument (rather than a differentiability argument) that x_0 is an attracting fixed point.

(b) Show that if $f(x)$ is a differentiable function with $f'(x_0) < 1$, then $f(x)$ indeed lies in the gray region specified.

(c) Show that if $f(x)$ is a differentiable function with $f'(x_0) > 1$, then x_0 is a repelling fixed point. Note that this is a little harder, and that it does not prove $f(x)$ will *never* come back to x_0, only that x_0 repels near neighbors.

5.1#6. Analyze the following functions under iteration, using your results from the previous exercises. First try to predict the behavior and then use *Analyzer* to confirm your predictions.

(a) $f(x) = x + \cos x$

(b) $f(x) = x + \sin x$

(c) $f(x) = \cos x + \dfrac{x^3}{2} - 1$

(d) $f(x) = \ln |x + 1|$

(e) $f(x) = x + e^{-x}$

(f) $f(x) = x - e^{-x^2}$

(g) $f(x) = -\left(e^x - 1 - \dfrac{x^2}{2} - \dfrac{x^3}{6} - \dfrac{x^4}{24} + x^5 \right)$

(h) $f(x) = e^x - 1 - \dfrac{x^2}{2} - 2x - x^3$

5.1#7. For the following functions, find the fixed points under iteration and classify them as attracting or repelling. Use the computer to confirm your results.

(a)° $f(x) = x^4 - 3x^2 + 3x$

(b) $f(x) = x^4 - \dfrac{7}{4}x^2 + \dfrac{1}{4}x$

5.1#8.

(a) Refer to Figures 5.1.6 and 5.1.7, iterating $x^2 - 0.7$ and $x^2 - 1.2$ respectively. Tell why, when they look at first glance so similar, they exhibit such different behavior under iteration.

(b) Referring again to Figure 5.1.7 and Example 5.5, iterating $x^2 - 1.2$, determine what happens between the two fixed points, since both are repelling. Try this by computer experiment.

5.1#9. Show that the multiplier m_n for a cycle order n is the same at all points of the cycle. That is, fill in the necessary steps in the text derivation.

5.1#10°. For Example 5.1.9 iterating $x^2 + c$,

(a) Find for which values of c there are, respectively, two fixed points, a single fixed point, and no fixed points.

(b) Sketch graphs for the three cases in part (a) showing how the diagonal line $x = y$ intersects the parabola $x^2 + c$ in three possible ways as the parabola moves from above the diagonal to below it.

(c) Find for which values of c there will be an attracting fixed point.

(d) Show that $c \in \left[-2, \frac{1}{4}\right]$, all orbits starting in the interval

$$I_c = \left[-\frac{1}{2}(1 + \sqrt{1 - 4c}), +\frac{1}{2}(1 + \sqrt{1 - 4c})\right]$$

are bounded, and all others are unbounded.

(e) Show that the orbit of 0 is bounded if and only if $c \in \left[-2, \frac{1}{4}\right]$.

The results of this exercise are used in Section 5.6, Example 5.6.2.

5.1#11. For iterating $x^2 + c$,

(a) verify the following facts about the period two cycle:

  (i) The periodic points of period 2 are real if $c < -\frac{3}{4}$.

  (ii) The multiplier m_2 of the period 2 cycle is $4x_0x_1 = 4(c + 1)$.

  (iii) The cycle of period 2 is attracting if and only if $-\frac{5}{4} < c < -\frac{3}{4}$.

(b) Illustrate the possibilities in part (a) graphically by sketching the various parabolas and the diagonal for each, then tracing out some appropriate iterations.

5.1#12. For each of the following functions, in the manner of Exercise 5.1#10b, analyze the various possible behaviors of the function under iteration for different values of the parameter α. That is, determine (graphically or numerically) which values of α will leads to different possible behaviors. Then for each α, sketch the curve, the diagonal, and the iterative behavior.

(a) $f(x) = 1 - (x - \alpha)^2$

(b) $f(x) = x^3 - x + \alpha$

5.1#13. For each of the following functions,

(a) $x^2 - 1.75$	(d) $x^2 - 1.30$	(g) $x^2 - 1.41739$
(b) $x^2 - 1.76$	(e) $x^2 - 1.35$	(h) $x^2 - 1.45$
(c) $x^2 + 1$	(f) $x^2 - 1.39$	(i) $x^2 - 1.69$

(i) Use *Analyzer* to make computer drawings for iterating them, and identify the order of any cycles that appear.

(ii) Use *DiffEq* to make time series drawings for iterating them, and add by hand the actual points. For cycles of an odd order or of no order, you may need to use a ruler to determine where the time series line bends, as in Example 5.1.10, or refer to a computer listing on the points on the trajectory.

(iii) Use *Cascade* to graph, with the diagonal, the nth iterate of a function that has a cycle order n. (The other purposes of the *Cascade* program will come up in Section 5.2.)

(iv) Match up the results of parts (i), (ii), and (iii) as in Examples 5.1.10 and 5.1.11. (If your scales are different for the different programs, your matching lines won't be horizontal and vertical, but you can still show the correspondences.)

5.1#14°. Match the numbered *Analyzer* iteration pictures with the lettered time series pictures:

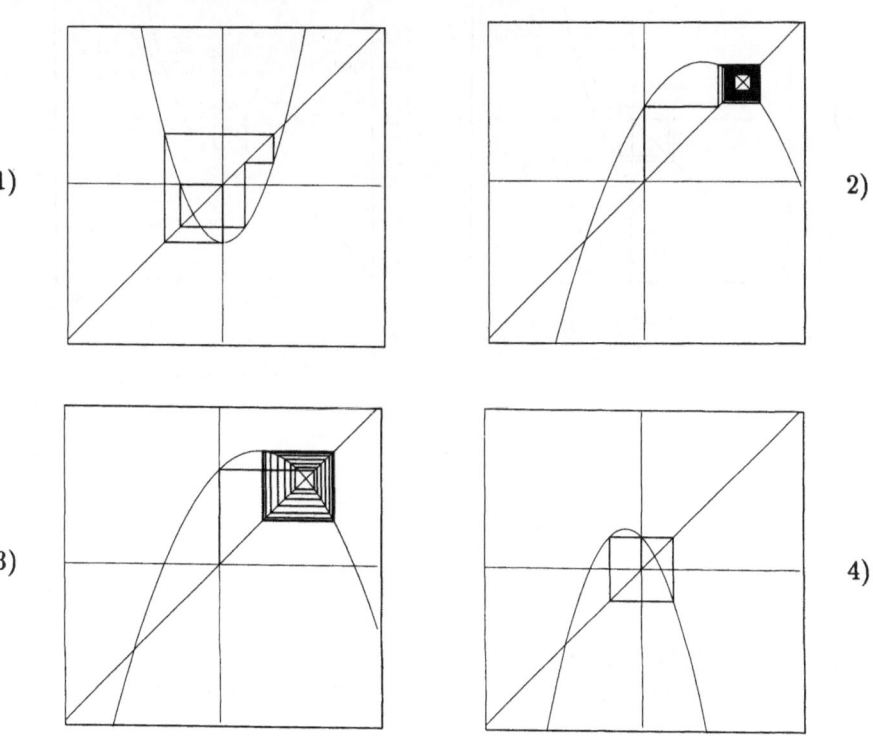

1)

2)

3)

4)

5)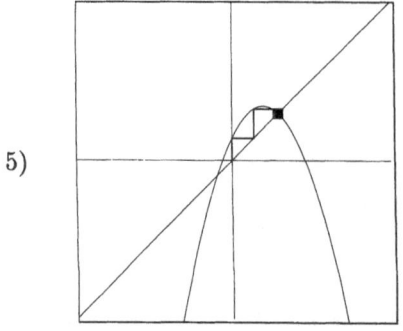

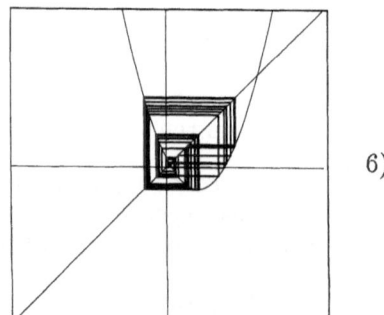 6)

7)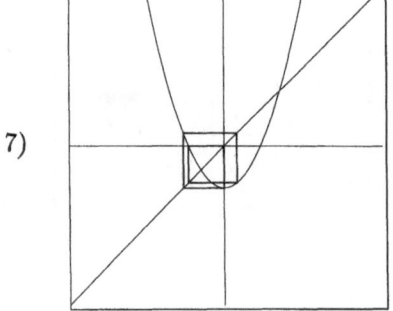

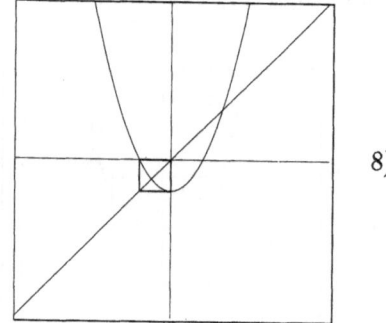 8)

a)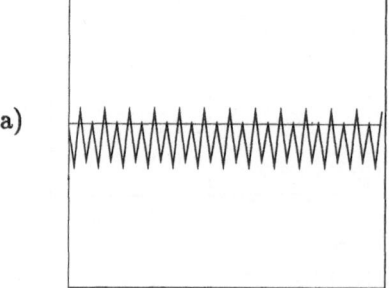

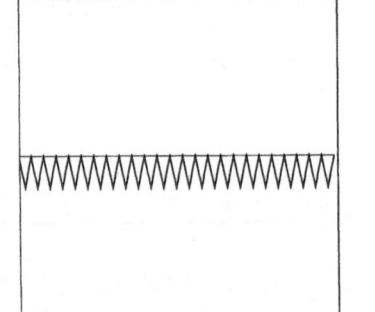 b)

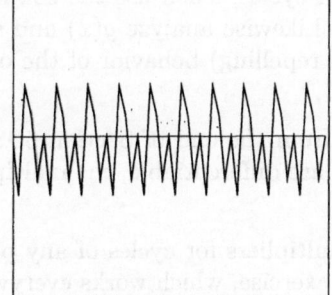

c)

d)

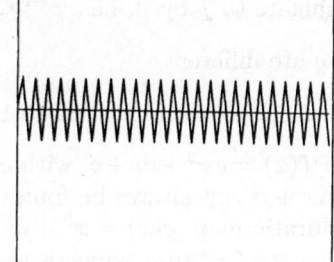

e)

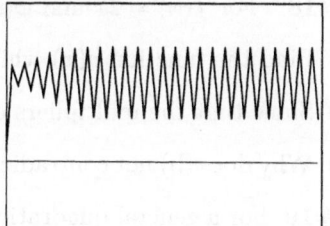

f)

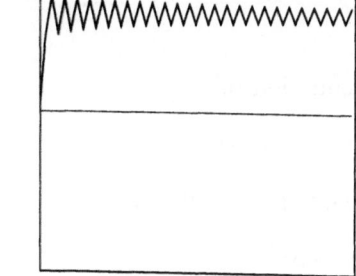

g)

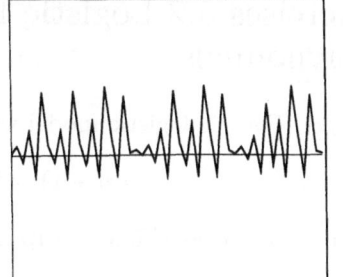

h)

5.1#15. For the function of Example 5.1.12, $f(x) = 2x - x^2$, analyze the iteration and identify fixed points and cycles. Then use the conjugate mapping $\varphi(x) = 1 - x$ to get $g(x) = x^2$. Likewise analyze $g(x)$ and show that in fact the qualitative (attracting or repelling) behavior of the orbits is preserved.

5.1#16. Prove the following statement from the end of Section 5.1: If a conjugate mapping φ and its inverse φ^{-1} are differentiable, the multipliers at corresponding points are equal.

5.1#17. For $f(x) = 2x^2 - 1$, find the multipliers for cycles of any period n. Hint: Use the results of the preceding exercise, which works everywhere except at the fixed points, where the conjugate mapping is not differentiable.

5.1#18°. For $f(x) = 2x$ and $\varphi(x) = x^3$,

(a) Find the function $g(x)$ which is conjugate to f by φ, i.e., $\varphi^{-1}f\varphi$.

(b) Show that the multipliers of f and g are different.

(c) Why does (b) not contradict the statement proved in Exercise 5.1#16?

5.1#19. For a general quadratic function $f(x) = ax^2 + bx + d$, with $a \neq 0$, show that a conjugate mapping $\varphi(x) = \alpha x + \beta$ can always be found such that $f(x)$ is conjugate to the simpler quadratic map, $g(x) = x^2 + c$. That is, show how to choose α, β, and c such that $\varphi \circ f = g \circ \varphi$, which is another way to state equation (3) in Section 5.1.7.

Exercises 5.2 Logistic Model and Quadratic Polynomials

5.2#1. For the logistic model we study equation (6)

$$q(n + 1) = (1 + \alpha)q(n) - \alpha q(n)^2 \qquad (6)$$

which may be considered as iterating the quadratic polynomial

$$Q(x) = (1 + \alpha)x - \alpha x^2.$$

(a) Show that the conjugate mapping $\varphi(x) = \delta x + \beta$ that transforms this quadratic to the simpler form $P(x) = x^2 + c$ (as in Exercise 5.1#19) can be written

$$\varphi(x) = \frac{(1 + \alpha)}{2\alpha} - \frac{x}{\alpha}.$$

(b) Show that the change of variables $x = (1 + \alpha)/2 - \alpha q$ in equation (6) is equivalent to the conjugation mapping you found in part (a).

5.2#2. Show that another way to look at equation (6) is as an application of Euler's method, with stepsize $h = 1$, to the differential equation $x' = \alpha(x - x^2)$ studied in Section 2.5.

5.2#3°. Show that $q(n)$ of equation (6) will be attracted to the stable population $q_s = 1$ if the fertility α is "small," $0 < \alpha \le 1$, and the initial population is "small," $0 < q(0) < 1 + \frac{1}{\alpha}$, as is the case with the corresponding differential equation.

5.2#4. (harder) Show that q_s of the preceding exercise is an attracting fixed point if and only if $0 < \alpha \le 2$.

5.2#5°. If $f: X \to X$ and $g: Y \to Y$ are two maps, and $\varphi: X \to Y$ is a mapping which is 1–1 and onto such that $\varphi \circ f = g \circ \varphi$, then *everything dynamical* about (X, f) gets carries into something with exactly the same dynamical properties for (Y, g). When φ is not 1–1 and onto, this is no longer quite true. Consider, for example,

$X =$ the unit circle in $\mathbb{C}$, $f(z) = z^2$;

$Y =$ the interval $[-1, 1]$, $g(x) = 2x^2 - 1$;

φ the mapping φ from X to Y given by $\varphi(e^{i\theta}) = \cos\theta$.

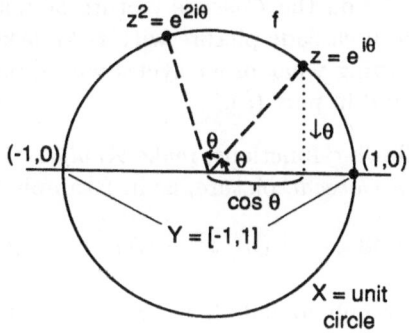

The accompanying figure shows two of the mappings: f (from the circle X to itself) and φ (from the circle X to the interval Y)

(a) Show that $\varphi \circ f = g \circ \varphi$.

(b) Show that φ is not 1–1.

(c) Find the fixed points of f and g. Do they correspond under φ?

(d) Show that both f and g have 2 periodic points of period exactly 2, and find them. Do they correspond under φ?

(e) Same as (d) for period three, in which case there are six points. Hint: in this case the periodic points for g are the numbers $\cos 2k\pi/7$, $k = 1, 2, 3$ and $\cos 2k\pi/9$, $k = 1, 2, 4$.

(f) Same as (d) for any period n, and describe how they correspond to the periodic points of g.

5.2#6.

(a) Make a *Cascade* picture and by locating "windows," find values of c that will give cycles orders 3, 4, and 5. Verify your results with *Analyzer* iteration pictures at those values.

(b) By suitable blowups of the cascade picture around values giving cycles orders 3 and 4, find different values of c that will give copies of order 12.

5.2#7.

(a) Use *Analyzer's* root-finding capability and Theorem 5.1 to find the values of c for which you will find real periodic points of order 3, 4, 5, and 7. That is, find the roots of equations like $((0+c)^2+c)^2+c=0$.

(b) Verify this fact experimentally by making either a time series or an *Analyzer* iteration picture for each value of c so found.

(c) Further verify your results by locating suitable "windows" at the given values of c on the *Cascade* picture or a suitable blowup of it. Mark the overall cascade picture with tickmarks on the vertical axis and labels showing what order cycles will appear for the particular values of c found in part (a).

5.2#8. For the following functions, make *Analyzer* iteration pictures and match them up to a *Cascade* picture, as in Example 5.14.

$$\text{(a) } x^2 - 1.48 \qquad \text{(b) } x^2 - 1.77 \qquad \text{(c) } x^2 - 1.76$$

5.2#9. Consider the polynomial $p(x) = 2x - cx^2$, where c is a number not equal to zero.

(a) What are the fixed points of $p(x)$?

(b) Show that $p(x)$ is conjugate to $q(x) = x^2$.

(c) What initial values are attracted under $p(x)$ to $1/c$?

(d) Suppose you knew π to 1000 significant digits. If you set $x_0 = 0.3$, how many iterations of $p(x)$ would you need to compute $1/\pi$ to 1000 digits? Compare this with the number of steps in long division necessary to achieve 1000 digits.

Remark. This is the algorithm actually used in computers to compute inverses to high precision.

Exercises 5.3 Newton's Method

5.3#1. Write the first five iterates of Newton's method for calculating the roots of

(a) $x^2 - 23 = 0$ starting at $x_0 = 5$;

(b) $x^3 + x - 1 = 0$ starting at $x_0 = 1$;

(c) $\sin x - \frac{x}{2} = 0$ starting at $x_0 = 1$.

5.3#2. Notice that if $x_0 = 1$, Exercise 5.3#1c does not converge very well under five steps of Newton's method. Describe why this is so, first graphing the function $\sin x - (x/2)$ with the *Analyzer* program, and then discussing the behavior of the iteration of Newton's method.

5.3#3°. Let P_b be the polynomial $z^3 - z + b$, and N_b the corresponding Newton's method.

(a) Show that there exists b such that $N_b(N_b(0)) = 0$.

(b) What happens to points z near 0 if you iterate Newton's method starting at z?

Hint: Compute $\dfrac{d}{dz}(N_b(N_b(z)))$.

5.3#4.

(a) Show that Newton's method for finding roots of $x^3 + ax + 1$ amounts to iterating $(2x^3 - 1)/(3x^2 + a)$, as in the Cascade picture of Figure 5.3.7.

(b) Show that any cubic polynomial $ax^3 + bx^2 + cx + d$ can be written in the form $x^3 + ax + 1$ if you make the appropriate change of variables, which you should state.

5.3#5. Show that if $p(z) = (z - a)(z - b)$ is a quadratic polynomial, with $a \neq b$, then the sequence

$$z, N_p(z), N_p(N_p(z)), \ldots$$

converges to a if $|z - a| < |z - b|$.
  Hint: Make the change of variables $u = (z - a)/(z - b)$.

5.3#6. Try to understand what Newton's method does for the polynomial $x^2 + 1$ on the real axis. Hint: Try setting $x = \arctan \theta$.

5.3#7. Refer to the end of Example 5.3.4, about using Newton's method to find square roots. For $x^2 - a = 0$, you will get

$$N_a(x) = \frac{1}{2}\left(x + \frac{a}{x}\right), \qquad x \neq 0, \quad a > 0.$$

(a) Draw the graph of N_a, showing that it is a hyperbola.

(b) Show that the square roots $\pm\sqrt{a}$ are superattracting fixed points for N_a.

(c) Show that for every positive x_0, Newton's method will converge to a positive root (i.e., $x_n \to \sqrt{a}$) and that for every negative x_0, Newton's method will converge to a negative root (i.e., $x_n \to -\sqrt{a}$).

(d) Show that $\pm\sqrt{a}$ are the only cycles.

5.3#8. Refer to Example 5.3.4 for using Newton's method to find the roots to $(x - a)(x - b) = 0$ (!!)

(a) Prove equation (15) in Example 5.3.4 (a "simple" computation with plenty of chances for errors).

(b) Verify that if $x > (a + b)/2$, you will converge to the larger root;

if $x_0 < (a + b)/2$, you will converge to the smaller root;

if $x_0 = (a + b)/2$, you will converge to no root at all.

5.3#9°. You might like to see how differently

$$T_a(x) = \frac{1}{2}\left(x - \frac{a}{x}\right) a > 0$$

behaves from the N_a of Exercise 5.3#7 (if under iteration you hit $x_n = 0$, just stop):

(a) Draw the graph of T_a (it is again a hyperbola). Show that there are no fixed points. But find a cycle of period 2 and show that the cycle is repelling by showing that $(T \circ T)'(x_0) > 1$.

(b) Why is $(T \circ T)'(x_0) = (T \circ T)'(x_1)$? Or, in general, why is for a periodic cycle, $x_0, x_1, \ldots, x_t = x_0$, $(T \circ \ldots \circ T)'(x_i)$ the same for $i = 0, 1, 2, \ldots, k - 1$?

Note: The functions $T_a(x)$ arise if you study Newton's iteration for $x^2 + a = 0$ with positive a. One can show that every complex number with *positive* real part converges to $\sqrt{a}$ under iteration with N_a, while every complex number with *negative* real part converges to $-\sqrt{a}$ under iteration. If x is purely imaginary, then $N_a(x)$ is purely imaginary with $\operatorname{Im} N_a(x) = T_a(\operatorname{Im} x)$.

5.3#10. Prove that each root r of an equation $f(x) = 0$ is a superattracting fixed point of Newton's equation, provided that $f'(r) \neq 0$ and $f''(r)$ exists.

5.3#11°.

(a) Let r be a root of $f(x) = 0$ and let $x = r + \varepsilon$, $\varepsilon \neq 0$. Derive an expression for $f(x - \varepsilon)$ using a Taylor polynomial with ε^2 as highest order ε-term. Assume f is twice differentiable.

(b) Assuming $f'(x_i) \neq 0$ and f is doubly differentiable, use your result from part (a) to show that

$$N_f(x_i) - r = \frac{f''(c)}{2f'(r)}(x_i - r)^2, \quad c \in \{[x_i - \varepsilon, x_i] \cup [x_i, x_i - \varepsilon]\}.$$

(c) Using the result from part (b), prove the assertion that if $x_i = r + \varepsilon$ and $|\varepsilon|$ is small enough, the number of correct decimal digits will approximately double after each iteration.

(Remember N correct decimal digits $\Rightarrow |\text{error}| \leq \frac{1}{2}10^{-N}$.)

(d) $\sqrt{3} = 1.732050808$. Using a calculator choose an initial $x > 0$ and iterate $N_3(x) = \frac{1}{2}(x + \frac{3}{x})$, Newton's equation for $x^2 - 3 = 0$. Indicate the doubling of the digit accuracy. Why is ε always "sufficiently small" in this case?

(e) Try $x = e^{-x}$ also. That is, find roots of $f(x) = x - e^{-x}$.

5.3#12. For each of the following equations, express as $f(x) = 0$; then find each $N_f(x)$ and analyze the iterations of each $N_f(x)$ using *Analyzer* (or the *Analyzer* component of the *Cascade* program with the Newton's method option). Then use *Analyzer* to graph each $f(x)$. Draw the linear approximations (tangents to the graph of $f(x)$) to demonstrate why solutions in a certain region converge to a certain root.

$\quad$ (a) $x^2 - x - 6 = 0$ $\qquad$ (b) $x^3 - x = 0$ $\qquad$ (c) $x = e^{-x}$.

5.3#13. Use the *Analyzer* component to the *Cascade* program with the Newton's method option to find the roots of each of the following equations:

(a) $x^4 + \frac{5}{2}x^3 - \frac{5}{2}x = 1$.

(b) $x^5 - x^4 + x^2 = 15x^3 - 38x - 24$.

(c) $2x^8 - 12x^7 + 19x^6 + 9x^5 - \frac{255}{8}x^4 - 9x^3 + 19x^2 + 12x + 2 = 0$.

5.3#14°. Let us consider a Newton's method problem in two variables: As you will see in Chapter 8, a singularity of a system of differential equations occurs when $x' = 0$ and $y' = 0$ simultaneously. If

$$x' = f(x, y) = (x - 1)^2 - y,$$

$$y' = g(x, y) = y + \frac{x^2}{2} - \frac{1}{2},$$

and you try to find a singularity by Newton's method, taking as an initial guess $(x_0, y_0) = (0,0)$, find (x_1, y_1). Then find (x_2, y_2).

5.3#15. Let f be a differentiable function near x_0, with a *degenerate root* at x_0, in the sense that at x_0, f has the asymptotic expansion (as explained in the Appendix)

$$f(x_0 + \xi) = A\xi^k + o(\xi^k)$$

for some $k > 1$.

(a) Show that N_f has an attracting fixed point at x_0, and compute the multiplier of N_f at x_0.

(b) If you try to solve $x^k = 0$ by Newton's method, starting at $x_0 = 1$, how many iterations will be required to find the root to within 10^{-6}?

Exercises 5.4 Numerical Methods as Iterative Systems

5.4#1. Considering Euler's method as an iterative system for an autonomous differential equation $x' = f(x)$, we derived equation (17) $F_h(x) = x + hf(x)$.

(a) Derive our subsequent statement that the corresponding iterative system for midpoint Euler is

$$F_h(x) = x + hf\left(x + \frac{h}{2}f(x)\right).$$

(b) More of a challenge is to do the same for Runge–Kutta:

$$x + \frac{h}{6}\left(f(x) + 2f\left(x + \frac{h}{2}f(x)\right) + 2f\left(x + \frac{h}{2}f\left(x + \frac{h}{2}f(x)\right)\right)\right.$$
$$\left. + f\left(x + hf\left(x + \frac{h}{2}f\left(x + \frac{h}{2}f(x)\right)\right)\right)\right).$$

5.4#2°. Consider $x' = -\alpha x (\alpha > 0)$, the equation of Examples 5.4.1 and 5.4.2.

(a) Analyze the iteration of the function resulting from midpoint Euler,

$$F_h(x) = x + h\left(-\alpha\left(x + \frac{h}{2}(-\alpha x)\right)\right) = \left(1 - \alpha h + \frac{\alpha^2 h^2}{2}\right)x,$$

to show that all solutions tend monotonically to 0 as $t \to \infty$ if $0 < h < 2/\alpha$, to ∞ if $h > 2/\alpha$.

(b) Analyze the iteration of the function resulting from Runge–Kutta,

$$F_h(x) = \left(1 - \alpha h + \frac{h^2 \alpha^2}{2} - \frac{h^3 \alpha^3}{6} + \frac{h^4 \alpha^4}{24}\right)x.$$

Find the range of values of h for which the orbits tend to 0 as $t \to \infty$, and show that numerical instability sets in for $h > k$, where $k > 2/\alpha$.

5.4#3. Consider $x' = x - x^2$, the equation of Example 5.4.3.

(a) Show that $F_h(x) = x - h(x - x^2)$ can be made equivalent to any quadratic polynomial by a proper choice of h and a change of variables $u = px + q$.

(b) Show that numerical solutions will never blow up by considering

$$x' = x - x^2 = (1 - x)x = \alpha x \quad \text{if} \quad \alpha = 1 - x,$$

thus treating the differential equation as in Example 5.4.1.

5.4#4. Consider $x' = -tx$, the equation of Example 5.4.4.

(a) Show that the condition for numerical stability of solutions is

$$1 - \frac{h^2}{2} + t_n\left(\frac{h^3}{4} - h\right) + \frac{h^2 t_n^2}{2} < 1,$$

and that it occurs for $t_n <$ approximately $2/h$.

(b) Explain the behavior of solutions to this differential equation by considering it as a case of $x' = -\alpha x$, of Example 5.4.1.

5.4#5. Consider the equation $v' = -2\sqrt{t}v$, which occurs in the discussion of Example 5.4.5.

(a) As in the preceding exercises, analyze the behavior of its solutions.

(b) Apply implicit Euler to this equation, with $t_n = nh$, and show that v_{n+1} goes monotonically to zero, regardless of the stepsize h. Explain why this means that computer drawings of solutions to $x' = x^2 - t$ would show no jaggies.

5.4#6°. For $x' = x^2 - t^2$, with stepsize $h = 0.03$, calculate at what value of t the Runge–Kutta solutions will break down into jagged junk. Refer back to the opening pictures of Section 5.4 and compare the results of your computer experiments on the second.

5.4#7. For each of the following differential equations, use stepsize $h = 0.03$ and calculate where the solutions will lose their numerical stability. Make computer drawings to show these results.

$$\text{(a) } x' = \sqrt{1 - tx} \qquad \text{(b) } x' = x - t^2 \qquad \text{(c) } x' = x^2 + t^2$$

Exercises 5.5 Periodic Differential Equations

5.5#1.

(a) Prove direct from Definition 5.5.2 of the one period later mapping P_{t_0} that $P_{t_0+T} = P_{t_0}$, a useful exercise in clarifying the roles of the t_0's and x_0's.

(b) Prove that for any s, t the period later mapping P_t is conjugate to P_s. Hint: Consider the map $Q_{s,t}(x)$ that associates to x the value of the solution to $x' = f(t, x)$ through (s, x) at time t.

5.5#2. To provide some understanding of Figure 5.5.3 for the differential equation $x' = \sin t \cos(x + t)$,

(a) Using the computer program *DiffEq*, print out a large direction field (default window is fine) with a number of solutions.

(b) Mark this printout with vertical lines at $t = -2\pi, 0, 2\pi$. For these three values of t_0, start solutions at the same x_0's and see that solutions indeed follow the same shape for either of the three t_0's. That is, for every initial condition $(0, x_0)$, try $(-2\pi, x_0)$ and $(2\pi, x_0)$.

(c) Make an appropriate P_{t_0} drawing by hand, showing which groups of solutions in the direction field go to the nearly horizontal sections of P_{t_0}, and which go to the nearly vertical sections, as in Figure 5.5.3.

(d) Show how the attracting fixed points of P_{t_0} correspond to attracting periodic solutions of the differential equation; look for and mark other corresponding features.

5.5#3. Find the one period later mapping P_{t_0} for the following periodic differential equations, with $t_0 = 0$ and $T = 2\pi$: (Remember that to keep the computing time finite, you must adjust your x_0 range to bracket just the values for which the solution makes it all the way across from t_0 to $t_0 + T$.) For equations (a) and (b) the direction fields are pictured in Figure 5.5.1. Make printouts of P_{t_0} and of the direction fields, and analyze, as in Exercise 5.5#1.

(a) $x' = (2 + \cos t)x - \dfrac{1}{2}x^2 - 1$

(b) $x' = (2 + \cos t)x - \dfrac{1}{2}x^2 - 2$

(c) $x' = (x^2 + 1)\cos t$

(d) $x' = (x^2 - 1)\cos t$.

5.5#4. Give a discussion comparing the preceding Exercises 5.1#3a and #3b, using the following guides. You can then look ahead to Example 5.5.6 to confirm and extend your comparison for $x' = (2 + \cos t)x - \frac{1}{2}x^2 - a$ in general. Proceed as follows.

(a) Show that if $x = u(t)$ is a solution, then $x = u(t + 2\pi)$ is also a solution.

(b) With the computer program *DiffEqSys*, make a computer drawing for $a = -1$ showing the following behavior: There are exactly *two* periodic solutions

$$u_i(t + 2\pi) = u_i(t) \qquad i = 1, 2.$$

Identify a funnel around one of the periodic solutions and an antifunnel around the other.

(c) There is *no* periodic solution for $a = -2$. By experimenting with the computer find the value to two significant digits of the constant "a" between -1 and -2, which separates the two behaviors: where there exists a periodic solution and where there exists no periodic solution.

(This is a repeat of Exercise 2.4–2.5#2, now that your perspective has a periodic point of view.)

5.5#5. Solve explicitly and give a discussion comparing Exercises 5.5#3c and #3d. These examples illustrate the theorem of the next exercise.

5.5#6°. For *separable* differential equations, sometimes the one period later mapping P_{t_0} will be the identity map, a segment of the diagonal, as in Exercise 5.5#3c, and sometimes not, as in Exercise 5.5#3d. The key is an additional "if and only if" hypothesis giving the following theorem:

Consider a separable and periodic differential equation $x' = f(t)g(x)$ with $f(t) = f(t + T)$ and $g(x)$ differentiable. Suppose $g(a) = g(b) = 0$ with $a < b$. Then the one period later mapping P_{t_0} is the identity on $[a, b]$ if and only if

$$\int_0^T f(t)dt = 0.$$

Prove this result. Hint: Show that if a and b are consecutive zeroes of g, and $c \in (a, b)$, then

$$\int_c^x \frac{du}{g(u)}$$

is a monotone function on (a, b), tending to positive or negative ∞ as $x \to a$ and $x \to b$.

5.5#7. Consider the differential equation

$$x' = (x^2 + 1)\cos t,$$

which is periodic with period 2π.

(a) Solve the differential equation by separation of variables to show that the solution $\gamma_x(t)$ with initial condition $\gamma_x(0) = x$ is

$$\gamma_x(t) = \tan(\sin t + \tan^{-1}(x)).$$

(b) Show that for this differential equation, the one period later mapping is

$$P_0(x) = \gamma_x(2\pi) = \tan[\sin(2\pi) + \tan^{-1} x] = x,$$

which is the identity *where P_0 is defined.*

(c) Show that $\gamma_x(t)$ only represents a solution of the differential equation over the entire interval $0 \le t \le 2\pi$ if $|\sin t + \tan^{-1} x| < \pi/2$, which corresponds to $|\tan^{-1} x| < (\pi/2) - 1$, so

$$P_0(x) = x \quad \text{if} \quad |x| < \tan\left(\frac{\pi}{2} - 1\right) \approx .64209,$$

and otherwise P_0 is undefined.

(d) Using the computer program *DiffEq*, draw some solutions to the differential equation. Then using the program *ID Periodic Equations*, draw the graph of P_0 as discussed in part (c). Show how your two drawings relate to one another.

5.5#8°. For a linear differential equation $x' = p(t)x + q(t)$ that is periodic in t with period T, prove that P_{t_0} is a linear function $P_{t_0} = ax + b$, for appropriate a and b.

5.5#9. Corroborate the statement of Example 5.5.3 for the differential equation

$$x' = (x^2 - 1)(\cos t + 1)$$

that if

$$x_0 > (e^{4\pi} + 1)/(e^{4\pi} - 1),$$

then the formula

$$\gamma_{x_0}(t) = \frac{(x_0 + 1) + (x_0 - 1)e^{2(t+\sin t)}}{(x_0 + 1) - (x_0 - 1)e^{2(t+\sin t)}}, \tag{25}$$

does not correspond to a genuine solution of the differential equation, but has a pole at some t, $0 < t < 2\pi$.

5.5#10. For Example 5.5.4, with the differential equation

$$x' = (\sin t + \alpha)x - \beta e^{\cos t}x^2,$$

verify the three statements explaining how the behavior of P_{t_0} changes according to whether, for $A = e^{-2\pi\alpha}$, we have $A > 1$, $A = 1$, or $A < 1$, as shown in Figure 5.5.5.

5.5#11. For Example 5.5.4, show with graphs how the statements explaining the behavior of P_{t_0} in terms of the values of the parameter $A = e^{-2\pi\alpha}$ translate into statements about the solutions to the differential equation $x' = (\sin t + \alpha)x - \beta e^{\cos t}x^2$.

5.5#12. Prove that any continuous function $g: \mathbb{R} \to \mathbb{R}$ which is onto and monotone increasing is the period-function $g = P_\pi$ for some differential equation periodic of period π.

   Hint: the idea is to draw curves in $\mathbb{R}^2$ joining every point $(0, x)$ to the point $(\pi, g(x))$, which are all disjoint, and fill up $[0, \pi] \times \mathbb{R}$. In order to guarantee that these curves, continued periodically to $[\pi, 2\pi] \times \mathbb{R}$, and so on, form differentiable curves in $\mathbb{R}^2$, we will require that they be horizontal at the endpoints. One way to do this is to consider, for each x, the graph of the function

$$\gamma_x(t) = \frac{g(x) + x}{2} - \frac{g(x) - x}{2}\cos(t).$$

   (a) Show that the graph of γ_x does join $(0, x)$ to $(\pi, g(x))$.

   (b) Show that every point (t, x) with $0 \le t \le \pi$ is on the graph of γ_x for some unique $x \in \mathbb{R}$.

   Consider the function $f(t, x)$, defined on $[0, \pi] \times \mathbb{R}$, which associates to (t, x) the slope of the curve through that point at that point.

   (c) If f is extended to $\mathbb{R}^2$ so as to be periodic of period π in t, show that it is a continuous function on $\mathbb{R}^2$.

   (d) Show that the period map P_π for the differential equation $x' = f(t, x)$ is exactly g.

5.5#13. The proof of Proposition 5.5.5 shows that the sequence $P_{t_0}^{\circ n}$ is monotone and bounded, so must tend to some limit. Show that the limit must be in the interval $[x_i, x_{i+1}]$, and must be a fixed point of P_{t_0}.

5.5#14. Predict and verify the implications of Figure 5.5.7 for the solutions to the differential equation $x' = \cos(x^2 + \sin 2\pi t) - a$ from Example 5.5.7.

5.5#15. Show that $x' = \sin(1/x) + \frac{1}{2}\sin t$ has infinitely many periodic solutions.

5.5#16. Show that $x' = \sin(x^2 - t)$ has a unique periodic solution, unstable in the forward direction, stable in the backward direction.

5.5#17. Consider the differential equation $x' = \cos x + \cos t$.

(a) Use *DiffEq* to make a computer picture of slopes and solutions.

(b) Show that the region $t \geq 0$, $0 \leq x \leq \pi$ is a weak funnel, and that the region $t \geq 0$, $\pi \leq x \leq 2\pi$ is a weak antifunnel.

(c) (harder) Show that the solution in the antifunnel is unique. Hint: You will need to show that such a solution satisfies $\pi + \varepsilon < x(t) < 2\pi - \varepsilon$ for some $\varepsilon > 0$, and use the theorem proved in Exercise 4.7#3 for the case of an antifunnel that does not narrow.

(d) Show that the solution in the antifunnel is periodic of period 2π.

(e) Show that there is a unique periodic solution in the funnel.

Exercises 5.6 Iterating in $\mathbf{c}^1$

5.6#1. Show that any quadratic polynomial can be conjugated to $z^2 + c$ by an affine map $az + b$. (This is shown for real z in Exercise 5.1#19.)

5.6#2. Try iterating $z^2 - \frac{1}{2}$, $z^2 - \frac{3}{2}$, $z^2 + 1$ and $z^2 - 3$ for various complex starting points z_0. Unless you have a program that will do complex arithmetic, you must do these by hand. That means you can multiply or add digits by calculator, but you will have to combine the proper terms by the rules of complex arithmetic before proceeding to the next step. What can you tell or suspect about the boundedness of the orbits in each case? Your results should fit the information to be obtained from the Mandelbrot set.

5.6#3. In the manner of Example 5.1.9, let us look at the periodic orbits of period *three* for $x^2 + c$. The order 3 analog of equation (2) is

$$((O^2 + c)^2 + c)^2 + c = 0,$$

which has two real solutions for c and two additional but complex solutions, $c = -0.1226 \pm 0.7499i$. This means that quadratic polynomials of form $x^2 + c$ for any of these four values of c, when iterated from $x_0 = 0$, produce a cycle of order three; one of these is actually a fixed point. We cannot directly graph this iteration as we did in Figure 5.1.10, since it involves complex numbers. But you can and should confirm by direct algebraic calculation that for $c = -.1226 \pm 0.7499i$ iterating from $z_0 = 0$ under $z^2 + c$ creates a cycle order three. Unless you have a program that will do complex arithmetic, you must do these by hand. That means you can multiply or add digits by calculator, but you will have to combine the proper terms by the rules of complex arithmetic before proceeding to the next step.

5.6#4°. Show that for $z^2 + c$ the following is a practical test for knowing whether an orbit is unbounded: whenever an iterate exceeds $|c| + 1$ in absolute value, the successive iterates will go off toward infinity.

5.6#5. Consider the polynomial $z^2 + 2$. As stated in Example 5.6.8, show that all points z with $|z| \geq 2$ escape to ∞.

5.6#6. You can learn a lot about the dynamics of complex iteration by examining the iteration of $z^2 - 1$. Although the resulting picture is symmetric (because z gives the same result as $-z$), you will see that the dynamics are certainly *not* symmetric.

(a) Calculate the iteration for various complex starting points z_0. For example, try $z_0 = 0, 1, -1, 1.5, i, -i, 1+i, 2i$. (See the notes regarding calculation in Exercise 5.6#2 above.)

(b) You also have the option of performing these operations geometrically: squaring means to square the absolute value and double the polar angle; adding or subtracting a real number amounts to a horizontal translation.

(c) Calculate the fixed points.

(d) Show that your results fit with Figure 5.6.2, the Julia set for $z^2 - 1$.

5.6#7. Confirm that there exist 2^{n-1} values of c for which $z^2 + c$ iterates from $z_0 = 0$ to a cycle of order n.

5.6#8. For $f(z) = z^2$.

(a) Find the periodic points of periods 2 and 3.

(b) Show that the periodic points of period dividing p are $e^{2k\pi i/2^p - 1}$.

5.6#9.

(a) Prove that every complex quadratic polynomial has $(2^p - 2)$ *complex* periodic points of period exactly p, counting multiplicity, for every *prime* number p.

(b) How many periodic points of period 4 does a complex quadratic polynomial have?

(c) How many periodic points of period 6 does a complex quadratic polynomial have?

(d) In general, how many periodic points of period k does a complex quadratic polynomial have?

5.6#10. Consider the question of Example 5.6.1. For how many *real* values of c is 0 periodic of period p, for some prime number p? The answer is $(2^{p-1} - 1)/p$ for all primes $p \neq 2$. That fact is too difficult to prove here, but you should verify it for $p = 3, 5, 7, 11$. That is, use the program *Analyzer* to graph the orbit of 0 under $x^2 + c$ and count the roots, using blowup when you need a clearer picture. (Two notes: (i) The function you wish to graph is a function of c; set $x_0 = 0$ and then change c to x for entering in *Analyzer*. (ii) To avoid problems with the blowup, keep the lefthand endpoint less than -2, and the vertical heights on the order of ± 1.)

5.6#11°. Prove that K_{-2}, the filled-in Julia set for $z^2 - 2$, is simply the interval $[-2, 2]$. For purely real $z = a$ or for purely imaginary $z = bi$, it is straightforward to find when the orbits are unbounded. But for $z = a + bi$, you will find that a different approach is required. We suggest the following:

(a) Show that for every point z in the complex plane outside the interval $[-2, 2]$, there is a unique point ξ with $|\xi| < 1$ such that $\xi + 1/\xi = z$. Hint: Observe that $\xi + \frac{1}{\xi} = z$ is a quadratic equation for ξ in terms of z, and that the two solutions are inverses of each other.

(b) Show that the map $\varphi(\xi) = \xi + 1/\xi$ conjugates P_{-2} to P_0, as shown in this diagram:

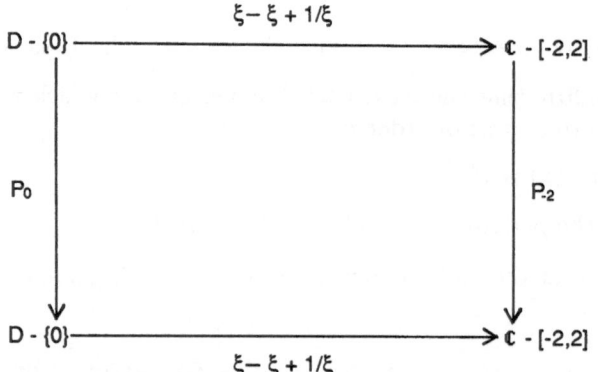

(c) Derive the result from the fact that if $|\xi| < 1$ then $\xi^{2^n} \to 0$ as $n \to \infty$.

5.6#12. The roots of the polynomial equation $P_c^{\circ n}(0) = 0$ are those values of c for which 0 is m-periodic, where m divides n. In Exercise 5.6#7 you have confirmed that there are always $k_1 = 2^{n-1}$ values of c that are roots of $P_c^{\circ n}(0) = 0$; this includes fixed points and cycles of lower order.

(a) For each of the first four possible values of n, find the c-values that are n-periodic and those that are exactly n-periodic. For each n locate these c-values on the Mandelbrot set and show that they confirm the following statement.

> *For any n, each value of c that is a root of $P_c^{\circ n}(0) = 0$ lies within a unique "ball" (called a* **component***) of the Mandelbrot set.*

(b) Verify by experiment with another point in the "2-ball," such as $(-1.1, 0)$, and another point in the topmost "3-ball" that the orbit of 0 confirms the following statement:

> *All other points of each* **component** *of the Mandelbrot set, have the property that* 0 *is not n-periodic, but is* **attracted** *to a cycle of period a.*

If you are lucky enough to have access to a graphics program that follows orbits, fine; otherwise you can do this part and the next just with algebraic calculation.

(c) Pick a point in another ball or baby Mandelbrot set and determine what the periodicity of the orbit will be. This should enable you to appreciate the fact that there exists a whole combinatorial litany of what you will find where on the Mandelbrot set; that is, what kinds of cycles the orbit of 0 will be attracted to, for the different values of c at the different points within M. Now it is time to go looking at the references, listed at the end of this volume.

5.6#13. For the original Cantor set:

(a) Show that it satisfies the defining requirements: that it is closed and bounded, totally disconnected, and without isolated points.

(b) Show that the endpoints of the deleted intervals can be written two different ways in base 3 decimals, so that they can have the proper decimal representation to remain in the Cantor set.

Appendix. Asymptotic Development

Asymptotic development or *asymptotic expansion* is the technical way of saying that one function looks like another function, and since in this book we will be constantly concerned with making such statements, we save an appendix for precisely describing the rules of the game.

A1. Equivalence and Order

The idea we wish to pursue in general is

> *how to express any function in terms of powers of x, or powers of $(x - x_0)$.*

The key property is the notion of *equivalence* of two functions near a point x_0. Actually, we will frequently be concerned with functions defined only for $x > 0$ and the behavior near 0 or ∞, so we will only require that functions be defined on a one-sided neighborhood of a point x_0.

Most often, it will be behavior near infinity that matters. But the change of variables

$$y = \frac{1}{x - x_0} \quad \text{or} \quad x = \frac{1}{y} + x_0$$

will transform a right x-neighborhood of x_0 into a y-neighborhood of $+\infty$; i.e., if we define

$$g(y) = f\left(\frac{1}{y} + x_0\right),$$

then the behavior of g near $+\infty$ will be the same as the behavior of f on a right-neighborhood of x_0.

For the purposes of defining terms, we shall use *right*-neighborhoods.

Definition A1.1. For two functions f and g defined to the right of x_0, they are *equivalent* at x_0 if

$$\lim_{x \downarrow x_0} \frac{f(x)}{g(x)} = 1.$$

This equivalence is noted $f \approx_{x_0} g$, or $f \approx g$ if $x \to \infty$.

This is not the only possible definition of equivalence, and it does not necessarily correspond to functions behaving "like each other." For instance, the functions

$$\sin(x) \quad \text{and} \quad \sin(x + e^{-x})$$

behave very much like each other near $+\infty$ (it would be hard to distinguish them numerically for $x > 20$) but they are not equivalent in the sense above. This is because the values of x for which they vanish do not coincide (although they are very close), so the ratio ricochets near these values, being zero at some x and undefined at others between the zeroes.

Remark. Definition A1.1 is only reasonable for functions which have constant sign on the right side of x_0; a different definition of asymptotic development would be required for functions which oscillate.

Associated with equivalence, there is a natural notion of *order* on functions defined on a right neighborhood of x_0.

Definition A1.2. Near but to the right of x_0, a function f is of *higher order* than a function g if

$$\lim_{x \downarrow x_0} \frac{f(x)}{g(x)} = 0.$$

If this is true as $x \to \infty$, a function f is said to be of *lower order* than a function g.

There is in general no reason to think, given two functions, that either is of higher order than the other; most often functions are not comparable.

We are now ready to proceed with the task of defining an asymptotic expansion.

A2. Technicalities of Defining Asymptotic Expansion

SETTING UP A SCALE

As we noted early in the last section, there is no loss of generality in describing only functions in a neighborhood of $+\infty$, and we will do that in the rest of this appendix, except where noted otherwise.

The next task is to describe our *scale functions,* a set S of functions (defined near $+\infty$) which can be considered to be known and which can be ranked in a linear order. The particular selection we will make is somewhat arbitrary, and the reader may at will discard some (fewer functions will then be describable) or add some (at the cost of being familiar with some rather elaborate functions).

Definition A2.1. Our set S of *scale functions* will be the functions

$$x^\alpha (\ln x)^\beta e^{P(x)}$$

where

$$P(x) = c_1 x^{\gamma_1} + \ldots + c_n x^{\gamma_n}$$

with $\gamma_1 > \ldots > \gamma_n > 0$, and with α, β, and the c_i arbitrary real numbers.

These functions are the ones which we will consider *known* near infinity, and in case you don't know them, the following may help:

Theorem A2.2. *The family S of scale functions has the following properties:*

(a) *Every function in S is positive near infinity.*

(b) *The family S is closed under products, and under raising to any real power; in particular it is closed under quotients.*

(c) *Every element of S tends to 0 or ∞ as $x \to \infty$ except the constant function 1.*

Proof. Statements (a) and (b) are clear upon reflection. Statement (c) is an elaborate way of saying that exponentials dominate powers, which themselves dominate logarithms.

Indeed, first suppose that $P(x) = 0$ and $\alpha = 0$. Then the scale function is

$$(\ln x)^\beta,$$

which clearly tends to 0, 1, or ∞, if $\beta < 0$, $\beta = 0$ or $\beta > 0$.

Now suppose that $P(x) = 0$ and $\alpha \neq 0$. Then the scale function is

$$x^\alpha (\ln x)^\beta,$$

and it is easy to see that this tends to infinity if $\alpha > 0$, and to 0 if $\alpha < 0$.

Finally if $P(x) \neq 0$, say

$$P(x) = c_1 x^{\gamma_1} + \ldots + c_n x^{\gamma_n}$$
$$= x^{\gamma_1}(c_1 + c_2 x^{\gamma_2 - \gamma_1} + \ldots + c_n x^{\gamma_n - \gamma_1}),$$

we need to show that the term $e^{c_1 x^{\gamma_1}}$ dominates, so that the function tends to 0 or infinity when $c_1 < 0$ and when $c_1 > 0$ respectively. This is left to the reader. $\square$

Actually, these properties are the only properties of a scale we will require, and the reader may add or delete functions at will so long that they are preserved. Please note that they imply that any two functions in the scale are comparable: in fact, either two functions are equal, or one is of lower order than the other. In this way the scale guarantees a linear ordering of the known functions.

PRINCIPAL PARTS

This section includes two rather nasty examples of theoretical interest; the more practical examples will be given in Sections A3 and A4.

Definition A2.3. A function $f(x)$ defined near $+\infty$ has *principal part* $cg(x)$ if $g \in S$, c is some constant, and $cg(x)$ is equivalent to f at $+\infty$. Then we can write $f \approx g$ as

$$f(x) = cg(x) + \text{lower order terms.} \tag{1}$$

Other ways of expressing that f is asymptotic or equivalent to $cg(x)$ are

$$f(x) - cg(x) \ll g(x), \tag{1a}$$

and

$$f(x) = cg(x) + o(g(x)). \tag{1b}$$

The "little o" notation $o(g(x))$ is in common usage, both to describe the set of all functions of lower order than $g(x)$ *and,* as in this case, to refer to some particular member of that set.

Another notation, "big O", which must be carefully distinguished from the "little o", is in common usage, especially by computer scientists. Upper case $O(g(x))$ means of order *at most* the same as $g(x)$—as opposed to lower case $o(g(x))$ meaning of order *strictly less* than that of $g(x)$. In this text we shall stick with $o(g(x))$.

A principal part already says a great deal about the function; and frequently knowing just the principal part is a major theorem, as the following somewhat remarkable example illustrates:

Example A2.4. Consider the function

$$\pi(x) = \text{number of prime numbers} \leq x.$$

It is not at all clear that this function should even have a principal part, but Legendre and Gauss, on the basis of numerical evidence conjectured that it would be $x/\ln x$. Indeed, the fact that

$$\pi(x) \approx x/\ln x,$$

was proved, about 1896, by the French mathematician Hadamard and the Belgian mathematician de la Vallée-Poussin. The result, called the Prime Number Theorem, is still considered a hard result, too hard for instance for a first graduate course in number theory.   ▲

ASYMPTOTIC EXPANSION

We are finally ready to finish the theoretical presentation. In many cases we will want more information about a function than just its principal part, and there is an obvious way to proceed, namely to look for a principal part for the difference, $f(x) - cg(x)$, which is of lower order than $g(x)$. That is, we now write

$$f(x) - cg(x) = c_2 g_2(x) + o(g_2(x)).$$

This procedure can sometimes be repeated.

Definition A2.5. For a function $f(x)$ an *asymptotic expansion*, or *asymptotic development*, occurs when we can write

$$f(x) = c_1 g_1(x) + c_2 g_2(x) + \ldots + c_n g_n(x) + o(g_n(x))$$

with the $g_i(x) \in S$.

Definition A2.5 means that for each $i = 1, \ldots, n$ we have

$$f(x) - \sum_{j=1}^{i} a_j g_j(x) \in o(g_i(x)).$$

Thus each term is a principal part of what is left after subtracting from f all the previous terms. Such an expansion is called "an asymptotic expansion to n terms."

A3. First Examples; Taylor's Theorem

We are now ready to state more precisely and prove Taylor's Theorem:

Theorem A3.1. *If $f(x)$ is n times continuously differentiable in a neighborhood of x_0, then*

$$f(x) = f(x_0) + f'(x_0)(x - x_0) + \ldots + \frac{1}{n!} f^{(n)}(x_0)(x - x_0)^n + o((x - x_0)^n).$$

Proof. This theorem is proved by repeated applications of l'Hôpital's rule:

$$\lim_{x \to x_0} \frac{f(x) - \left[f(x_0) + f'(x_0)(x - x_0) + \ldots + \frac{1}{n!} f^{(n)}(x_0)(x - x_0)^n \right]}{(x - x_0)^n}$$

$$= \lim_{x \to x_0} \frac{f'(x) - \left[f'(x_0) + f''(x_0)(x - x_0) + \ldots + \frac{1}{(n-1)!} f^{(n)}(x_0)(x - x_0)^{n-1} \right]}{n(x - x_0)^{n-1}}$$

$$\vdots$$

$$= \lim_{x \to x_0} \frac{f^{(n)}(x) - f^{(n)}(x_0)}{n!} = 0. \qquad \square$$

At $x_0 = 0$, a few standard Taylor expansions are

$$e^x = 1 + x + \frac{1}{2!}x^2 + \ldots + \frac{1}{n!}x^n + o(x^n)$$

$$\sin x = x - \frac{1}{3!}x^3 + \ldots + \frac{(-1)^n}{(2n+1)!}x^{2n+1} + o(x^{2n+1})$$

$$\cos x = 1 - \frac{1}{2!}x^2 + \ldots + \frac{(-1)^n}{(2n)!}x^{2n} + o(x^{2n+1})$$

$$\ln(1+x) = x - \frac{1}{2}x^2 + \ldots + \frac{(-1)^{n-1}}{n}x^n + o(x^n)$$

$$(1+x)^\alpha = 1 + \alpha x + \frac{\alpha(\alpha-1)}{2!}x^2 + \ldots + \frac{\alpha(\alpha-1)\ldots(\alpha-n+1)}{n!}x^n$$
$$+ o(x^n).$$

Note carefully that *these Taylor series are asymptotic developments in a neighborhood of* 0. It is true that e^x looks quite a bit like 1 and even more like $1 + x$, and so on, near 0. But it does not look the least bit like 1 near infinity (or any other point except 0). In fact, the asymptotic development of e^x near infinity is itself; it is an element of the scale and hence already fully developed.

Similarly, the function $e^{1/x}$ is an element of the scale at 0 and fully developed there, although *at infinity* we have

$$e^{1/x} = 1 + 1/x + (1/2)1/x^2 + \ldots + (1/n!)1/x^n + o(1/x^n).$$

It is quite important to keep the notions of Taylor series and asymptotic development separate, and we will show this with the following example, actually studied by Euler:

Example A3.2. Consider the function

$$f(x) = \int_0^\infty \frac{e^{-t}}{1+xt}dt.$$

This function is well defined for $0 \le x < \infty$. It has an asymptotic development at 0 which we shall obtain as follows:

First, observe that near 0, we have a geometric series (attainable by long division):

$$\frac{1}{1+xt} = 1 - xt + (xt)^2 - \ldots + (-1)^n(xt)^n + \frac{(-xt)^{n+1}}{1+xt}. \qquad (2)$$

This is simply the formula for summing a finite geometric sequence.

Second, we will need to know that

$$\int_0^\infty t^n e^{-t}dt = n!. \qquad (3)$$

This is easy to show by induction on n and is left to the reader.

Third, multiply each term of (2) by e^{-t}, then, remembering that x is independent of t and hence can be brought outside the integral, we use (3) to integrate each term with respect to t from 0 to ∞, yielding

$$\int_0^\infty \frac{e^{-t}dt}{1+xt} = 1 - x + 2!\,x^2 - 3!\,x^3 + \ldots + (-1)^n n!\,x^n$$

$$+ (-1)^{n+1}x^{n+1}\int_0^\infty \frac{t^{n+1}e^{-t}}{1+xt}dt$$

$$= 1 - x + 2!\,x^2 - 3!\,x^3 + \ldots + (-1)^n n!\,x^n + o(x^n).$$

Let us now examine the remainder, to show that it is in fact $o(x^n)$. Since $1 + xt \geq 1$, we see that the remainder is *bounded* by

$$(n+1)!\,x^{n+1} \in o(x^n).$$

However, this power series

$$\sum (-1)^n n!\,x^n$$

does not converge for any $x \neq 0$!   ▲

You should think of what Example A3.2 means. Here we have a series, and the more terms you take, the "better" you approximate a perfectly good function. On the other hand, the series diverges, the terms get larger and larger, and the more terms you take, the more the sum oscillates wildly. How can these two forms of behavior be reconciled? The key point is the word "better." It is true that the nth partial sum is "better" than the $n - 1^{\text{th}}$, in the sense that the error is less than x^n near zero. But nobody says how close to zero you have to be for this to occur, and because of the $(n + 1)!$ in front, it only happens for x's very close to zero, closer and closer as n becomes large. So we are really only getting better and better approximations to the function on smaller and smaller neighborhoods of zero, until at the end we get a perfect approximation, but only at a single point.

Also, Example A3.2 should show that even if a function has an asymptotic expansion with infinitely many terms, there is no reason to think that the series these terms form converges, or, if it converges, that it converges to the function. Somehow, these convergence questions are irrelevant to asymptotic expansions. In particular, the authors strongly object to the expression "asymptotic series," which is common in the literature. *It is almost always wrong to think of an asymptotic expansion as a series; you should think rather of the partial sums.*

A4. Operations on Asymptotic Expansions

For nice functions, the previous section gave a way of computing asymptotic expansions. However, even if a function is perfectly differentiable, the computation of the successive derivatives that are needed in Taylor's formula is usually quite unpleasant, it is better to think of the function as made up from simpler functions by additions, multiplications, powers and compositions, and then to compute the first several terms of the asymptotic expansion the same way. We begin with several examples, before developing the theory which makes it work.

Example A4.1. Consider the function

$$f(x) = \sin(x + \tan x),$$

we will find an expansion near zero (where the function, on the right side of zero, is *not* oscillatory) with precision x^4. First

$$\tan x = \frac{\sin x}{\cos x} = \frac{x - \frac{x^3}{6} + o(x^4)}{1 - \frac{x^2}{2} + \frac{x^4}{24} + o(x^5)}$$

$$= \left(x - \frac{x^3}{6} + o(x^4)\right)\left\{1 + \left(\frac{x^2}{2} - \frac{x^4}{24}\right) + \left(\frac{x^2}{2} - \frac{x^4}{24}\right)^2 + o(x^4)\right\}$$

$$= \left(x - \frac{x^3}{6} + o(x^4)\right)\left\{1 + \left(\frac{x^2}{2} + \frac{5}{24}x^4\right) + o(x^4)\right\}$$

$$= x + \frac{x^3}{3} + o(x^4).$$

Now we can write

$$\sin(x + \tan x) = \sin\left(2x + \frac{x^3}{3} + o(x^4)\right)$$

$$= 2x + \frac{x^3}{3} + o(x^4) - \frac{1}{6}\left(2x + \frac{x^3}{3} + o(x^4)\right)^3 + o(x^4)$$

$$= 2x - x^3 + o(x^4). \quad \blacktriangle$$

Example A4.2. Consider the function $f(x) = \sqrt{(x+1)} - \sqrt{x}$ near ∞. We find $\sqrt{(x+1)} = \sqrt{x}\sqrt{(1 + 1/x)}$, and since $1/x$ is small near ∞, we can develop

$$\sqrt{1 + \frac{1}{x}} = 1 + \frac{1}{2!}\frac{1}{x} - \frac{1}{2!}\frac{1}{4}\frac{1}{x^2} + o\left(\frac{1}{x^2}\right)$$

leading to

$$\sqrt{x+1} - \sqrt{x} = \frac{1}{2}x^{-1/2} - \frac{1}{8}x^{-3/2} + o(x^{-3/2}). \quad \blacktriangle$$

Example A4.3. Consider the behavior near $x = 0$ of the function

$$f(x) = \frac{x \ln |x|}{1 + e^x}.$$

We find

$$\frac{x \ln |x|}{1 + e^x} = \frac{x \ln |x|}{2\left(1 + \frac{x}{2} + \frac{x^2}{4} + o(x^2)\right)}$$

$$= \frac{1}{2} x \ln |x| \left\{ \left(1 - \frac{x}{2} - \frac{x^2}{4} + o(x^2)\right) + o(x^2) \right\}$$

$$= \frac{x \ln |x|}{2} - \frac{x^2 \ln |x|}{4} - \frac{x^3 \ln |x|}{8} + o(x^3 \ln |x|). \qquad \blacktriangle$$

Example A4.4. Now for something nastier. How does

$$f(x) = (1 + x)^{1/x}$$

behave near ∞? Write

$$f(x) = (1 + x)^{1/x} = e^{(\ln(1+x))/x}.$$

Then we need to write

$$\ln(1 + x) = \ln x + \ln\left(1 + \frac{1}{x}\right) = \ln x + \frac{1}{x} - \frac{1}{2x^2} + o\left(\frac{1}{x^2}\right)$$

so that

$$\frac{\ln(1 + x)}{x} = \frac{\ln x}{x} + \frac{1}{x^2} - \frac{1}{2x^3} + o\left(\frac{1}{x^3}\right).$$

Since this function goes to 0 as $x \to \infty$, we can apply the power series of the exponential at 0 to get

$$f(x) = 1 + \frac{\ln x}{x} + \frac{1}{x^2} - \frac{1}{2x^3} + o\left(\frac{1}{x^3}\right) +$$

$$\frac{1}{2}\left[\frac{\ln x}{x} + \frac{1}{x^2} - \frac{1}{2x^3} + o\left(\frac{1}{x^3}\right)\right]^2 + o\left(\frac{(\ln x)^3}{x^3}\right)$$

$$= 1 + \frac{\ln x}{x} + \frac{(\ln x)^2}{2x^2} + \frac{1}{x^2} + \frac{(\ln x)}{x^3} + \frac{(\ln x)^3}{x^3} + o\left(\frac{(\ln x)^3}{x^3}\right).$$

In this case the result was not really obvious; it wasn't even clear that $f(x)$ tended to 1. $\blacktriangle$

A5. Rules of Asymptotic Development

It should be clear from the examples of the last section that there are two aspects to asymptotically developing functions: one is combining the parts found so far, and the other is recognizing what can be neglected at each step. Here are the rules.

1. *Addition.* You can add two asymptotic developments terms by term, neglecting the terms of higher order than the lower of the two precisions, and getting a development of that precision.

2. *Multiplication.* If

$$f_1(x) = c_1 g_1(x) + \ldots + c_n g_n(x) + o(g_n(x))$$
$$f_2(x) = b_1 h_1(x) + \ldots + b_m h_m(x) + o(h_m(x)),$$

   then you get an asymptotic expansion of f_1, f_2 by multiplying together the expressions above, collecting terms of the same order, and neglecting all terms of higher order than

$$g_1(x)h_m(x) \quad \text{or} \quad g_n(x)h_1(x).$$

The development obtained has precision equal to the lower of these two precisions.

3. *Powers* (and in particular *inverses*). This is a bit more subtle. The idea is to factor out the principal part of the function to be raised to a power, i.e. to write

$$f(x) = c_1 g_1(x)\underbrace{\left[1 + \left(\frac{c_2}{c_1}\right)\left(\frac{g_2}{g_1}\right) + \ldots\right]}_{\varphi} \tag{4}$$

so that we can apply the formula for $(1 + \varphi)^\alpha$ to the second factor. We get

$$(f(x))^\alpha = c_1^\alpha g_1^\alpha \left(1 + \alpha(\varphi) + \frac{\alpha(\alpha - 1)}{2}(\varphi)^2 + \ldots\right).$$

If the Taylor expansion of $(1 + \varphi)^\alpha$ is carried out to m terms, the precision of the development (4) is

$$g_1^\alpha \left(\frac{g_2}{g_1}\right)^m.$$

Here we stop the list of rules. A good exercise for the reader is the problem, very analogous to the Taylor series, of developing

$$\ln f \quad \text{and} \quad e^f,$$

or rather of formulating conditions under which it can be done (this is not always possible without enlarging the scale).

References

FOR DIFFERENTIAL EQUATIONS IN GENERAL

Artigue, Michèle and Gautheron, Véronique, *Systèmes Différentiels: Étude Graphique* (CEDIC, Paris, 1983).

Dieudonné, Jean, *Calcul Infinitésimal* (Hermann, Paris, 1968). This great book has deeply influenced the mathematical substance of our own volume; it is a valuable resource for anyone wanting a more mathematical treatment of differential equations. On page 362 is the Fundamental Inequality which we have expanded in Chapter 4.

Hirsch, Morris W. and Smale, Stephen, *Differential Equations, Dynamical Systems, and Linear Algebra* (Academic Press, 1974). This is the first book bringing modern developments in differential equations to a broad audience. Smale, a leading mathematician of the century and the only Fields medal winner who has worked in dynamical systems, also has profoundly influenced the authors.

Simmons, George F., *Differential Equations, with Applications and Historical Notes* (McGraw Hill, 1972). We have found this text particularly valuable for its historical notes.

Sanchez, David A., Allen, Richard C. Jr., and Kyner, Walter T., *Differential Equations, An Introduction* (Addison-Wesley, 1983). In many ways this text is close to our own treatment.

FOR NUMERICAL METHODS (CHAPTER 3)

Forsythe, George E. et al., *Computer Methods for Mathematical Computations* (Prentice-Hall, Inc., NJ, 1977). An interesting implementation of Runge–Kutta which has most of the benefits of the linear multistep methods can be found in Chapter 6.

Gear, C. William, *Numerical Initial Value Problems in Ordinary Differential Equations* (Prentice-Hall, Inc., NJ, 1971). A good summary of "what to use when" is Chapter 12.

Henrici, P., *Discrete Variable Methods in Ordinary Differential Equations* (John Wiley & Sons, 1962).

Moore, Ramon, *Methods and Applications to Interval Analysis* (SIAM Series in Applied Mathematics, 1979); helpful for error approximation. In particular, for appropriate cases this approach can even make Taylor series the method of choice.

Noye, J., *Computational Techniques for Differential Equations* (North-Holland, 1984), pp. 1–95; an excellent survey. An extensive bibliography of further references is provided therein.

FOR ITERATION IN THE COMPLEX PLANE (CHAPTER 5.6)

Blanchard, Paul, "Complex analytic dynamics on the Riemann sphere," *Bull. Amer. Math. Soc.* **11** (1984), pp. 85–141.

Devaney, Robert, *An Introduction to Chaotic Dynamical Systems* (Benjamin Cummings, 1986); a text detailing the mathematics.

Dewdney, A.K., "A computer microscope zooms in for a look at the most complex object in mathematics" (*Scientific American,* August 1985); color illustrations.

Peitgen, H-0. & Richter, P.H., *The Beauty of Fractals* (Springer-Verlag, 1986); especially the chapter by Adrien Douady, "Julia Sets and the Mandelbrot Set"; color illustrations.

Peitgen, H-O. & Saupe, D., *The Science of Fractals* (Springer-Verlag, 1988); concentrating on computer algorithms for generating the beautiful pictures; color illustrations.

FOR COMPUTER EXPERIMENTATION WITH MANDELBROT AND JULIA SETS (CHAPTER 5.6)

Munafo, Robert, *SuperMandelZoom* for the Macintosh; black and white, for all Macs; public domain; very useful documentation. Send disk to 8 Manning Dr., Barrington, RI 02806.

Parmet, Marc, *The Game of Fractals* (Springer-Verlag, 1988); written for the MacII with optional color, but does not transmit to postscript for laser printing.

Write to Artmatrix, P.O. Box 880, Ithaca, NY 14851-0880 for listings of other software for the Mandelbrot set, including color programs, for Macintosh, IBM, and other personal computers; Artmatrix also markets educational slide sets and zoom videos.

FOR ASYMPTOTIC DEVELOPMENT (APPENDIX)

Erdelyi, Arthur, *Asymptotic Expansions* (Dover, 1956).

See also Dieudonné, op.cit.

Answers to Selected Problems

Since many of the problems in this book are unlike those in other differential equations texts, solutions to a selected subset of the problems are given here. It is hoped that this will provide the reader with useful insights, which will make all of the problems stimulating and tractable.

Solutions for Exercises 1.1

1.1#2.b. The isoclines for $x' = x^2 - 1$ are of the form $x = \pm\sqrt{m+1}$, with slope m. A slope field is shown below, with dashed lines for isoclines. Note that $-1 \leq m < \infty$.

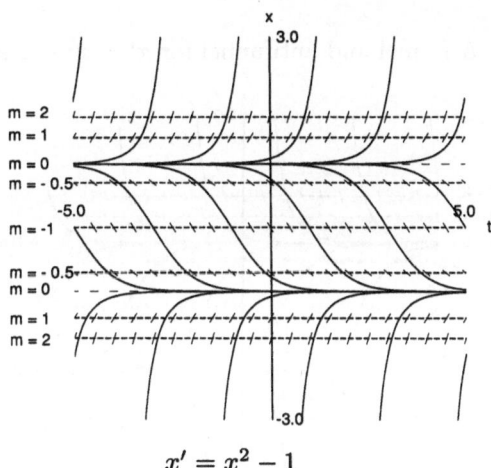

$$x' = x^2 - 1$$

Solutions above $x = 1$ and below $x = -1$ appear to become vertical. The functions $x = (1-Ce^{2t})/(1+Ce^{2t})$ are solutions, with $C = (1-x_0)/(1+x_0)$. Therefore, for $C > 0$, the solutions lie between $x = -1$ and $x = 1$; for $-1 < C < 0$, they are above $x = 1$; for $C < -1$, they lie below $x = -1$. The isocline $x = -1$ is a solution, but cannot be written as $(1-Ce^{2t})/(1+Ce^{2t})$ for any finite C.

1.1#4.iv. $1 - a$, $2 - h$, $3 - d$, $4 - e$, $5 - b$, $6 - c$, $7 - f$, $8 - g$.

1.1#5. $a - 4$ (or 8), $b - 6$, $c - 1$, $d - 2$, $e - 3$, $f - 5$. Graph 7 could be $x' = x + 1$.

1.1#9. If t is replaced by $-t$ in the equation $x' = f(t, x)$, we get $d(x)/d(-t) = -x'(-t) = f(-t, x(-t))$, so symmetry about the x-axis will result from the transformed equation $x' = -f(-t, x)$. Similarly, symmetry about the t-axis comes from replacing x by $-x$. This leads to the new equation $x' = -f(t, -x)$. For symmetry about the origin, replace x by $-x$ and t by $-t$, and the new equation will be $x' = f(-t, -x)$. None of the other sign variants lead to symmetric graphs, in general.

1.1.#13. Implicit differentiation of $x' = x^2 - 1$ gives $x'' = 2xx' = 2x(x^2 - 1)$, so that inflection points can only occur where $x = 0, 1,$ or -1. Note in the graph for **Exer. 1.1#2.b** the solutions do have inflection points at $x = 0$. No solutions pass through $x = 1$ or -1 except the two constant solutions, which do satisfy $x'' = 0$ at every value of t. It can also be inferred that all solutions above $x = 1$ are everywhere concave up, those below $x = -1$ are everywhere concave down, and those between $x = -1$ and $x = 1$ are concave down where $x > 0$ and concave up where $x < 0$.

Solutions for Exercises 1.2–1.4

1.2–1.4#3.c. A funnel and antifunnel for $x' = x^2 - 1$ are shown below.

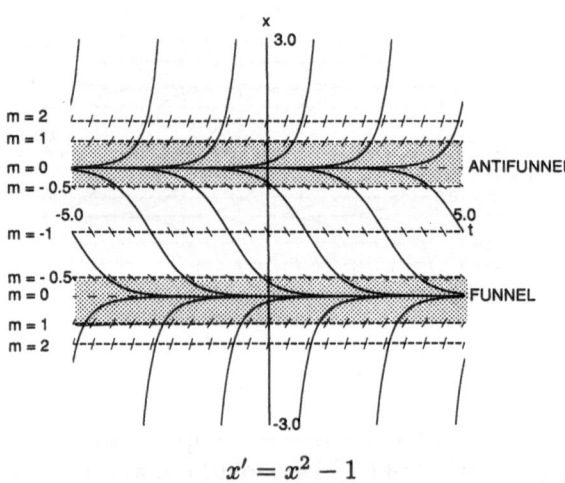

$$x' = x^2 - 1$$

1.2–1.4#4. The computer pictures are provided in the text, as Figure 5.4.6. It shows three different sets of computer solutions for $x' = x^2 - t$, with stepsize $h = 0.4$. The three pictures correspond to three different numerical methods for computing the solutions. You will learn all about these in Chapter 3. Your program *DiffEq* uses the third method.

1.2–1.4#9. (a) In the equation $x' = f(t, x) = x - x^2$, f does not depend explicitly on t. This means that the isoclines are horizontal lines $x = $ constant. Therefore, to check that $|x| \leq 1/2$ is an antifunnel, we need only show that x' is (strictly) positive for $x = 1/2$ and (strictly) negative for $x = -1/2$. Similarly, $1/2 \leq x \leq 3/2$ is a funnel because $x' < 0$ on the isocline $x = 3/2$.

(b) A narrowing antifunnel containing $x(t) = 0$ can be obtained by using any monotonically decreasing curve $\gamma(t)$, with $\lim \gamma(t) = 1$ as $t \Rightarrow -\infty$ and $\lim \gamma(t) = 0$ as $t \Rightarrow \infty$, as the upper fence. The lower fence can be any monotonically increasing curve $\delta(t)$ with $\lim \delta(t) = 0$ as $t \Rightarrow \infty$. For example, let $\delta(t) = -e^{-t}$ and $\gamma(t) = (\pi/2 - \tan^{-1} t)/\pi$. See drawing a:

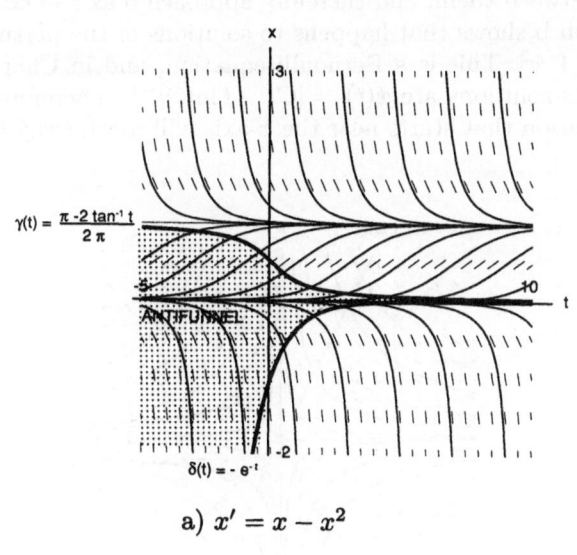

a) $x' = x - x^2$

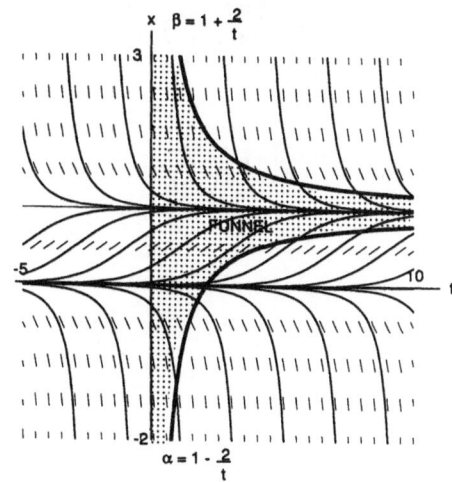

b) $x' = x - x^2$

A narrowing funnel containing $x(t) = 1$ can be constructed by using $\beta(t) = 1 + a/t$, $a > 0$ as an upper fence, and $\alpha(t) = 1 - a/t$ as a lower fence. To make $|\beta - \beta^2| > |\beta'|$, we need $|(1 + a/t) - (1 + a/t)^2| > |-a/t^2|$. If $a > 1$, this condition is satisfied for all $t > 0$. The curve $\alpha = 1 - a/t$ will be a lower fence if $|a/t - a^2/t^2| > |a/t^2|$. This holds whenever $t > a + 1$. Graph b shows $\alpha(t) = 1 - 2/t$ and $\beta(t) = 1 + 2/t$ bounding a narrowing funnel about $x = 1$.

1.2–1.4#15. For $x' = -x/t$, the isoclines are $x = -mt$. A slope field is shown in graph a.

   (a) Use $\beta(t) = t^{-1/2}$ and $\alpha(t) = -t^{-1/2}$ as upper and lower fences, respectively. Any solution that starts between these two curves, for $t > 0$, will stay between them, and therefore approach 0 as $t \to \infty$.

   (b) Graph b shows that happens to solutions of the perturbed equation $x' = -x/t + x^2$. This is a Bernoulli equation, and in Chapter 2 you will find that its solutions are $x(t) = [Ct - t\ln(t)]^{-1}$. Therefore, any solution of this equation that starts near the x-axis will move away from it for any $C > 0$.

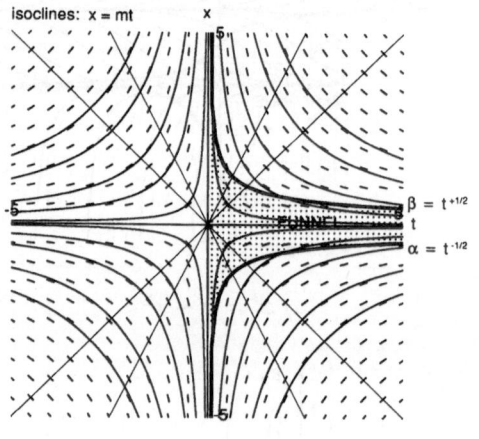

$$\text{a. } x' = -\frac{x}{t}$$

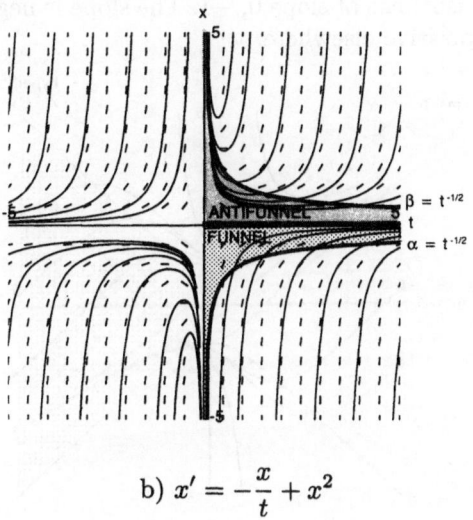

b) $x' = -\dfrac{x}{t} + x^2$

Solutions for Exercises 1.5

1.5#1.h. The isoclines for $x' = t(2 - x)/(t + 1)$ are of the form $x = (2-m)-m/t$. Note that $x = 2$ is a solution. A narrowing funnel around the solution $x = 2$ can be constructed by using the upper fence $\beta(t) = 2+e^{-0.5t}$ and lower fence $\alpha(t) = 2 - e^{-0.5t}$. Both of these satisfy the conditions for a strong fence, for all $t > 1$. A picture is shown below.

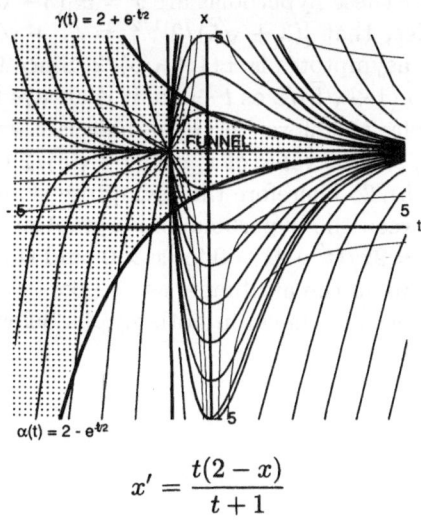

$$x' = \frac{t(2 - x)}{t + 1}$$

1.5#10. (a) If $x' = x^2/(t^2 +1)-1$, the isoclines of slope m satisfy $x^2/(t^2 + 1) = m + 1$, hence are hyperbolas of the form $x^2/(m + 1) - t^2 = 1$. The sketch shows the isoclines of slope $0, -1$. The slope is negative where $|x| < (t^2 + 1)^{1/2}$, and positive elsewhere.

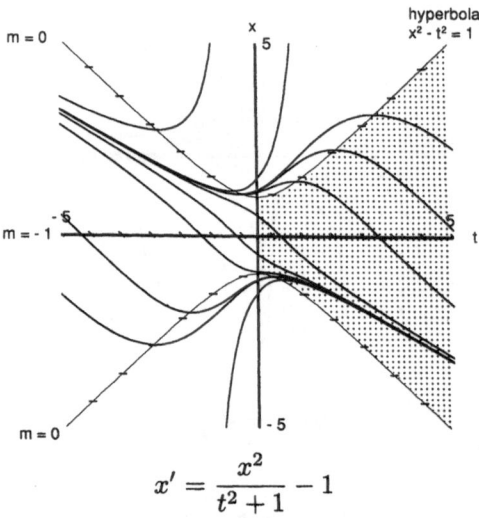

$$x' = \frac{x^2}{t^2 + 1} - 1$$

(b) The region between the isoclines for $m = 0$ is a funnel for $t > 0$; therefore, any solution satisfying $|u(0)| < 1$ stays inside that funnel for all $t > 0$. Every solution in this region is a monotonically decreasing function of t. These solutions cannot approach $-\infty$ faster than the lower fence $-(t^2 + 1)^{1/2}$, hence are defined for all $t > 0$.

(c) For slope $m = (1+\sqrt{5})/2$, the isoclines satisfy $x^2/[(3+\sqrt{5})/2]-t^2 = 1$. The asymptotes for these hyperbolas are $x = \pm[(3 + \sqrt{5})/2]^{1/2}t$. Check, by squaring both sides, that $[(3 + \sqrt{5})/2]^{1/2} = (1 + \sqrt{5})/2$. Therefore, the slope of these two asymptotes is $\pm(1 + \sqrt{5})/2$. Since the hyperbola's slope increases from 0 to $(1+\sqrt{5})/2$ as $t \to \infty$, it is always less than the slope x' along the hyperbola; therefore, the hyperbola is a lower fence. Along the line $x = [(1+\sqrt{5})/2]t$, the value of x' is $\{[(1+\sqrt{5})/2]t^2 - 1\}/(t^2 + 1)$, which is less than $(1+\sqrt{5})/2$. Therefore the asymptote is an upper fence and the shaded region between the hyperbola and asymptote is an antifunnel. The dispersion $\partial f/\partial x = 2x/(t^2 + 1) > 0$ for $x > 0$, and this implies that exactly one solution remains in the antifunnel as $t \to \infty$.

(d) Solution curves are shown in both graphs below.

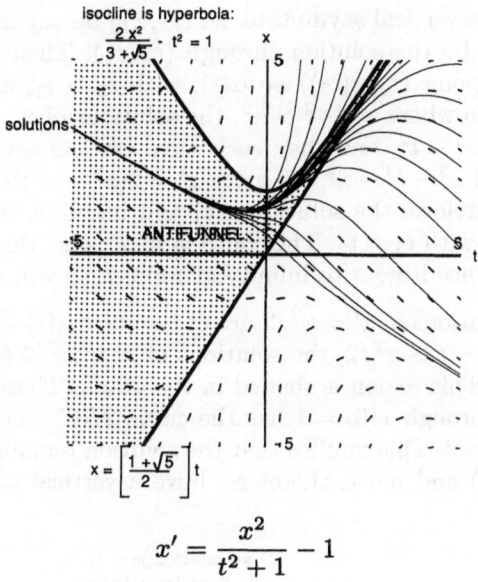

isocline is hyperbola:
$$\frac{2x^2}{3+\sqrt{5}} - t^2 = 1$$

solutions

ANTIFUNNEL

$$x = \left[\frac{1+\sqrt{5}}{2}\right] t$$

$$x' = \frac{x^2}{t^2+1} - 1$$

Solutions for Exercises 1.6

1.6#3. (a) Use the isoclines $\beta(t) = (t^2 - 1)^{1/3}$ as an upper fence and $\alpha(t) = (t^2 + 1)^{1/3}$ as a lower fence. On $x = \beta(t)$, $x' = -1$ and $\beta'(t) = 2t/[3(t^2 - 1)^{2/3}] > 0$ for $t > 1$. On $x = \alpha(t)$, $x' = 1$ and $\alpha'(t) = 2t/[3(t^2 + 1)^{2/3}] < 1$ for all $t > 0$. The shaded region in the graph below, between α and β, is a narrowing antifunnel along $x = t^{2/3}$. The dispersion $\partial f/\partial x = 3x^2 > 0$ implies that there exists a unique solution x^* which remains in the antifunnel.

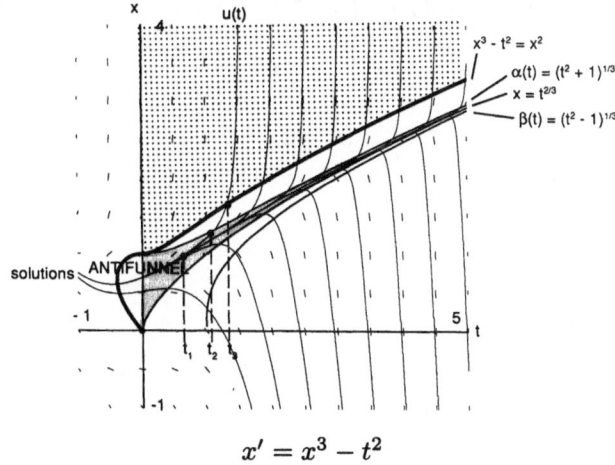

u(t)

$x^3 - t^2 = x^2$

$\alpha(t) = (t^2 + 1)^{1/3}$

$x = t^{2/3}$

$\beta(t) = (t^2 - 1)^{1/3}$

solutions ANTIFUNNEL

$$x' = x^3 - t^2$$

(b) To show that any solution in the antifunnel, above the exceptional solution x^*, has a vertical asymptote, let (t_1, x_1) be any initial point in this region. Let $u(t)$ be the solution through (t_1, x_1). Then u must leave the antifunnel at a point $(t_2, u(t_2))$ on $\alpha(t)$, with $t_2 > t_1$, and $u'(t_2) = 1$. In the shaded region where $x^3 - t^2 > x^2$, the solutions of $x' = x^2$ will be lower fences for $x' = x^3 - t^2$, and they each have vertical asymptotes. Implicit differentiation of $x^3 - t^2 - x^2 = 0$ gives $x' = 2(x^3 - x^2)^{1/2}/(3x^2 - 2x) \to 0$ as $x \to \infty$. Therefore, the solution $u(t)$ will intersect this curve at some point $(t_3, u(t_3))$ with $t_3 > t_2$. The solution of $x' = x^2$ through this point is a lower fence. Since it goes to infinity at finite t, so will $u(t)$.

1.6#6. The solutions of $x' = x^2/2$ are of the form $x(t) = 2/(C - t)$. In the region where $x^2 - t > x^2/2$, the solutions of $x' = x^2/2$ form a lower fence for $x' = x^2 - t$. This region is shaded in the graph. Therefore, the solution of $x' = x^2 - t$ through $x(0) = 1$ has the curve $2/(C - t)$ as a strong lower fence, where $C = 2$. This implies that the solution remains above the curve $\alpha(t) = 2/(2 - t)$ and must, therefore, have a vertical asymptote at some value of $t < 2$.

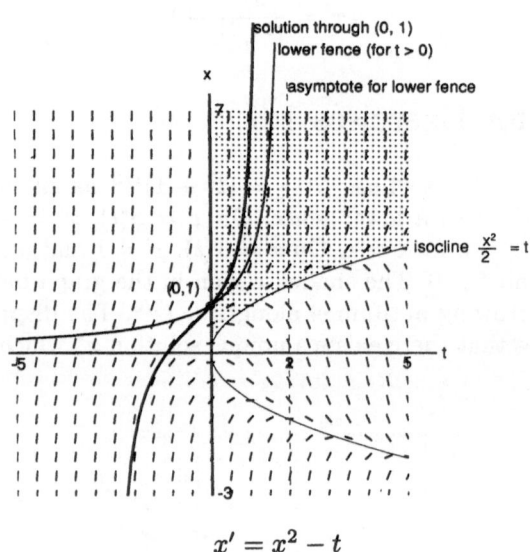

$$x' = x^2 - t$$

Solutions for Exercises 2.1

2.1#2. The answers are:

(a) $\ln(tx) + x - t = C$
(b) $(1+x)(1-t) = C$
(c) $(t+x)/(tx) + \ln(x/t) = C$
(d) $(x-a) = Ce^{1/t}$
(e) $x^{2a} = C(t-a)/(t+a)$

(f) $x = (t+C)/(1-Ct)$
(g) $\tan(\theta)\tan(\varphi) = C$
(h) $\sin^2(\theta) + \sin^2(\varphi) = C$
(i) $\tan(x) = C(1-e^t)^3$
(j) $t^2 + x^2 = t^2x^2 + C$

Solutions for Exercises 2.2–2.3

2.2–2.3#4. The answers are:

(a) $x = t(t+1)^2 + C(t+1)^2$
(b) $x = Ct^a + t/(1-a) - 1/a$
(c) $x = at + Ct(1-t^2)^{1/2}$
(d) $x = \sin(t) - 1 + Ce^{-\sin(t)}$

(e) $x = t^n(e^t + C)$
(f) $t^n x = at + C$
(g) $e^t x = t + C$
(h) $x = t^2(1 + Ce^{1/t})$

2.2–2.3#8. Assume a quadratic solution of the form $x(t) = \alpha + \beta t + \gamma t^2$. Then $(t^2+1)(2\gamma) + (2\gamma t + \beta)^2 + k(\alpha + \beta t + \gamma t^2) \equiv t^2$ implies that $t^2(2\gamma + 4\gamma^2 + k\gamma) + t(4\gamma\beta + k\beta) + (2\gamma + \beta^2 + k\alpha) \equiv t^2$, which leads to the three equations

$$2\gamma + 4\gamma^2 + k\gamma = 1$$
$$4\gamma\beta + k\beta = 0$$
$$2\gamma + \beta^2 + k\alpha = 0.$$

From the first of these equations the possible values for γ are $[-(2+k) \pm \{(2+k)^2 + 16\}^{1/2}]/8$. If $k = -2$, then $\gamma = \pm 1/2$. When $\gamma = 1/2$, β is arbitrary, $\alpha = (1+\beta^2)/2$, and there is a family of quadratic solutions of the form $x(t) = t^2/2 + \beta t + (1+\beta^2)/2$. When $k = -2$ and $\gamma = -1/2$, β is 0 and $\alpha = -1/2$, so there is one extra quadratic solution $x(t) = -(t^2+1)/2$. If $k \neq -2$, $\beta = 0$ and $x(t) = \gamma(t^2 - 2/k)$, where γ can have only the two distinct values shown in the quadratic formula above. This gives exactly two quadratic solutions, one opening upward and the other downward. Note that this method does not find any of the solutions that are not quadratics. But you can show that there are no other polynomial solutions of degree $n > 2$ because the middle term alone would then be of higher degree, and there would be nothing to match it up to.

2.2–2.3#10. The answers are:

(a) $x^2(t^2 + 1 + Ce^{t^2}) = 1$
(b) $[C(1-t^2)^{1/2} - a]x = 1$

(c) $a^2x^3 = Ce^{at} - a(t+1) - 1$
(d) $x = [\tan(t) + \sec(t)]/[\sin(t) + C]$

Solutions for Exercises 2.4–2.5

2.4–2.5#2. (a) If $u(t)$ satisfies $du/dt = (2 + \cos(t))u - u^2/2 + \alpha$, then $u(t + 2\pi) \equiv z(t)$ satisfies $dz/dt = (2 + \cos(t + 2\pi))z - z^2/2 + \alpha$, which is the same equation because of the periodicity of the cosine function.

(b) The graph below shows the two periodic solutions for $\alpha = -1$. The funnel around the upper solution and antifunnel around the lower one are shaded in.

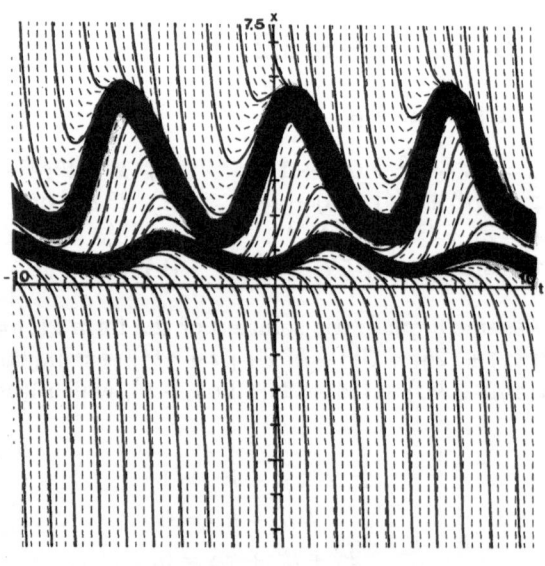

$$x' = (2 + \cos t)x - x^2/2 - 1$$

(c) The three graphs below show the slope fields for $\alpha = -1.37, -1.375,$ and -1.38. Notice that the two periodic solutions exist, but are very close together, for $\alpha = -1.37$. By the time α reaches -1.38 it is clear that no periodic solution can exist, since it would have to cross the trajectory shown in the third figure. This in turn would violate the existence and uniqueness theorem.

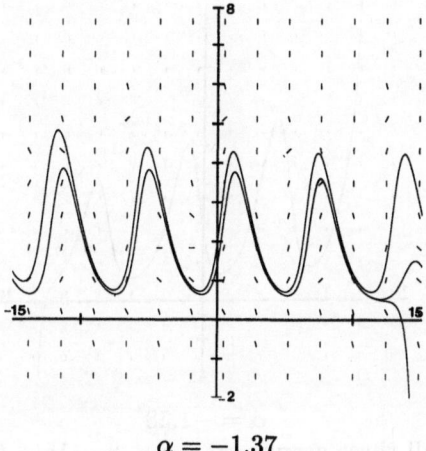

$$\alpha = -1.37$$

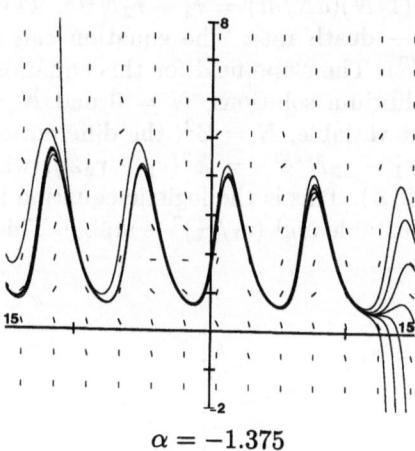

$$\alpha = -1.375$$

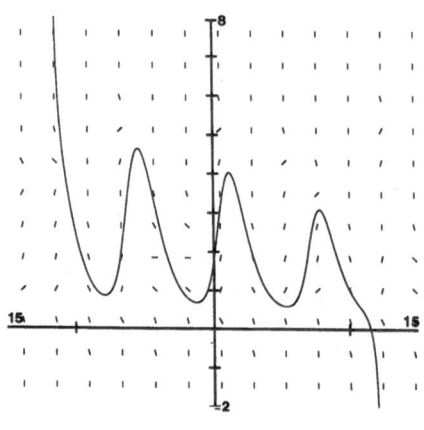

$$\alpha = -1.38$$

For all three graphs $-2 < x < 8$, $-15 < t < 15$.

2.4–2.5#6. The differential equation for N is obtained by setting the per capita growth rate $(1/N)(dN/dt) = r_1 - r_2 N^{1/2}$. This is the net growth rate, or birth rate $-$ death rate. The equation can then be written as $N' = N(r_1 - r_2 N^{1/2})$. The slope field for this equation is graphed below. There are two equilibrium solutions, $N = 0$ and $N = (r_1/r_2)^2$. With a change of dependent variable, $N = Z^2$, the differential equation becomes $2ZZ' = N' = N(r_1 - r_2 N^{1/2}) = Z^2(r_1 - r_2 Z)$, which can be written $Z' = Z(r_1/2 - (r_2/2)Z)$. This is the logistic equation in Z. Therefore, the solution $N = 0$ is unstable and $(r_1/r_2)^2$ is stable. This is indicated in the graph below.

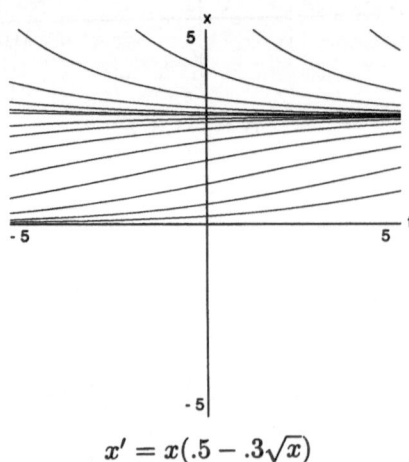

$$x' = x(.5 - .3\sqrt{x})$$

Solutions for Exercises 2.6

2.6#1. The answers are:

(a) $t^3/3 + xt - x^2 = C$ (d) $\ln(x/t) - tx/(t-x) = C$
(b) $2x^2 - tx + t^3 = C$ (e) not exact
(c) $x^4 = 4tx + C$ (f) $\ln(t+x) - t/(t+x) = C$

Solutions for Exercises 2.7

2.7#2.c. To solve $(1+t)x' - kx = 0$, with $x(0) = 1$, let $u(t) = 1 + a_1 t + a_2 t^2 + \cdots = \sum a_n t^n$. Then $u'(t) = \sum n a_n t^{n-1}$. If these sums are substituted into the differential equation, we have

$$(1+t) \sum_{n=1}^{\infty} n a_n t^{n-1} - k \sum_{n=0}^{\infty} a_n t^n \equiv 0.$$

Assuming the sum is convergent, we can write the first term as two sums:

$$\sum_{1}^{\infty} n a_n t^{n-1} + \sum_{0}^{\infty} n a_n t^n - k \sum_{0}^{\infty} a_n t^n \equiv 0.$$

A change of index on the first sum allows us to write

$$\sum_{0}^{\infty} (n+1) a_{n+1} t^n + \sum_{0}^{\infty} n a_n t^n - k \sum_{0}^{\infty} a_n t^n \equiv 0,$$

and this gives a recurrence relation for the coefficients, $(n+1)a_{n+1} = (k-n)a_n$. Therefore, $a_0 = 1$, $a_1 = k a_0 = k$, $a_2 = (k-1)a_1/2 = (k-1)k/2!$, and in general $a_n = k(k-1)\ldots(k-n+1)/n!$. Therefore, the series solution is $u(t) = 1 + kt + k(k-1)t^2/2! + \ldots = (1+t)^k$ (the Binomial Series), and converges for $|t| < 1$.

2.7#7.c. If the series $\sum a_n t^n$ is substituted into the equation for $x(t)$, you will obtain the recurrence relation $a_{n+2} = -(n-1)a_n/[(n+1)(n+2)]$. The initial conditions determine $a_0 = 1$ and $a_1 = 0$. Note that with $a_1 = 0$, the recurrence relation implies that all a_n, n odd, will be 0. The series solution is $x(t) = 1 - t^2/2! + t^4/4! - 3t^6/6! + 5 \cdot 3t^8/8! - \cdots$ with nth term given by

$$a_n t^{2n} = (-1)^n 1 \cdot 3 \cdot 5 \cdots (2n-3) t^{2n}/(2n)!, \text{ for } n \geq 2.$$

This is an alternating series, with terms monotonically decreasing to zero, which means that the error made by stopping with the nth term is less in absolute value than the $(n+1)$st term. Therefore $x(0.5) \cong 7/8$, with absolute error less than $(1/2)^4/4! \cong 0.0026$.

Solutions for Exercises 2. Miscellaneous Problems

2.Misc#2.(ii). The answers are:

(a) $x^2 + 2tx - t^2 = C$

(b) $t^2 + 2tx = C$

(c) $\ln(t^2 + x^2)^{1/2} - \tan^{-1}(x/t) = C$

(d) $1 + 2Cx - C^2t^2 = 0$

(e) $(t+x)^2(2t+x)^3 = C$

(f) $t\exp[(s/t)^{1/2}] = C$,
   or $s = t[\ln(C/t)]^2$

Solutions for Exercises 3.1–3.2

3.1–3.2#1. (a) The first two steps in the approximate solutions of $x' = x$, $x(0) = 1$, with stepsize $h = 1$, are shown below:

x	Euler	Midpoint	$R - K$	Exact Sol.
0	1.0	1.0	1.0	1.0
1	2.0	2.5	2.7083333	2.71828183
2	4.0	6.25	7.33506945	7.38905610

(b) The analytical solution is $x(t) = e^t$, with the values shown in the table above.

(c) Graphical solution:

THREE METHODS FOR NUMERICALLY APPROXIMATING SOLUTIONS TO X' = X
STARTING AT $(t_o, x_o) = (0,1)$
WITH STEPSIZE: h = 1
FOR TWO STEPS.

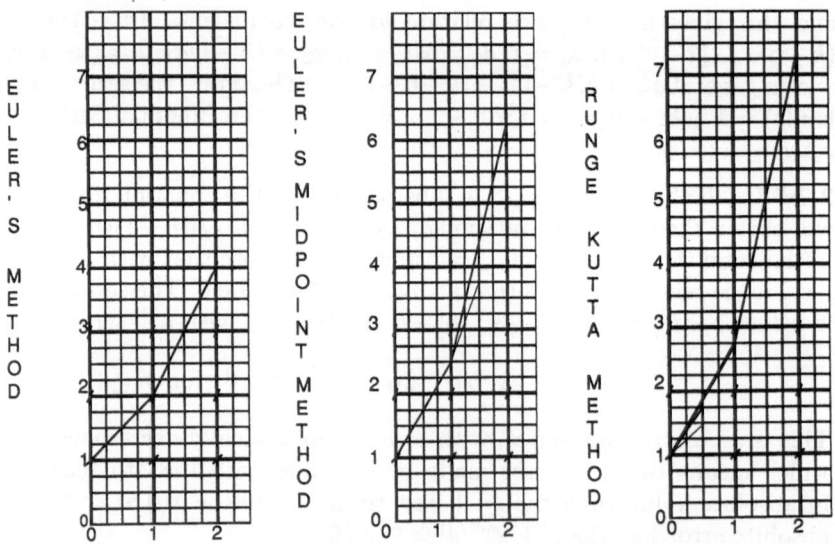

3.1–3.2#7. (a) This method uses the formula $x_{n+1} = x_n + hm$, where m is the average of the slopes at the two ends of the interval. See the drawing below.

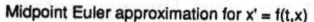

Midpoint Euler approximation for x' = f(t,x)

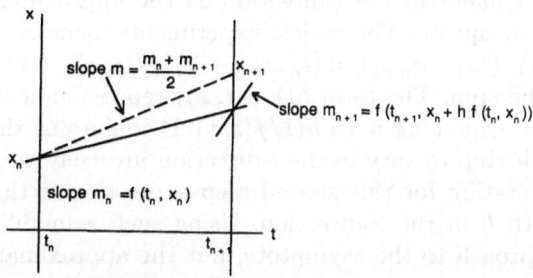

slope $m = \dfrac{m_n + m_{n+1}}{2}$

slope $m_{n+1} = f(t_{n+1}, x_n + h f(t_n, x_n))$

slope $m_n = f(t_n, x_n)$

$x_{n+1} = x_n + h m$

m = average of slope at two ends of the interval

(b) Let $dx/dt = g(t)$. Then $x_{n+1} = x_n + h[g(t_n) + g(t_{n+1})]/2$. This is the Trapezoidal Rule for integration, which approximates the area under the curve between $t = t_n$ and t_{n+1} by the area of the trapezoid having sides of height $g(t_n)$ and $g(t_{n+1})$.

3.1-3.2#8. The Euler and Runge–Kutta approximations of $x(1)$, where $x' = x^2$ and $x(0) = 2$, are shown below. The stepsize used is $h = 0.1$.

t	x (Euler)	$x(R-K)$	x (exact)
0	2.0	2.0	2.0
0.1	2.4	2.499940	2.5
0.2	2.976000	3.332946	3.333333
0.3	3.861658	4.996857	5.0
0.4	5.352898	9.930016	10.0
0.5	8.218249	82.032516	∞
0.6	14.972210	1.017404×10^{-12}	-10.0
0.7	37.388917	overflow	-5.0
0.8	177.1820		-3.333333
0.9	3316.5292		-2.5
1.0	1103253.14		-2.0

(b) Looking at the results of the Euler approximation, you might think it was just increasing exponentially. After running the Runge–Kutta approximation, it appears that something is quite wrong.

(c) The analytic solution can be found by separation of variables. $dx/x^2 = dt \rightarrow -1/x = t + C \Rightarrow x = 2/(1 - 2t)$, where $x(0) = 2$. Note that this solution is undefined at $t = 0.5$. It has a vertical asymptote there. This was not obvious from either approximation.

(d) A smaller constant step h in the t-direction will not help, because eventually the distance from the approximate solution to the asymptote will be smaller than h; then the next step simply jumps across the asymptote without even knowing it is there.

(e) The only hope for not crossing the asymptote would be a step-size that gets smaller in the t-direction as the approximate solution approaches the asymptote. The easiest experiments focus on Euler's method for $x' = f(t, x)$, $(t_{n+1}, x_{n+1}) = (t_n, x_n) + h(1, f(t, x))$, and change the expression for the step. The term $h(1, f(t, x))$ represents a step length h in the t-direction. Changing it to $h(1/f(t, x), 1)$ maintains the proper slope, but forces each step to vary in the t-direction inversely as slope. An alternative interpretation for this second step expression is that the step has constant length h in the x-direction. Using such a modified step greatly delays the approach to the asymptote, but the approximate solution nevertheless eventually crosses it (e.g., with $h = 0.1$ at $x \approx 78.7$), because the approximate solution eventually gets closer to the asymptote than the horizontal component of the step.

(f) A possible refinement of the ideas in (e) is to force the Euler step to have length h in the direction of the slope, by using

$$h(1, f(t, x))/\sqrt{1 + f^2(t, x)}.$$

This actually works better at first, but an approximate solution with this step also eventually crosses the asymptote, a little higher up (for $h = 0.1$ at $x \approx 79.7$).

Our conclusion is that the step variations described in (e) and (f) definitely are much better than the original Euler method, but they fail to completely remove the difficulty. We hope you have seen enough to suggest directions for further exploration.

3.1–3.2#10. (a) $x' = 4 - x$, $x(0) = 1$. The implicit Euler formula gives $x_{i+1} = x_i + h(4 - x_{i+1})$. This is easily solved for x_{i+1} to give $x_{i+1} = (x_i + 4h)/(1 + h)$. Therefore, with $h = 0.1$, we can find $x_1 = 1.272727$, $x_2 = 1.520661$, and $x_3 = 1.746056$. This compares well with the exact solution $x(t) = 4 - 3e^{-t}$, which gives $x(0.3) = 1.777545$.

(b) Here, $x_{i+1} = x_i + h(4x_{i+1} - x_{i+1}^2)$, which can be written in the form $hx_{i+1}^2 + (1 - 4h)x_{i+1} - x_i = 0$. The quadratic formula can be used to obtain $x_{i+1} = (4h - 1)/(2h) \pm [(1 - 4h)^2 + 4hx_i]^{1/2}/(2h)$. This gives the values $x_1 = 1.358899$, $x_2 = 1.752788$, and $x_3 = 2.150522$. The analytic solution $x(t) = 4e^{4t}/[3(1 + e^{4t}/3)]$, so the exact value of $x(0.3)$ is 2.101301.

(c) The implicit Euler formula is $x_{i+1} = x_i + h(4x_{i+1} - x_{i+1}^3)$. In this case we must solve the cubic equation

$$hx_{i+1}^3 + (1 - 4h)x_{i+1} - x_i = 0$$

for x_{i+1} at each step. The solution of $0.1x_1^3 + (0.6)x_1 - 1 = 0$ is $x_1 \cong 1.300271$; and continuing, $x_2 \cong 1.548397$, $x_3 \cong 1.725068$. In this case the analytic solution $x(t) = [4/(3e^{-8t} + 1)]$ gives $x(0.3) \cong 1.773210$.

(d) For this last equation, the implicit Euler formula is $x_{i+1} = x_i + h\sin(t_{i+1}x_{i+1})$. Here, the equation at each step must be solved by something like Newton's method. For $x_1 = 1 + 0.1\sin(0.1x_1)$ we get $x_1 \cong 1.010084$; and continuing, $x_2 \cong 1.030549$, $x_3 \cong 1.061869$.

Solutions for Exercises 3.3

3.3#2. (a) For the midpoint Euler method, Table 3.3.1 gives

N	h	$E(h)/h^2$
512	1/256	2.4558
1024	1/512	2.4594
2048	1/1024	2.4612
4096	1/2048	2.4621
8192	1/4096	2.4626
16384	1/8192	2.4628

so that C_M appears to have a value of approximately 2.463.

(b) To find the asymptotic development of the error, use $u_h(t) = (1 + h + h^2/2)^{t/h}$. Then $u_h(2) = e^{\ln(u_h(2))} = e^{(2/h)\ln(1+h+h^2/2)}$, and using the Taylor series for $\ln(1+x)$, $\ln(1+h+h^2/2) = (h+h^2/2) - (h+h^2/2)^2/2 + (h+h^2/2)^3/3 - \cdots = h - h^3/6 + \mathcal{O}(h^4)$. Therefore

$$e^2 - u_h(2) = e^2 - e^{2-h^2/3+\mathcal{O}(h^3)} = e^2 - e^2[1 + (-h^2/3 + \mathcal{O}(h^3))$$
$$+ [-h^2/3 + \mathcal{O}(h^3))]^2/2! + \cdots] = e^2 - e^2[1 - h^2/3 + \mathcal{O}(h^3)]$$
$$= e^2h^2/3 + \mathcal{O}(h^3) = C_M h^2 + \mathcal{O}(h^3).$$

The value $C_M = e^2/3 \cong 2.46302$.

(c) The exact value of C_M compares very well with the experimental value found in part (a).

3.3#6. (a) The Taylor series can be computed from the successive derivatives of f, as

$$x' = (x+t)^{-1}, \quad x'' = -(x+t)^{-2}(x'+1) = -(x+t)^{-2}[(x+t)^{-1}+1]$$

or $x'' = -(x+t)^{-3} - (x+t)^{-2}$. The successive derivatives can be written as polynomials in $(x+t)^{-1}$. If we let $z = (t_0 + x(t_0))^{-1}$, the 6th degree Taylor polynomial is

$$x(t_0 + h) = x(t_0) + hz - h^2(z^3 + z^2)2! + h^3(3z^5 + 5z^4 + 2z^3)/3!$$
$$- h^4(15z^7 + 35z^6 + 26z^5 + 6z^4)/4!$$
$$+ h^5(105z^9 + 315z^8 + 340z^7 + 154z^6 + 24z^5)/5!$$
$$- h^6(945z^{11} + 3465z^{10} + 4900z^9 + 3304z^8 + 1044z^7 + 120z^6)/6!$$
$$+ \mathcal{O}(h^7).$$

To find $x(0.1)$, with $x(0) = 1.0$, for example, this polynomial must be evaluated with $h = 0.1$ and $z = 1/(0.0 + 1.0) = 1$.

(b) The equation is a linear differential equation for t, as a function of x. That is $dt/dx = x + t$, or $t'(x) - t(x) = x$. This gives the result $t(x) = -x - 1 + Ce^x$. With $x(0) = 1$, $0 = -1 - 1 + Ce \Rightarrow C = 2/e$.

(c) The table below compares the Taylor series approximation and the Runge–Kutta approximation with the exact solution. Note that the Runge–Kutta solution is more accurate, and definitely easier to compute, especially if the program is already available on your computer.

t	exact sol.	Taylor series	$R - K$
0	1.0	1.0	1.0
0.1	1.09138791	1.09138403	1.09138895
0.2	1.16955499	1.16955068	1.16955629

(d) When $x(0) = 0$, the solution is undefined.

(e) The table for $x(0) = 2$:

t	exact sol.	Taylor series	$R - K$
0	2.0	2.0	2.0
0.1	2.04822722	2.04822721	2.04822733
0.2	2.09326814	2.09326813	2.09326816

The Taylor series is more accurate when $x(0) = 2$, because the value of z is smaller, so the series converges more quickly.

3.3#8. (a) The exact solution of $x' = f(t)$ is of course

$$u(t) = \int_{t_0}^{t} f(s)ds;$$

Euler's method gives the lefthand Riemann sum approximation to the integral as

$$u_h(t_1) = \sum_{i=0}^{n-1} hf(s_i) + (t_1 - s_n)f(s_n),$$

where we have set $s_i = t_0 + ih$, and n is the number of whole increments of h which fit in $[t_0, t_1]$. Then

$$E_h = \sum_{i=0}^{n-1} \int_{s_1}^{s_{i+1}} (f(s) - f(s_i))ds + \int_{s_n}^{t} (f(s) - f(s_n))ds.$$

Taylor's theorem says that there exists a function $c_i(s)$ with $s_i \le c_i(s) \le s$ such that

$$|f(s) - f(s_i) - f'(s_i)(s - s_i)| \le \frac{|f''(c_i(s))|}{2}(s - s_i)^2,$$

so that

$$\left| \int_{s_i}^{s_{i+1}} (f(s) - f(s_i))ds - \int_{s_1}^{s_{i+1}} f'(s_i)(s - s_i)ds \right| \le \sup|f''|\frac{h^3}{6}.$$

Make a similar argument for the last term of the Riemann sum, and evaluate the second integral on the left, and sum, to get

$$\left| E(h) - \frac{h^2}{2} \sum_{i=0}^{n} f'(s_i) \right| \le \sup|f''|(t_1 - t_0)\frac{h^2}{6}.$$

The second term within the absolute value is

$$\frac{h}{2} \times \text{a Riemann sum for } \int_{t_0}^{t_1} f'(s)ds.$$

An argument just like the one above gives the error:

$$\frac{h}{2}|h\sum_{i=0}^{n} f'(s_i) - \int_{t_0}^{t_1} f'(s)ds| \le \sup|f''|(t_1 - t_0)\frac{h^2}{4}.$$

Putting the two estimates together, and calculating the integral of the derivative of f, this leads to

$$\left| E(h) - \frac{h}{2}(f(t_1) - f(t_0)) \right| \le \sup|f''|(t_1 - t_0)\frac{h^2}{2}.$$

(b) Part (a) was relatively straightforward: write everything down explicitly and apply Taylor's theorem. This part is harder. To simplify notation, let us suppose that $t_1 = t_0 + (n+1)h$. Then the Euler approximation u_h to the solution of $x' = g(t)x$ with $x(t_0) = x_0$ is given at time t_1 by the product

$$\log u_h(t_1) = (1 + hg(s_0))(1 + hg(s_1)) \cdots (1 + hg(s_n))x_0.$$

Since x_0 is a factor of $u(t_1)$ and $u_h(t_1)$, we can set $x_0 = 1$.
Use logarithms to study the product:

$$\log u_h(t_1) = \sum_i \log(1 + hg(s_i)) = \sum_i (hg(s_i) - \frac{h^2}{2}(g(s_i))^2 + O(h^3)).$$

The first term in the sum is a Riemann sum for $\int_{t_0}^{t_1} g(s)ds$, and we saw in part (a) that

$$\sum_i hg(s_i) = \int_{t_0}^{t_1} g(s) - \frac{h}{2}\int_{t_0}^{t_1} g'(s)ds + O(h^2).$$

The next term is a Riemann sum for $(h/2)\int_{t_0}^{t_1}(g(s))^2 ds$.

Putting this together, we find

$$\log u_h(t_1) = \int_{t_0}^{t_1} g(s)ds - \frac{h}{2}\left(\int_{t_0}^{t_1} g'(s)ds + \int_{t_0}^{t_1} (g(s)^2 ds) + O(h^2)\right).$$

Now exponentiate, remembering that $e^{ah+O(h^2)} = 1 + ah + O(h^2)$:

$$u_h(t_1) = e^{\int_{t_0}^{t_1} g(s)ds}\left(1 - \frac{h}{2}\left(\int_{t_0}^{t_1} g'(s)ds + \int_{t_0}^{t_1} (g(s))^2 ds\right) + O(h^2)\right).$$

Finally, we find that

$$E(h) = u(t_1) - u_h(t_1) = \frac{h}{2}e^{\int_{t_0}^{t_1} g(s)ds}\left(\left(\int_{t_0}^{t_1} (g'(s))ds + (g(s)^2 ds) + O(h^2)\right)\right).$$

(c) For the equation $x' = x$ we have

$$\log u_h(t) = \log(1+h)^{t/h} = \frac{t}{h}\left(h - \frac{h^2}{2} + O(h^3)\right) = t - \frac{ht}{2} + O(h^2).$$

Exponentiating leads to

$$e^t - u_h(t) = \frac{h}{2}(te^t) + O(h^2);$$

this agrees with part (b), since $\int_0^t 1^2 ds = t$.

(d) If $g(t) = -t$, then $\int_0^t (g'(s) + g^2(s))ds = 0$ when $t = \sqrt{3}$. The program *numerical methods* indicates that over $0 < t \le \sqrt{3}$, Euler's method for the differential equation $x' = -x$ converges with order 3. We don't know why it doesn't converge with order 2, as one would expect.

Solutions for Exercises 3.4

3.4#1. The table below shows the results of the analysis of the errors when the equation $x' = x^2 \sin(t)$ is approximated by Euler's method, with (a) 18 bits rounded down, and (b) 18 bits rounded round. The solution was run from $t = 0$ to $t = 6.28$. The exact solution $x(t) = [7/3 + \cos(t)]^{-1}$, if $x(0) = 0.3$. The value used for $x(6.28)$ was 0.3000004566.

	(b) Rounded round		(a) Rounded down	
N	Error	order	Error	order
4	0.16463688		0.16463783	
8	0.07148865	1.2035	0.07149055	1.2035
16	0.0401814	0.8312	0.0408524	0.8311
32	0.02201011	0.8684	0.02202919	0.8672
64	0.0116551	0.9172	0.01169327	0.9137
128	0.00601891	0.9534	0.00609329	0.9404
256	0.00306251	0.9748	0.00321319	0.9232
512	0.00154236	0.9896	0.00184372	0.8014

	(b) Rounded round		(a) Rounded down	
N	Error	order	Error	order
1024	0.00080421	0.9395	0.00136497	0.4337
2048	0.00037125	1.1152	0.00160149	−0.2305
4096	0.00020340	0.8681	0.00262001	−0.7102
8192	0.00009849	1.0462	0.00488594	−8991
16384	0.000052718	0.9017	0.00965431	−0.9825

(c) Notice that the errors in the rounded round calculation are still decreasing, although the order is beginning to fluctuate further away from 1.0000. For the rounded down case, the order of $\cong 1$ is seen for N between 64 and 256. It goes to $\cong -1.0$ at $N = 16384$. The smallest error in the 2nd column occurs at $N = 1024$.

Solutions for Exercises 4.1–4.2

4.1–4.2#3. At any time t, $r(t)/h(t) = \tan 30° = 1/\sqrt{3}$. The area of the cross-section A at height h is $\pi r^2 = \pi h^2(t)/3$. From equation (3) in Section 4.2, $h'(t) = -\sqrt{2g}(a/A)\sqrt{h} = -3a\sqrt{2g}\, h^{-3/2}/\pi$. We are given the constant area $a = 5 \times 10^{-5} m^2$, and $g = 9.8$. The differential equation for h is $h'(t) = -Kh^{-3/2}$, where $K = \sqrt{19.6} \times 15 \times 10^{-5}/\pi \cong 2.114 \times 10^{-4}$. Solving, by separation of variables, gives $h(t) = [-2.5Kt + C]^{2/5}$. At time $t = 0$, $h(0) = 0.1 m$. This makes $C = (0.1)^{5/2} \cong 0.00316$. To find the time T when $A = (\pi/3)h^2(t) = a$, set $h(T) = 6.91 \times 10^{-3}$. Then $T \cong 6$ seconds.

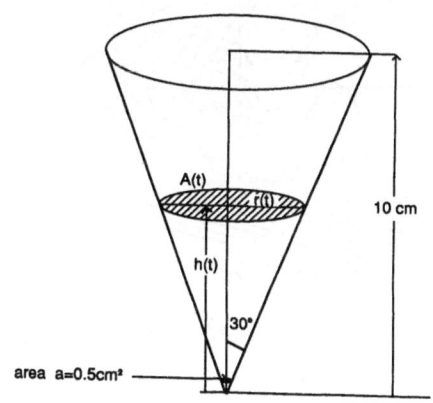

Solutions for Exercises 4.3

4.3#2. For the equation $x' = \cos(x^2 + t)$, $\partial f/\partial x = -2x\sin(x^2 + t)$. To have $|\partial f/\partial x| \geq 5$ requires $|x\sin(x^2 + t)| \geq 2.5$. This can only occur when $|x| \geq 2.5$. The narrow triangular regions where the inequalities $\partial f/\partial x \geq 5$ and $\partial f/\partial x \leq -5$ are satisfied are shown in the graph below. These are found by graphing the functions apart or come together much more rapidly than in the part of the plane where $|x| < 2.5$.

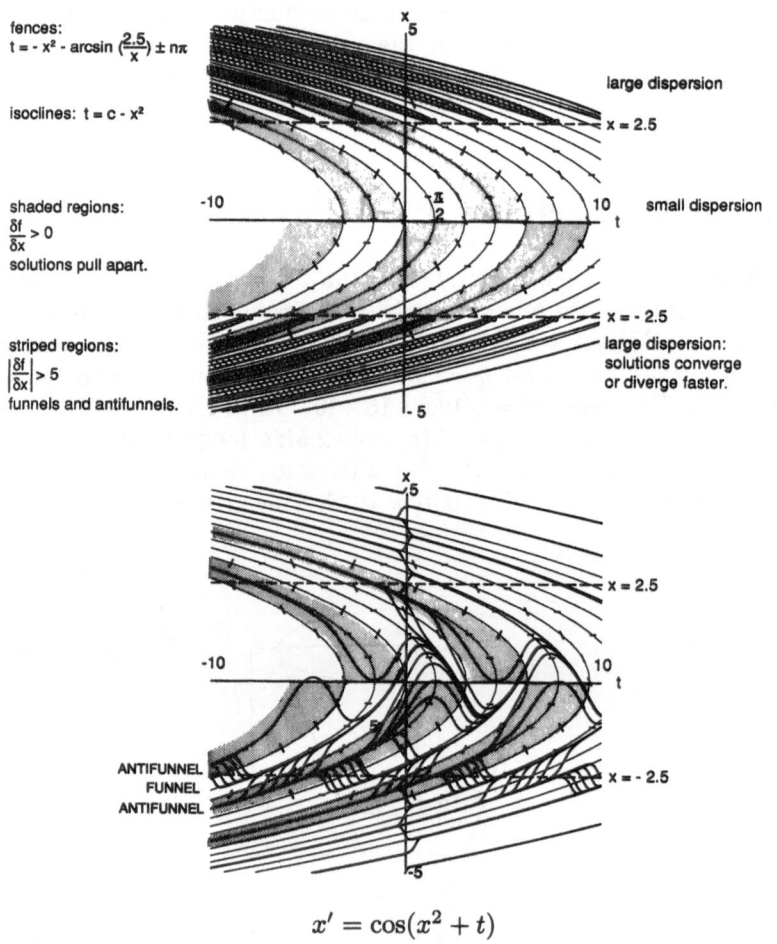

fences:
$t = -x^2 - \arcsin\left(\frac{2.5}{x}\right) \pm n\pi$

isoclines: $t = c - x^2$

shaded regions:
$\frac{\delta f}{\delta x} > 0$
solutions pull apart.

striped regions:
$\left|\frac{\delta f}{\delta x}\right| > 5$
funnels and antifunnels.

large dispersion

$x = 2.5$

small dispersion

$x = -2.5$

large dispersion:
solutions converge
or diverge faster.

ANTIFUNNEL
FUNNEL
ANTIFUNNEL

$x = 2.5$

$x = -2.5$

$$x' = \cos(x^2 + t)$$

Solutions for Exercises 4.4

4.4#4. Let $u(t)$ be the solution of $x' = f(t, x)$, with $u(0) = 0$, and $u(5) = 17$. Assume f has Lipschitz constant $K = 2$. If $x(t)$ satisfies $x' = f(t, x) + g(t, x)$, with $|g(t, x)| < 0.1$, then

$$|u(t) - x(t)| \leq 0.03\, e^{2|t - t_0|} + (0.1/2)\left(e^{2|t - t_0|} - 1\right).$$

Therefore, $|u(5) - x(5)| \leq 0.03\, e^{10} + 0.05(e^{10} - 1) \leq 1763$. This says that $x(5)$ can lie between 17 ± 1763 (not a very useful estimate).

Solutions for Exercises 4.5

4.5#1. (a) The differential equation $x' = (|x|^{1/2} + k)$ can be solved explicitly by solving it in each of the regions $x > 0$ and $x < 0$, by separation of variables. For example, for $x > 0$, $x' = \sqrt{x} + k$, or $dx/(\sqrt{x} + k) = dt$. This can be integrated to get $2\sqrt{x} - 2k \ln(\sqrt{x} + k) = t + C$. A similar result for $x < 0$ leads to the solution $2|x|^{1/2} - 2k \ln(|x|^{1/2} + k) = t + C$, which holds for all x. If $k \neq 0$, there is a unique solution for any initial condition. Notice that $x = 0$ is not a solution of the equation if $k \neq 0$.

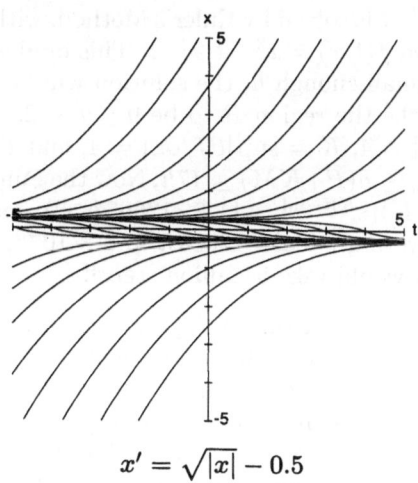

$$x' = \sqrt{|x|} - 0.5$$

(b) This differential equation does not satisfy a Lipschitz condition in any region including $x = 0$, because $\partial f/\partial x \Rightarrow \infty$ as $x \Rightarrow 0^+$ or $x \Rightarrow 0^-$. That implies that $|f(t, 0) - f(t, x)|$ can be made greater than $K|x|$ for any constant K. This problem is meant to show you that a Lipschitz condition is sufficient, **but not necessary**, for uniqueness of solutions.

4.5#5. (a) If $x(t) = Ct - C^2/2$, then $tx' - (x')^2/2 = tC - C^2/2 = x$.
   (b) If $x(t) = t^2/2$, then $tx' - (x')^2/2 = t \cdot t - t^2/2 = t^2/2 = x$.

(c) The line $x = Ct - C^2/2$ intersects $x = t^2/2$ where $t^2/2 = Ct - C^2/2$; that is, where $(t - C)^2/2 = 0$. Therefore the point of intersection is $(C, C^2/2)$. The slope of the parabola at this point is C, so the line is tangent there. This gives two solutions through every point outside the parabola; there are none inside. To see that this is true, let $(\bar{t}, \bar{x})$ be an arbitrary point in the plane. Set $\bar{x} = C\bar{t} - C^2/2$ and solve for the constant C. By the quadratic formula, there will be two solutions if $\bar{x} < \bar{t}^2/2$, one solution if $\bar{x} = \bar{t}^2/2$, and none otherwise.

(d) In the text, a differential equation is defined to be an equation of the form $x' = f(t, x)$, *solved for x' in terms of x and t.* If we try to put *this* equation in that form, we get $x' = t \pm (t^2 - 2x)^{1/2}$. In the region $(t^2 - 2x) > 0$, i.e., outside the parabola of equation $x = t^2/2$, this gives *two* differential equations; both are easily seen to be Lipschitz, and the two straight lines tangent to the parabola from any point outside it are the solutions to those two equations through that point. Along the parabola, the equation is not defined in a neighborhood of any point of the parabola, so the existence and uniqueness theorem says nothing.

Solutions for Exercises 4.6

4.6#1.a. If $x' = x^2 - t$ is solved by Euler's Method, with $x(0) = 0$, the value of the slope function $f(t, x) = x^2 - t > -t$. This implies that x will remain above $-t^2/2$. For small enough h, the solution will be strictly negative for $t > 0$, so we can take the region R to be $0 \le t \le 2$, $-2 \le x \le 0$. In this region $M = \sup |f| = 4$, $K = \sup |\partial f/\partial x| = 4$, and $P = \sup |\partial f/\partial t| = 1$. This implies that $\varepsilon_h \le h(P + KM) \le 17h$. Now the Fundamental Inequality gives $E(h) \le (\varepsilon_h/K)(e^{K|t-t_0|} - 1) \cong 12665\,h$. To make $x(2)$ exact to 3 significant digits, we must make $12665\,h < 5 \times 10^{-3}$, which requires that $h < 4 \times 10^{-7}$. This would take 5 million steps!!

4.6#3. (a) At each step, the initial slope is $-x_i$, and therefore with $h < 1$, you cannot reach 0 in one step. The Runge–Kutta slope $(m_1 + 2(m_2 + m_3) + m_4)/6$ will be negative, but smaller in absolute value than x_i, because m_2, m_3, and m_4 are each less than m_1 in absolute value (i.e. $0 > m_1 = -x_i$, $0 > m_2 = -(x_i - hx_i/2) > m_1$, etc.). With $h < 1$, you will never go down to zero, and the entire argument can be repeated for each subsequent step.

(b) Look at the Runge–Kutta approximation:

$$x_{i+1} = x_i + (h/6)(m_1 + 2(m_2 + m_3) + m_4), \text{ where}$$
$$m_1 = f(x_i) = -x_i$$
$$m_2 = f(x_i + hm_1/2) = -x_i + (h/2)x_i$$
$$m_3 = f(x_i + hm_2/2) = -x_i + (h/2)x_i - (h^2/4)x_i$$
$$m_4 = f(x_i + hm_3) = -x_i + hx_i - (h^2/2)x_i + (h^3/4)x_i.$$

Combining these gives

$$x_{i+1} = x_i[1 - h + h^2/2 - h^3/6 + h^4/24].$$

If we let

$$u_h(t) = x_i[1 - (t - t_i) + (t - t_i)^2/2 - (t - t_i)^3/6 + (t - t_i)^4/24],$$

in the interval between t_i and t_{i+1}, the slope error will be

$$\begin{aligned}
\varepsilon_h &= |u_h'(t) - f(t, u_h(t))| \\
&= |x_i[-1 + (t - t_i) - (t - t_i)^2/2 + (t - t_i)^3/6] + u_h(t)| \\
&= |x_i(t - t_i)^4/24| \le (t - t_i)^4/24 \le h^4/24,
\end{aligned}$$

since $0 < x_i \le 1$, and $t - t_i < h$.

The Fundamental Inequality now gives

$$|u_h(t) - e^{-t}| \le (\varepsilon_h/K)(e^{K|t-t_0|} - 1) \le \varepsilon(e^t - 1),$$

since $K = \sup |\partial f/\partial x| = 1$ and $t_0 = 0$. With $\varepsilon = h^4/24$, we get

$$|u_h(t) - e^{-t}| \le (e^t - 1)h^4/24 = C(t)h^4.$$

(c) To make $|u_h(1) - e^{-1}| < 5 \times 10^{-6}$, it is sufficient to take

$$h < [(24 \times 5 \times 10^{-6})/(e - 1)]^{1/4} \cong 0.092.$$

A numerical computation with $h = 0.10$ gives $u_h(1) = 0.3678798$, which is exact to 6 significant digits, and with $h = 0.25$ you get $u_h(1) = 0.367894$, a drop to 4 significant digits ($e^{-1} = 0.36787944$ to 8 decimal places).

Solutions for Exercises 4.7

4.7#1. In the graph below, a funnel between $x + t = 3\pi/4$ and $x + t = \pi$ is shaded in on the slope field for $x' = \cos(t + x)$. The antifunnel is between $x + t = -\pi$ and $x + t = -3\pi/4$.

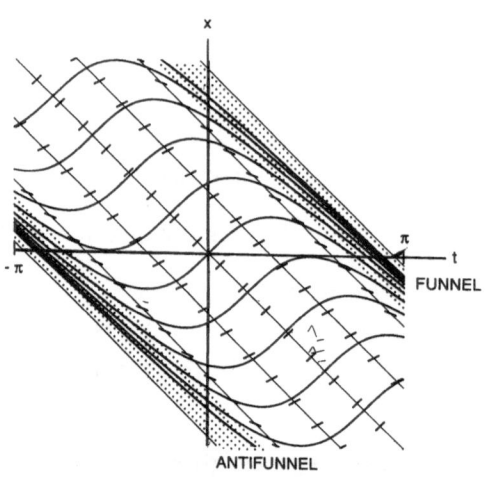

$$x' = \cos(t + x)$$

(b) Both the funnel and antifunnel are weak, since the fences $x + t = \pm\pi$ are isoclines of slope -1. The function $f(x) = \cos(t+x)$ satisfies $|\partial f/\partial x| \leq 1$ for all x and t, so f satisfies a Lipschitz condition everywhere on $\mathbb{R}^2$. By Theorem 4.7.1 and Corollary 4.7.2 this implies that both fences $x = -t \pm \pi$ are nonporous, and that any solution that enters the funnel stays in the funnel. Furthermore, Theorem 4.7.3 implies that there exists a solution that remains in the antifunnel for all t.

(c) One solution that stays in the antifunnel is $x = -\pi - t$.

(d) Notice that the funnel can be defined by $x = \pi - t$ and $x = \pi - t - \varepsilon$ for any $\varepsilon > 0$. Therefore, again by Corollary 4.7.2, any solution between $x = -\pi - t$ and $x = \pi - t$ will approach $x = \pi - t$ as closely as desired as $t \Rightarrow \infty$. It is also true that any line $x = -\pi - t + \varepsilon$ can be used to form the lower fence of the funnel; therefore all solutions to the right of $x = -\pi - t$ leave the antifunnel, so that the only solution that stays in it for all t is $x = -\pi - t$. A nice finale to these arguments is suggested by John Hubbard: the curve $\alpha(t) = \pi - t - t^{-1/3}$ can be shown to form the lower fence for a *narrowing* funnel about the trajectory $x = \pi - t$. The slope $\alpha'(t) = -1 + t^{-4/3}/3$. The slope of the vector field on α is given by

$$x' = \cos(t + x) = \cos(t + \pi - t - t^{1/3}) = \cos(\pi - t^{-1/3}) \approx -1 + t^{-2/3}/2$$

for t large, and this is therefore greater than α'.

Solutions for Exercises 5.1

5.1#4. (a) The required Taylor Series expansion is

$$f(x_0 + h) = f(x_0) + f'(x_0)h + f''(x_0)h^2/2! + \mathcal{O}(h^3).$$

Given $f(x_0) = x_0$, $f'(x_0) = 1$, and $f''(x_0) = \alpha^2 > 0$, this becomes

$$f(x_0 + h) = x_0 + h + (\alpha^2/2)h^2 + \mathcal{O}(h^3).$$

If $x_0 + h$ is any nearby point (i.e., with $\alpha^2 h^2/2$ small relative to h), it will map into a point further away from x_0 if h is positive, but closer to x_0 if h is negative. See graph below for $f(x) = x^3/3 + 2/3$.

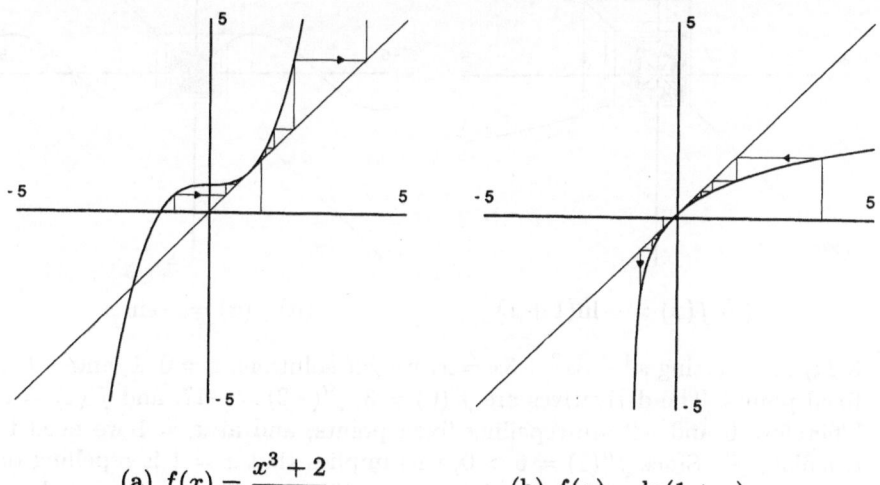

$$\text{(a) } f(x) = \frac{x^3 + 2}{3} \qquad\qquad \text{(b) } f(x) = \ln(1 + x)$$

(b) If $f''(x_0) = -\alpha^2 < 0$, then $f(x_0 + h) = x_0 + h - \alpha^2 h^2/2 + \mathcal{O}(h^3)$, and the iterates will be attracted to x_0 on the right and repelled on the left. See right-hand graph above for $f(x) = \ln(1 + x)$.

(c) If $f''(x_0) = 0$, behavior at x_0 depends on the first nonzero derivative. For example, if $f(x_0 + h) = x_0 + h + f^{(k)}(x_0)h^k/k! + \mathcal{O}(h^{k+1})$, then there are 4 cases, depending on whether k is even or odd, and whether $f^{(k)}(x_0)$ is positive or negative.

(d) If $f'(x_0) = -1$, the series is

$$f(x_0 + h) = x_0 - h + f''(x_0)h^2/2! + f'''(x_0)h^3/3! + \mathcal{O}(h^4).$$

The iterates oscillate about x_0, so convergence depends on whether $f(f(x_0 + h))$ is closer to, or further from x_0 than $x_0 + h$. Applying the Taylor Series to $f(f(x_0 + h))$ gives

$$f(f(x_0 + h)) = x_0 - [-h + f''h^2/2 + f'''h^3/6 + \cdot]$$
$$+ (f''/2)[-h + f''h^2/2 + \cdots]^2 + (f'''/6)[-h + f''h^2 + \cdots]$$
$$+ \cdots = x_0 + h\{1 - [(f'')^2/2 + f'''/3]h^2\} + \mathcal{O}(h^4).$$

Therefore, convergence depends on the sign of $[(f''(x_0)^2/2 + f'''(x_0)/3]$. Two examples are shown below. For $f(x) = -\ln(1+x)$, $(f''(0))^2/2 + f'''(0)/3 = -1/6 < 0$, and x_0 can be seen to be repelling. In the case of $f(x) = -\sin(x)$, the sum of the derivatives is $1/3 > 0$ and the point $x = 0$ is attracting.

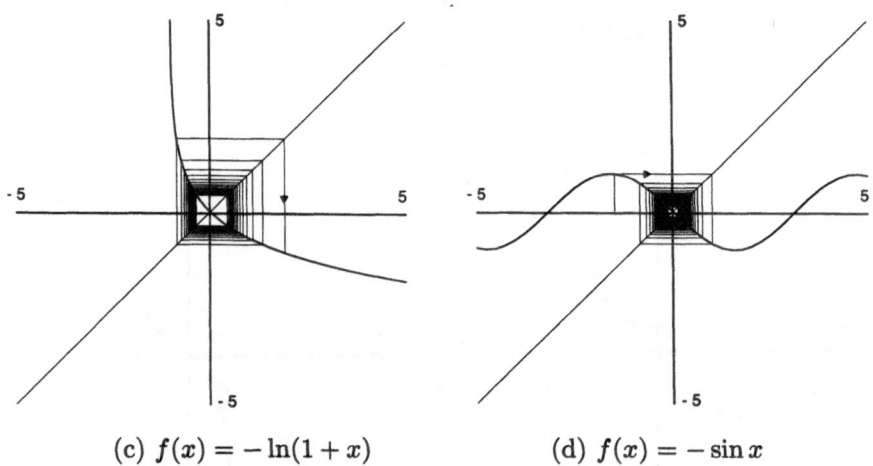

(c) $f(x) = -\ln(1 + x)$ (d) $f(x) = -\sin x$

5.1#7.a. Setting $x^4 - 3x^2 + 3x = x$, we get solutions $x = 0$, 1, and -2 as fixed points. The derivatives are $f'(0) = 3$, $f'(-2) = -17$, and $f'(1) = 1$. Therefore 0 and -2 are repelling fixed points, and at $x = 1$ we need to calculate f''. Since $f''(1) = 6 > 0$, this implies that $x = 1$ is repelling on the right and attracting on the left. For verification, see the graph below.

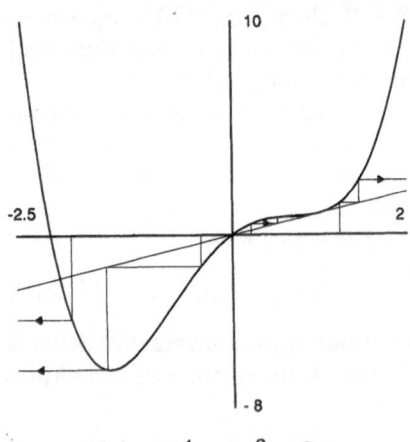

$$f(x) = x^4 - 3x^2 + 3x$$

5.1#10. (a) The fixed points of $f(x) = x^2 + c$ must satisfy $x^2 + c = x$, or $x = 1/2 \pm (1 - 4c)^{1/2}/2$. Therefore, there are 2 real fixed points if $c < 1/4$, one fixed point if $c = 1/4$, and none if $c > 1/4$.

(b) The graphs below show how the parabola intersects the line $y = x$ in each of the three different cases in (a).

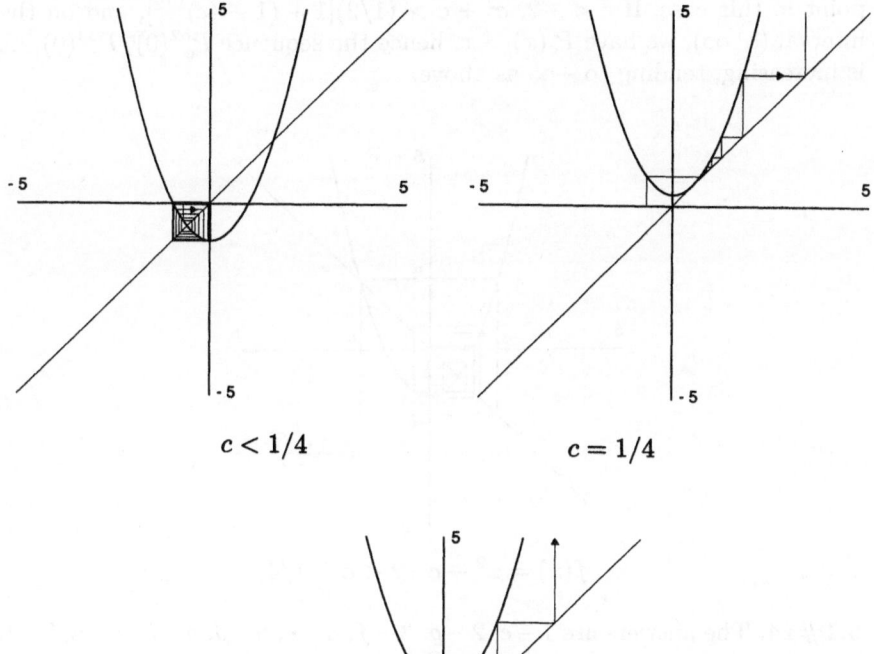

$$c < 1/4 \qquad\qquad c = 1/4$$

$$c > 1/4$$
$$f(x) = x^2 + c$$

(c) A fixed point x_0 is attracting if $|f'(x_0)| = |1 \pm (1 - 4c)^{1/2}| < 1$. If the plus sign is used, this inequality is never true. With the minus sign, $|1 - (1 - 4c)^{1/2}| < 1$ implies $-3/4 < c < 1/4$.

(d) Let α denote $[1/2 + (1 - 4c)^{1/2}/2]$. Since $P_c(x) = x^2 + c$ has no local maxima, the maximum of $x^2 + c$ on the interval $I_c = [-\alpha, \alpha]$ is realized at an end point; since $P_c(-\alpha) = P_c(\alpha) = \alpha$, the maximum is in the interval

I_c. The minimum is $c = P_c(0)$; so $P_c(I_c) \subset I_c$ precisely if $c \le 1/4$ (for I_c to exist at all) and if $c \ge -(1/2)[1 + (1 - 4c)^{1/2}]$ or $1 + 2c \ge -(1 - 4c)^{1/2}$, which occurs precisely if $c \ge -2$ (and, of course, $c \le 1/4$).

(e) If $c \in [-2, 1/4]$, the result is already proved in (d), since $0 \in I_c$. For $c > 1/4$, the orbit of 0 is strictly increasing since $P_c(x) > x$ for $x \ge 0$ and $c > 1/4$, so it must increase to ∞ or a fixed point; by (a) there is no fixed point in this case. If $c < -2$, $c^2 + c > (1/2)[1 + (1 - 4c)^{1/2}]$, and on the interval (α, ∞), we have $P_c(x) > x$, hence the sequence $P_c^{o2}(0)$, $P_c^{o3}(0), \ldots$ is increasing, tending to $+\infty$ as above.

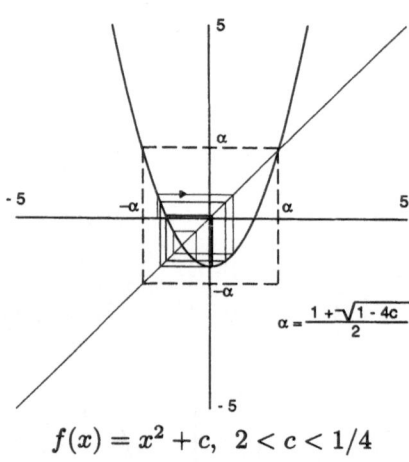

$$f(x) = x^2 + c, \quad 2 < c < 1/4$$

5.1#14. The answers are $1 - c$, $2 - g$, $3 - f$, $4 - e$, $5 - d$, $6 - h$, $7 - a$, $8 - b$.

5.1#18. (a) For $\varphi(x) = x^3$, $\varphi^{-1}(x) = x^{1/3}$, and

$$g(x) = \varphi^{-1}(f(\varphi(x))) = \varphi^{-1}(2x^3) = 2^{1/3}x.$$

(b) The functions f and g are both linear, so that their multipliers are 2 and $2^{1/3}$ respectively.

(c) The fact that the multipliers are not equal does not contradict the statement proved in **Ex. 5.1#16**, because the hypothesis that "φ^{-1} is differentiable" is not satisfied at $x = 0$.

Solutions for Exercises 5.2

5.2#3. For the map $q(n + 1) = (1 + \alpha)q(n) - \alpha(q(n))^2$, let x_n denote $q(n)$. Then the equation becomes $x_{n+1} = (1 + \alpha)x_n - \alpha x_n^2 = f(x_n)$. The fixed points of this map are points x_s such that $f(x_s) = x_s$, or $(1 + \alpha)x_s - \alpha x_s^2 = x_s$. This gives $x_s = 0$ or 1. The derivative is $f'(x) = (1 + \alpha) - 2\alpha x$, so $f'(0) = 1 + \alpha$ and $f'(1) = 1 - \alpha$. Therefore, for $0 < \alpha \le 1$, $q_s = 1$ will be an attracting fixed point of the map, with positive derivative. Moreover, for $x \in [0,1]$, we have $x < f(x) \le 1$, so any sequence x_0, $x_1 = f(x_0)$, $x_2 = f(x_1), \ldots$ with $x \in [0,1]$ is strictly increasing and bounded by 1. Hence the sequence must tend to a fixed point, which must be 1 (rather than 0).

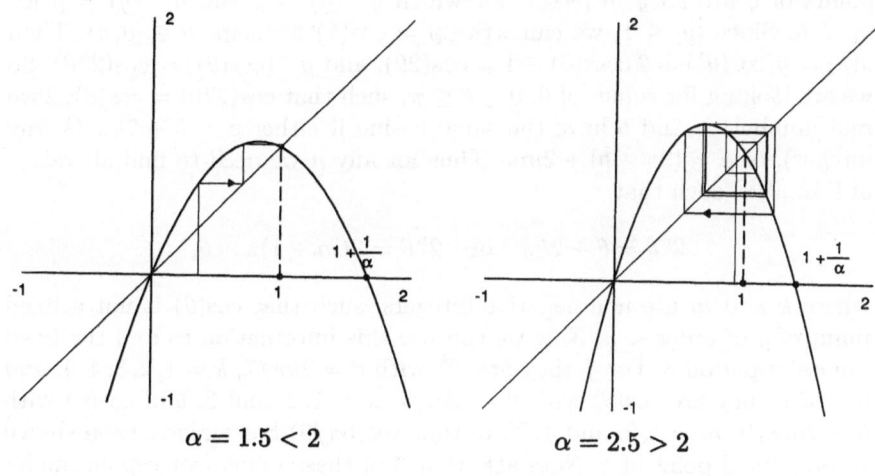

$$\alpha = 1.5 < 2 \qquad\qquad \alpha = 2.5 > 2$$

$$f(x) = (1 + \alpha)x - \alpha x^2$$

5.2#5. Let $f(z) = z^2$, for z on the unit circle X, $g(y) = 2y^2 - 1$ for $y \in [-1,1] = Y$, and $\varphi(e^{i\theta}) = \cos(\theta)$ be the map from X to Y.

(a) To show that $\varphi \circ f = g \circ \varphi$, we simply calculate:

$$\varphi \circ f(e^{i\theta}) = \varphi(e^{2i\theta}) = \cos(2\theta),$$

and

$$g \circ \varphi(e^{i\theta}) = g(\cos(\theta)) = 2\cos^2(\theta) - 1 = \cos(2\theta).$$

(b) The function φ is not 1–1 since $\varphi(e^{i\theta}) = \varphi(e^{i(2\pi-\theta)})$.

(c) For $e^{i\theta}$ to be a fixed point of f, we need $e^{2i\theta} = e^{i\theta}$, or $e^{i\theta} = 1$. Therefore 1 is the only fixed point in X, under f. For fixed points of g, it is necessary that $2y^2 - 1 = y$, so that $y = 1$ or $-1/2$. We see that the fixed points $x = e^{0i}$ and $y = 1$ correspond under the map φ, but the point $-1/2$ does not correspond to a fixed point of f. Note that $-1/2$ is $\cos(2\pi/3)$.

(d) The period 2 points of f must satisfy $f^{\circ 2}(z) = f(z^2) = z^4 = z$. That is, $z(z^3 - 1) = 0$. Therefore, the period 2 points of f are $e^{2\pi i/3}$ and $e^{4\pi i/3}$. The point $z = 1$ is a fixed point. Period 2 points of g must satisfy $g(g(y)) = y$, or $2(2y^2 - 1)^2 - 1 - y = 0$. This polynomial is divisible by $2y^2 - y - 1$, so the period 2 points are roots of $4y^2 + 2y - 1 = 0$; that is, $y = -1/4 \pm \sqrt{5}/4$. Note that these can also be written as $\cos(4\pi/3)$ and $\cos(4\pi/5)$. Once again, one of the period 2 points of f corresponds under φ to a period 2 point of g, but the other does not.

(e and f) At this point it becomes imperative to look at the map φ and see what it is telling us. For any n, a period n point of f is a point $e^{i\theta}$ such that $f^{on}(e^{i\theta}) = e^{i\theta}$; that is, $e^{2^n \theta i} = e^{\theta i}$, or $(2^n - 1)\theta$ must be an integral multiple of 2π. Therefore the period n points of f are all $e^{i\theta}$ such that $\theta = [2\pi/(2^n - 1)]k$ for positive integers k such that the points are distinct and are not period m points for some $m < n$. The period n points of g are all y in $[-1, 1]$ for which $g^{on}(y) = y$ and $g^{om}(y) \neq y$ for $m < n$. Since $|y| \leq 1$, we can write $y = \cos(\theta)$ for some $\theta \in [0, \pi]$. Then $g(y) = g(\cos(\theta)) = 2\cos^2(\theta) - 1 = \cos(2\theta)$, and $g^{on}(\cos(\theta)) = \cos(2^n\theta)$. So we are looking for values of θ, $0 \leq \theta \leq \pi$, such that $\cos(2^n\theta) = \cos(\theta)$. Two real numbers a and b have the same cosine if either $a = b + 2k\pi$ (k any integer), or $a = (2\pi - b) + 2m\pi$. Thus for any n we need to find all values of θ in $[0, \pi]$ such that

$$2^n\theta = \theta + 2k\pi \quad \text{or} \quad 2^n\theta = 2(m + 1)\pi - \theta,$$

where k and m are non-negative integers, such that $\cos(\theta)$ is not a fixed point of g of order $< n$. Now we can use this information to find the fixed points of period 3. For f they are $e^{i\theta}$ with $\theta = 2k\pi/7$, $k = 1, 2, 3, 4, 5$, and 6. For g they are $\cos(\theta)$ with $\theta = 2k\pi/7$, $k = 1, 2$, and 3; and $\cos(\theta)$ with $\theta = 2m\pi/9$, $m = 1, 2$, and 4. Note that $\cos(6\pi/9)$ has already been shown to be a fixed point of g. Note also that 3 of these points correspond under the map φ, but the other 3 do not. This will be the general pattern for all n.

Solutions for Exercises 5.3

5.3#3. If P_b is the polynomial $z^3 - z + b$, then the corresponding Newton's method formula is

$$N_b(z) = z - (z^3 - z + b)/(3z^2 - 1) = (2z^3 - b)/(3z^2 - 1).$$

Therefore, $N_b(N_b(0)) = N_b(b) = (2b^3 - b)/(3b^2 - 1)$. This will be 0 if, and only if, $b(2b^2 - 1) = 0$, or $b = 0$, $\pm 2^{-1/2}$. If $b = 0$, 0 is also a fixed point of the map N_b but for $b = 1/\sqrt{2}$, for example, 0 is a point on a period 2 cycle. Furthermore, $(d/dz)(N_b(N_b(z)) = N_b'(N_b(z))N_b'(z)$, and $N_b'(z) = (6z^4 - 6z^2 + 6bz)/(3z^2 - 1) = 0$ at $z = 0$. This implies that the period 2 cycle will be superattracting. If you start the Newton's method for a root of $z^3 - z + (1/\sqrt{2})$ near 0 it will rapidly begin to cycle, rather than converge to a root.

5.3#9. The graph of T_a is shown below. To show that there are no fixed points, set $T_a(x) = (x - a/z)/2 = x$. This gives $x^2 = -a$, and with $a > 0$, there are no real solutions. A period 2 point must satisfy $T_a(T_a(x)) = x$. This leads to the equation $[(x/2 - a/(2x)) - a/(x/2 - a/(2x)) = x$, or

$3x^4 + 2ax^2 - a^2 = 0$. This can be factored to give $(x^2 + a)(3x^2 - a) = 0$, and since $a > 0$, the only real roots are $x = \pm\sqrt{a/3}$. To show that the 2-cycle is repelling, look at $(d/dx)(T \circ T)(x) = (d/dx)(T(T(x)) = T'(T(x))T'(x) = T'(\sqrt{a/3})T'(-\sqrt{a/3}) = 4 > 1$.

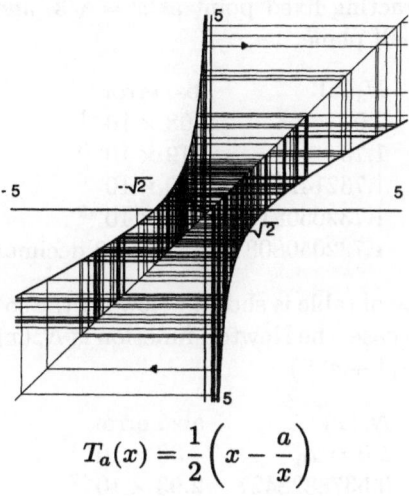

$$T_a(x) = \frac{1}{2}\left(x - \frac{a}{x}\right)$$

(b) By the Chain Rule, the derivative of $(T \circ T \circ \cdots \circ T) = (T^{on})$ is $T'(T^{on-1}(x))\,(d/dx)[T^{on-1}(x)]$ and by induction on n, this is $T'(T^{on-1}(x))$ $T'(T^{on-2}(x)) \cdots T'(T(x))T'(x)$. If x belongs to an n-cycle, this is just the product of the derivative of T at each point in the cycle, and is the same no matter which point of the cycle you start at.

5.3#11. Assume $f(r) = 0$ and $x = r + \varepsilon$.

(a) Then $f(x - \varepsilon) = f(r) = 0$, but by Taylor's Theorem with remainder, $f(x - \varepsilon) = f(x) - \varepsilon f'(x) + \varepsilon^2 f''(c)/2$ where c lies between x and $x - \varepsilon$. Therefore, $0 = f(x) - \varepsilon f'(x) + \varepsilon^2 f''(c)/2$. If $f'(x) \not\equiv 0$ we can divide by f' and get $f(x)/f'(x) = \varepsilon - \varepsilon^2 f''(c)/(2f'(x))$.

(b) Applying (a) to the function $N_f(x_i)$, gives

$$\begin{aligned} N_f(x_i) - r &= x_i - f(x_i)/f'(x_i) - r \\ &= x_i - [\varepsilon - \varepsilon^2 f''(c)/(2f'(x_i))] - r \\ &= (x_i - \varepsilon - r) + \varepsilon^2 f''(c)/(2f'(x_i)) \\ &= \varepsilon^2 f''(c)/(2f'(x_i)). \end{aligned}$$

Setting $\varepsilon = x_i - r$ gives the desired result.

(c) Part (b) implies that if we take an initial guess $x_i = r + \varepsilon$, then the next iteration $N_f(x_i)$ will satisfy the inequality

$$|N_f(x_i) - r| \le \varepsilon^2 |f''(c)/(2f'(x_i))|.$$

If x_i is close enough to the root r, $f'(r) \neq 0 \Rightarrow f'(x_i) \neq 0$, and $|f''(c)/(2f'(x_i))|$ will be less than some positive constant B. Therefore, if $\varepsilon = \mathcal{O}(10^{-n})$, then the error at the next step will be $\mathcal{O}(10^{-2n}) = \mathcal{O}(\varepsilon^2)$.

(d) The table below shows successive iterates of $N_3(x) = (x + 3/x)/2$, with $x_0 = 1$. Any initial value $x_0 > 0$ will converge to $\sqrt{3}$ because the map N_3 has a superattracting fixed point at $x = \sqrt{3}$, and it is the only fixed point in the right-half plane.

step	$N_3(x)$	abs. error
1	2.0	2.68×10^{-1}
2	1.75	1.79×10^{-2}
3	1.732142857	9.20×10^{-5}
4	1.732050810	2.43×10^{-9}
5	1.732050808	exact to 9 decimal places

(e) The same type of table is shown below for the solution of the equation $x - e^{-x} = 0$. In this case, the Newton function is $N_e(x) = x - (x - e^{-x})/(1 + e^{-x}) = e^{-x}(1 + x)/(1 + e^{-x})$.

step	$N_e(x)$	abs. error
1	$1.0 = x_0$	4.33×10^{-1}
2	0.5378828427	2.93×10^{-2}
3	0.5669869914	1.56×10^{-4}
4	0.5671432860	4.41×10^{-9}
5	0.5671432904	exact to 9 decimal places

5.3#14. Let $x' = f(x, y) = (x - 1)^2 - y$ and $y' = g(x, y) = y + x^2/2 - 1/2$. The zeros of x' and y' can be seen to lie on two parabolas which intersect at the points $(1, 0)$ and $(1/3, 4/9)$. To solve the nonlinear system by Newton's method, we must solve the linear system:

$$f(x_0 + h, y_0 + k) = f(x_0, y_0) + f_x(x_0, y_0)h + f_y(x_0, y_0)k = 0$$
$$g(x_0 + h, y_0 + k) = g(x_0, y_0) + g_x(x_0, y_0)h + g_y(x_0, y_0)k = 0$$

for the increments h and k. The matrix of partial derivatives is

$$M = \begin{bmatrix} \partial f/\partial x & \partial f/\partial y \\ \partial g/\partial x & \partial g/\partial y \end{bmatrix} = \begin{bmatrix} 2(x - 1) & -1 \\ x & 1 \end{bmatrix}.$$

Using the initial guess $(x_0, y_0) = (0, 0)$, Newton's method leads to the system

$$M|_{(0,0)} \begin{bmatrix} h \\ k \end{bmatrix} = \begin{bmatrix} -f(0, 0) \\ -g(0, 0) \end{bmatrix},$$

which gives

$$\begin{bmatrix} -2 & -1 \\ 0 & 1 \end{bmatrix} \begin{bmatrix} h \\ k \end{bmatrix} = \begin{bmatrix} -1 \\ 1/2 \end{bmatrix}.$$

Therefore, $(x_1, y_1) = (x_0+h, y_0+k) = (1/4, 1/2)$ is the next approximation. The second approximation is found by repeating the process with (x_0, y_0) replaced by (x_1, y_1). This leads to the system

$$\begin{bmatrix} -3/2 & -1 \\ 1/4 & 1 \end{bmatrix} \begin{bmatrix} h \\ k \end{bmatrix} = \begin{bmatrix} -1/16 \\ -1/32 \end{bmatrix},$$

having solution $(h, k) = (3/40, -1/20)$. Therefore (x_2, y_2) is $(13/40, 9/20) = (0.325, 0.45)$ which is quite close to the solution $(1/3, 4/9)$.

Solutions for Exercises 5.4

5.4#2. (a) Let $w = \alpha h$. Then $F_h(x) = [1 - w + w^2/2]x$, and we need to see where $|1 - w + w^2/2| < 1$. This is a parabola with minimum $1/2$ at $w = 1$, and remains below 1 for $0 < w < 2$. Therefore, the iterates tend to 0 if $0 < \alpha h < 2$, i.e., for $h < 2/\alpha$. If $h > 2/\alpha$, x will be multiplied for a value > 1 at each step, and the iterates will tend to ∞.

(b) For Runge–Kutta $F_h(x) = [1-w+w^2/2-w^3/6+w^424]x$, where again we have let $w = \alpha h$. It can be shown that the fourth degree polynomial $p(w) = 1 - w + w^2/2 - w^3/6 + w^4/24$ has a unique minimum greater than 0 at $w \cong 1.596$, and ranges between 0 and 1 for $0 < w < 2.78529356$. Therefore, for $h < 2.78529356/\alpha$, the iterates will tend to 0, and for h larger than that, they will tend to ∞. This shows that the Runge–Kutta method has a slightly larger interval of stability than Euler's method.

5.4#6. Theoretically, we know that the solutions of $x' = x^2 - t^2$ must enter the funnel and approach $-t$ as $t \Rightarrow \infty$. If a solution starts at a point near the funnel, so that $x(t) = -t + w(t)$, with $w(t)$ small, then

$$w'(t) = x'(t) + 1 = [-t + w(t)]^2 - t^2 + 1 = -2tw(t) + [w(t)]^2 + 1;$$

and for large t the approximation $w'(t) = -2tw(t)$ is reasonable. The Runge–Kutta step for the differential equation $x' = ax$ was shown in problem 5.4$2(b) to be $x_{n+1} = x_n(1 - ah + a^2h^2/2 - a^3h^3/6 + a^4h^4/24)$. It was shown there that the polynomial $1 - z + z^2/2 - z^3/6 + z^4/24$ has values between 0 and 1 for $0 < z < 2.7853$, and is greater than 1 for $z > 2.7853$. Therefore, with $h = 0.03$, this multiplier will be < 1 for $a < 2.7853/0.03 \cong 92.8$. For the linearized equation $w' = -2tw$, this means that the solution will be stable as long as $2t < 92.8$, or $t < 46.4$. A numerical Runge–Kutta solution programmed on an IBM PC appeared to converge to $x(t) = -t$ until t reached about 49. At that point it was clear that there was some kind of instability in the method. Results using the graphics program *DiffEqSys* are shown below. The trajectory starting at the dot appears to stay in the funnel for $t < 46$. At that point it seems to head off on a spurious trajectory, and finally breaks down completely into jagged junk. The solution between the dashed lines is correct; that in the shaded regions is not.

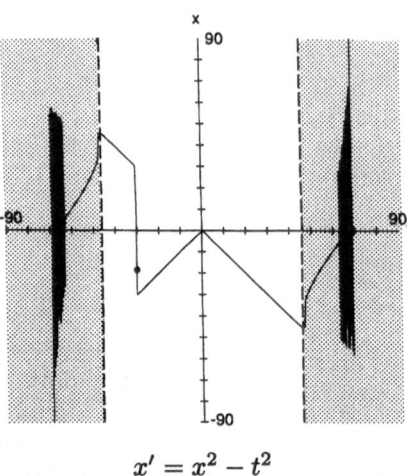

$$x' = x^2 - t^2$$

Solutions for Exercises 5.5

5.5#6. Choose $c \in [a, b]$, and consider

$$G(x) = \int_c^x \frac{du}{g(u)} \quad \text{for } x \in [a, b].$$

Then $G(x)$ is monotone (increasing if $g(x) > 0$ for $x \in [a, b]$, and decreasing if $g(x) < 0$), and tends to $\pm\infty$ as $x \to a$ or $x \to b$. In particular, the expression

$$G(x(t)) = \int_c^{x(t)} \frac{du}{g(u)} = \int_{t_0}^t f(s)ds$$

defines $x(t)$ implicitly as a function of t, and this function with $x(t_0) = c$ is the solution of our differential equation. If

$$\int_0^T f(s)ds = \int_{t_0}^{t_0+T} f(s)ds = 0$$

this means that $G(x(t_0 + T)) = 0 = G(c)$, so $x(t_0 + T) = c$, and the one-period later mapping is the identity on $[a, b]$.

5.5#8. Let $x' = p(t)x + q(t)$ be a periodic linear differential equation. Then we can solve the equation by variation of parameters and get

$$x(t) = A(t)x(t_0) + B(t), \quad \text{where} \quad A(t) = \exp\left\{\int_{t_0}^t p(\tau)d\tau\right\}$$

and

$$B(t) = A(t) \int_{t_0}^{t} [1/A(s)]q(s)ds.$$

Then $x(t_0+T) = A(t_0+T)x(t_0) + B(t_0+T)$ which is of the form $ax(t_0)+b$, with constants a and b.

Solutions for Exercises 5.6

5.6#4. If z satisfies $|z| > |c| + 1$, then $|f(z)| = |z^2 + c| \geq |z^2| - |c| > (|c|+1)^2 - |c| = 1 + |c| + |c|^2$. On the next iteration,

$$|f^{\circ 2}(z)| > (1 + |c| + |c|^2)^2 - |c| > 1 + |c| + 3|c|^2,$$

and by induction on n, $|f^{\circ n}(z)|$ can be seen to be $> 1 + |c| + K_n|c|^2$ where $K_n = 2K_{n-1} + 1$, $K_1 = 1$. Since the sequence $K_n \Rightarrow \infty$ with n, the orbit for z will be unbounded.

5.6#11. (a) Clearly, if ξ is a solution of $\xi + 1/\xi = z$, then by the quadratic formula, ξ is either $\xi_1 = [z + (z^2 - 4)^{1/2}]/2$ or $\xi_2 = [z - (z^2 - 4)^{1/2}]/2$. Furthermore, these two values of ξ are reciprocals; so exactly one of $|\xi_1|$ or $|\xi_2|$ is < 1, or else $|\xi_1| = 1$. In this latter case, $z = \xi_1 + 1/\xi_1 = e^{i\theta} + e^{-i\theta} = 2\cos\theta$ for some θ, and z lies in the closed interval $[-2, 2]$.

(b) To show that $\varphi = \xi + 1/\xi$ conjugates P_{-2} to P_0, we need only check that $P_{-2}\varphi(\xi) = \varphi P_0(\xi)$ for any ξ in the unit disk. Simply compute:

$$P_{-2}(\varphi(\xi)) = P_{-2}(\xi + 1/\xi) = (\xi + 1/\xi)^2 - 2 = \xi^2 + 1/\xi^2,$$

and $\varphi(P_0(\xi)) = \varphi(\xi^2) = \xi^2 + 1/\xi^2$.

(c) For any $\xi \in D$ (the unit disk), the orbit of ξ under P_0 converges to the fixed point 0, since $|\xi| < 1$. This implies that the corresponding point $z = \Phi(\xi)$ converges to $\Phi(0)$ under iteration by P_{-2}. But $\Phi(0)$ is the point at ∞. Therefore the orbit of z is unbounded.

Index